D0914179

THE REACTION PATH IN CHEMISTRY:
CURRENT APPROACHES AND PERSPECTIVES

Understanding Chemical Reactivity

Volume 16

The Reaction Path in Chemistry: Current Approaches and Perspectives

edited by

Dietmar Heidrich

Fakultät für Chemie und Mineralogie,
Institut für Physikalische und Theoretische Chemie,
Universität Leipzig,
Leipzig, Germany

KLUWER ACADEMIC PUBLISHERS
DORDRECHT / BOSTON / LONDON

A C.I.P. Catalogue record for this book is available from the Library of Congress.

ISBN 0-7923-3589-9

Published by Kluwer Academic Publishers,
P.O. Box 17, 3300 AA Dordrecht, The Netherlands.

Kluwer Academic Publishers incorporates
the publishing programmes of
D. Reidel, Martinus Nijhoff, Dr W. Junk and MTP Press.

Sold and distributed in the U.S.A. and Canada
by Kluwer Academic Publishers,
101 Philip Drive, Norwell, MA 02061, U.S.A.

In all other countries, sold and distributed
by Kluwer Academic Publishers Group,
P.O. Box 322, 3300 AH Dordrecht, The Netherlands.

Printed on acid-free paper

Printed in the Netherlands

040496-12760C

To Katja and Marit

TABLE OF CONTENTS

Introduction

The so-called reaction path (RP) with respect to the potential energy or the Gibbs energy ("free enthalpy") is one of the most fundamental concepts in chemistry. It significantly helps to display and visualize the results of the complex microscopic processes forming a chemical reaction. This concept is an implicit component of conventional transition state theory (TST). The model of the reaction path and the TST form a qualitative framework which provides chemists with a better understanding of chemical reactions and stirs their imagination. However, an exact calculation of the RP and its neighbourhood becomes important when the RP is used as a tool for a detailed exploring of reaction mechanisms and particularly when it is used as a basis for reaction rate theories above and beyond TST. The RP is a theoretical instrument that now forms the "theoretical heart" of "direct dynamics". It is particularly useful for the interpretation of reactions in common chemical systems.

A suitable definition of the RP of potential energy surfaces is necessary to ensure that the reaction theories based on it will possess sufficiently high quality. Thus, we have to consider three important fields of research:

- Analysis of potential energy surfaces and the definition and best calculation of the RPs or - at least - of a number of selected and chemically interesting points on it.

- The further development of concrete versions of reaction theory beyond TST which are applicable for common chemical systems using the RP concept.

- The investigation of chemical reactions with RP concepts and theories based on it.

These fields are treated in this book. The present volume begins with an "Introduction to the Nomenclature and Usage of the RP concept" (D.Heidrich). The brief chapter tries to find a common language and better understanding between chemists when they are dealing with the RP and its applications.

Then, four contributions are grouped together (those from P.G.Mezey, X.Chapuisat, T.Iwai and A.Tachibana, and from W.Quapp) which are related to important aspects or new developments of the reaction path theory.

The next block of chapters (by T.Helgaker, K.Ruud and P.R.Taylor, by W.Quapp, O.Imig, and D.Heidrich, and by G.Seifert and K.Krüger) is dedicated to the mathematical and quantum chemical techniques of calculating points on the RP. The last of these chapters includes a review of applications of the density functional theory in chemistry.

The following two chapters (written by D.G.Truhlar, and by A.D.Isaacson) are devoted to the description and application of current versions of reaction theories based on the RP concept ("direct dynamics" methods) for the ground state.

Finally, in the contribution of A.L.Sobolewski and W.Domcke, the "state of art" in treating the RP in excited states (of hydrogen transfer processes) are brought into focus. The last contribution by V.Engel gives us insights into the RP regions gained by modern time-resolved femtosecond spectroscopy and reviews literature in this field.

I asked the authors of the chapters to use clear and instructive introductions as well as more illustrations. So, in part, the book aims at a deepening of theoretical insights for the average chemist. If our book contributes to a better understanding of the RP concept and to encouraging theoreticians to fill or further develop the existing grey zones between quantum mechanic scattering theory and trajectory calculations, on the one hand, and conventional TST, on the other hand, then it has served its purpose.

I would like to thank the authors for their valuable contributions. Special personal thanks are dedicated to Prof.P.G.Mezey and Prof.D.G.Truhlar for their stimulating discussions and friendly support. The interdisciplinary "Naturwissenschaftlich-Theoretisches Zentrum" (NTZ) of the Leipzig University has greatly promoted our activities. Finally, I am very grateful to Dr.W.Quapp for his support in preparing the book and Margo Eastlund and Mike Cloud for corrections in our English text.

                                        Dietmar Heidrich
                                        Leipzig
                                        March 1995

AN INTRODUCTION TO THE NOMENCLATURE
AND USAGE OF THE REACTION PATH CONCEPT

DIETMAR HEIDRICH
Institut für Physikalische und Theoretische Chemie der Universität Leipzig
Fakultät für Chemie und Mineralogie
D-04109 L e i p z i g
Germany

Our thinking about chemical reactions has benefitted greatly from the model of energy profiles as functions of so-called "reaction coordinates" (RC). In textbooks, such profiles are often used to illustrate the main features of a chemical reaction, including its mechanism, in terms of transition state theory (TST)[1]. In his book "Potential Energy Hypersurfaces" (p. 86), P.G.Mezey [2] rightly describes the term "reaction coordinate" (RC) as a misleading one to characterize the reaction path (RP). Confusion results from the term "coordinate" which suggests that only one dominant internal coordinate describes the reaction path. The use of one selected "distinguished" coordinate (or "leading" coordinate) in higher-dimensional systems (by minimizing the other independent internal coordinates at fixed values of the distinguished coordinate) generally does not produce cross-sections through a PES (potential energy hypersurface) that uniquely correspond to minimum energy paths (MEP), see below. The chemist must - nolens volens - understand what a curve in the high-dimensional coordinate space taken as RP actually means (cf. RC in **Figure** 1)! The comparison of the RP to a normal mode resulting from Wilson's **FG** matrix [3] method may be helpful.

One possible means of avoiding misunderstandings with the term "RC" is to distinguish the reaction coordinate from the RP (= MEP). However, the term "RC" is a deeply entrenched term in chemistry and related natural sciences. It would be better to re-define the meaning of the term "reaction coordinate" (RC) so that it becomes intrinsically synonymous with the term "RP". This is done by understanding the RC as x-"coordinate" in the usual energy profile representation of a chemical reaction. "RC", then, does not mean the use of selected leading coordinate(s), but the use of a mathematically well-defined RP as a line in a high-dimensional coordinate space (e.g. that line given by steepest descent or gradient extremals - see below). Each point of the RC and the RP specifies a distinct geometry of all nuclei of the system along the MEP, and, in the case of the RC, presented on an axis. The occasional use of a one (or two) independent internal coordinate(s) of larger systems instead of a well-defined RP remains a special case of or an approximation to a RC and should be indicated. Used in this way, the term "reaction *coordinate*" can now be understood clearly, avoiding the confusion mentioned above.

The RP may be interpreted (**Figure** 1) as: **a.** a pure geometric entity, or, **b.** a

1

D. Heidrich (ed.), The Reaction Path in Chemistry:
Current Approaches and Perspectives, 1–10.
© 1995 *Kluwer Academic Publishers. Printed in the Netherlands.*

massweighted RP, a borderline case of whether or not it is a trajectory and, in this manner, connected with the dynamics of the reaction (see below). It seems that the concept of energy profile along a RP will outlive many theories that employ it because of its simplicity that helps to *visualize* the results of a "chemical reaction".

The concept of the reaction path can be further developed to form a suitable basis for advanced reaction rate theories. For chemically interesting systems (which often involve more than 4 atoms), reaction path methods provide powerful tools for the study of aspects of chemical dynamics. Hence, the first step of such theoretical treatments consists in a PES analysis, locating chemically interesting points and the chemically interesting paths between them. The field is not without misunderstandings and unsolved problems. Thus, the analysis of energy hypersurfaces in order to locate chemically interesting points and paths is an evolving field of great theoretical interest [2,4]. Additionally, the terms used to characterize points and paths of energy hypersurfaces

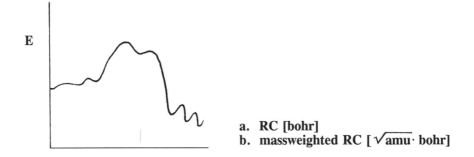

a. **RC [bohr]**
b. **massweighted RC [$\sqrt{\text{amu}} \cdot$ bohr]**

Figure 1. The "energy profile" of a chemical reaction with the "RC" representing the RP in the diagram

vary greatly in the literature partly because this field of research is an inter-disciplinary one. We must change this situation in the course of discussing approaches to the reaction path concept, its possibilities and shortcomings, as well as new horizons.

Terms relevant to the minimum energy path and their meaning

The reaction path (RP) is a projection of the valley path of the PES onto the coordinate space, i.e., it is a line in the multidimensional coordinate space connecting a particular minimum from the pool of minima representing the reactant system, with a minimum representing a metastable state or with a minimum from the product system along points of lowest potential energy. Hence, it describes the "minimum energy path" (MEP). As outlined above, *"reaction coordinate" (RC)* and *"reaction path" (RP)* are inherently synonymous to characterize the MEP. However, the terms RC and MEP have been introduced to illustrate microscopic [5,6] as well as macroscopic [6] energy sizes on the y-axis, i.e. potential energies as well as Gibbs energies ("free enthalpies"). We propose to take the term reaction path (RP) for potential energy surfaces (PES) only.

A RP (identical to terms such as "minimum potential energy path", or "potential energy RC") is a static path neglecting all kinetic energy terms and, thus, is fundamentally different from any "real" trajectory that forms a solution to the "classical" equations of motion. The RP may also be inadequate to represent quantum effects which are intrinsically nonlocal, but they may additionally be estimated by calculating "tunneling paths" [7]. The RP, however defined, cannot be interpreted as illustrating the stereochemical course of a chemical reaction (in which the trajectories hypothetically pass the neighbourhood of a transition structure). Whatever we use as a theoretical approach for approximating the progress of a chemical reaction (by semiclassical or quantum mechanical procedures), the reaction proceeds along one or more reaction channels, each representing a valley with 3N-7 walls, with N being the number of atoms). If more than one reaction channel is available with increasing energy, the mechanism of the reaction may be changed (cf.also P.G.Mezey, this book).

In summary, the reaction path does not have direct physical meaning. It is an artificial chemical instrument, a fiction of chemical thinking, but it is extremely valuable in overcoming the dimensionality dilemma. The structure of the reaction valley can be characterized by the RP itself, and by the frequencies of the transverse directions (i.e. by the steepness of the "walls of the valley"). Theories using these kinds of information represent RP concepts. This often works well because the trajectories, or wave-packets, representing the chemical system, are concentrated within the valleys.

The stationary points of the PES are, in mathematics and in quantum chemistry, also called critical points, cf. for instance [2]; however, the term "stationary" seems to be more suited. This term is mostly used in the presentation of electronic structure results. The chemically important stationary points are minima (related to the equilibrium structure), and saddle points of index 1 which form the transition structures of the transition state theory (TST). Only these types of stationary points are components of each RP regardless of how the RP is defined. The transition structure has one, and only one, imaginary frequency. As a rule, the equilibrium structure is characterized by at least one, but usually more, local minima, frequently close in energy, each minimum representing one or another of various conformers, complexes etc.

Low-lying saddle points may exist between different conformers, etc., that together characterize the reactant minimum. In an ideal case, the proper chemical reaction (breaking and formation of bonds) then leads over higher lying saddle points. Each of the saddle points of index 1 (index = number of imaginary frequencies) represents a mathematically defined point of highest energy along a minimum energy path of the PES. The synonymous term is "transition structure" (see [8]) which indicates the relation to TST. If the statistical part of the Eyring equation is calculated at this point, the transition structure determines the characteristics of a transition state. RPs are frequently degenerate by symmetry. In TST, the number of equivalent RPs is considered in context with the use of symmetry numbers in the rotational partition function of the saddle point structure. In mathematics, the term "saddle point" also includes saddle points of a higher index. These do occur in the course of geometry optimization. Saddle points of index 2 (two imaginary frequencies), or higher ones, may have importance for theoretical interpretations when the corresponding structures may change their PES characteristic by **a.** the choice of the level of theory, or by **b.** substituting one, or a set of atoms, for others (for instance comparing a series of structures of the same type). On the other hand, saddle points of index 2 (or higher indices) may be led back to a superposition of two or more saddle points of index 1, cf.[9]). Saddle points of index 2 are often called "hilltops".

Points of increasing interest are the nonstationary "branching and dissipation points" of valleys which are connected in a more complex way, with zeros of the Hessian matrix. With respect to the coordinates, the Hessian matrix is the second partial derivatives' potential energy matrix. At the minima, it corresponds to the "classical" force constant matrix [3]. The stationary points of a PES are "a priori" defined mathematically and related to physical and chemical entities. In other words, one can use any coordinate system to describe the geometry of the stationary points of the PES. However, this is not the case for points between stationary points, for instance those of a RP. Here, we must accept a suited primary coordinate system from which all transformations must be defined by means of differential geometry. Accepting the Cartesian coordinate system as the genuine one, the RP will change into a unique reaction feature, as proposed by W.Quapp and D.Heidrich in 1984 [10] (see also the chapter of W.Quapp in this book).

The "steepest descent" path was first established as a mathematically well-defined approach to the RP. Recently, it has also been termed the "gradient line" (better would be "gradient path") on a PES [11]. The path of steepest descent follows the negative gradient **g** from a saddle starting along the eigenvector direction of the imaginary frequency at the saddle point (because $\mathbf{g} = \mathbf{0}$ at the stationary point). The gradient is the vector containing all first partial derivatives of the energy with respect to the atomic coordinates. Then, the direction of the negative gradient goes downhill perpendicularly to the isopotential lines of a PES (**Figure** 2). Physically, the gradient is the vector of forces acting on the atoms of a chemical system and following this vector finally leads to a local or a global minimum of the energy.

Although it has been repeatedly discussed in the literature (cf. for instance [4a]), it is nevertheless necessary to characterize the distinguished-coordinate RP [12a] procedures (=coordinate driving procedures) once again. The idea of forcing the reactions by changes of one leading coordinate while optimizing the rest of the atomic coordinates, does not generally work (in the case of two leading coordinates, the RP must be derived from the two-dimensional cross-section of the PES). This procedure is, as already mentioned above, mathematically unsuitable, because of its possible arbitrariness in cutting the PES, i.e. the cross-sections do not lead along routes which are identical with minimum energy pathways on PES, even though it is a suitable or possible - in each case convenient - procedure for generating points near the MEP using selected examples [12]. Another question, also of considerable practical interest, is how such procedures may be utilized in order to obtain suited algorithms to find saddle points and approximate RPs.

The mass-weighted path of steepest descent in Cartesian coordinates has been termed the "intrinsic reaction coordinate" [13]. In the past, it was used by a number of scientists [14] to describe some kind of reaction dynamics. Since then, the path has been analyzed, particularly by Fukui et al. [13,15,16]. At present, it is well-accepted, and its calculation is, for instance, implemented in the GAUSSIAN program package [17], using a procedure established by Schlegel [18]. In the coordinate space, this path is measured in $\sqrt{\text{amu}} \cdot$ bohr (amu: atomic mass units, cf. **Figure** 1b) and forms the abscissa in the corresponding energy profiles.

STEEPEST DESCENT AND THE VALLEY FLOOR LINE (VFL)

The criteria for defining a VFL was first discussed by J.Panciř in 1975 [19]. The corresponding path is another that has a claim to be called the "minimum energy path". It has been studied more recently by means of a "gradient extremal path" (GE)[20]. It follows weakest ascent along the valley ground by determining the minima of the gradient norm along the isopotential lines. This requires setting the Hessian matrices at the points calculated. In mathematical terms, each point on a GE is an eigenvector of the Hessian matrix: $\mathbf{Hg} = \lambda\mathbf{g}$. In general (considering all extreme values of the gradient norm along the isopotential lines), gradient extremals do not follow only valley directions, but they also follow, for instance, ridges or flanks. We focus our interest on the gradient extremals (VF-GE) which follow valley floor lines (VFL)

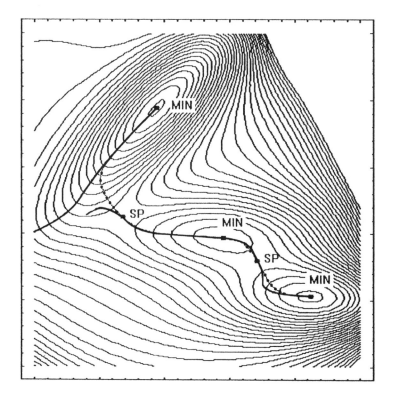

Figure 2. Steepest descent and gradient extremals (GE) using the Müller-Brown potential [23], ···· steepest descent path; ---- GE.

The VFL, defined in this way, does not always lead continuously to the saddle of interest. Discontinuities of the VF-GE indicate complex valley structures. Some algorithms have been proposed for GE following. They work in ascent as well as in descent [20]. The tracing of eigenvectors (EF) [21] along eigenvectors of suited Hessian eigenvalues in order to reach the saddle point have been used quite successfully [22]. However, the algorithm seems to be unsuitable for providing a mathematically unique path definition.

It should be noted that, in the literature, the term "reaction path" is often tacitly identified as the course of the steepest descent path. Please, note, that many workers, including authors of this book, use IRC or IRP (mass-weighted steepest descent paths) and MEP as synonymous. This simplification is no longer sufficient. It is evident (**Figure** 2) that steepest descent from the saddle and VFL-tracing are not identical. Steepest descent describes aspects of minimum energy paths of PES. It does not necessarily follow the "valley bottom" (-) whenever it passes the neighbourhood of it. We must note that there are no local criteria to decide whether or not a gradient line point is on the MEP (2-). Steepest descent has the advantage of representing the most direct and continuous line from any saddle to the minimum. However, steepest descent may change in a "hidden" way to trace the "ridge line" between a symmetrically branching RP (3-). Thus, steepest descent does not indicate bifurcation points and fundamental changes in the valley structure (emerging and disappearing valleys)(4-). Furthermore, the line perpendicular to the gradient cuts an energy profile with a minimum at the point on the steepest descent line. The line perpendicular to the VF-GE does not have this property, and thus, this fact forms a challenge to the theoreticians to make a path suited for developing a RP concept using the characteristics of the walls of the valley transverse to the path. Steepest descent reaches the reactant minimum along the smallest eigenvalue of the force constant matrix (if mass-weighted: along the normal mode of lowest frequency) [2,24]. This eigenvalue is, at least for larger chemical systems, not identical to the onset of the RP that we are interested in (5-). Thus, in theories beyond TST, steepest descent can be used to describe the RP only near the saddle, for instance, in order to find the actual position and numbers of transition states by Variational TST (VTST). If it is necessary that a RP represents a continuous line in the coordinate space to formulate theories beyond TST, use of the steepest descent may be tolerated. However, it should be used cautiously in critical regions, considering possible difficulties or breakdowns. It should also be discussed whether or not a RP can actually be, in each case, best characterized by a continuous curve.

We should keep in mind that steepest descent can be defined starting at any point of the PES, whereas GE are, by definition, restricted to a limited number of lines in the coordinate space. The possibilities of GE have not been shown yet for larger systems. Until now, they are mainly characterized by two-dimensional test potentials. However, GE are promising tools for analyzing the actual structure of a reaction channel or parts of the PES of larger systems. The valley structure of the reaction channel should be of considerable interest for classifying chemical reactions with respect to their vibrational excitation and dynamics.

A saddle in the valley direction may be reached by mode-selective vibrational excitation. The normal mode(s) of the minimum leading along the valley may reach a saddle by large amplitude motions. However, a saddle not lying in the valley path direction is reached by other types of trajectorial motion and should have a different dynamic characteristic. So, the weakest ascent (VF-GE) of the valley measured above the conformational and complex minima of the equilibrium structure may not lead to the desired saddle, or to any saddle at all (closing valley). This cannot be considered a disadvantage of the VF-GE algorithm, because, in such cases, the valley structure is actually characterized by two crossing valleys, and hence, by two minimum energy paths. GE and steepest descent may be further demonstrated by the following illustration in the two-parameter world of coordinates:

The course of least ascent should correspond to a stream bed path. It is the description of the move "uphill" which the front of the water level would follow in the mountains when it overflows gradually. One may imagine a valley which is continuously filled with water. If the water reaches the saddle located in the direction of the valley, the stream of water will eventually pass over the saddle. In this case, the steepest descent and VF-GE (valley path) are nearly identical or similar. However, if while mounting the valley bottom, a lower lying saddle occurs "at the side of the valley" in another dimension of the space, the stream suddenly goes to this side overcoming this barrier (cf. **Figure** 2). This means that two valleys come into contact with: the main valley, which may be closing or leading to a saddle, and a valley coming from the "side-saddle". The second valley ends suddenly when it has reached the main direction. The steepest descent does not indicate this valley structure, whereas the GEs draw two lines often crossing each other. It is evident that steepest descent and VF-GE may be nearly identical in weakly curved parts of the path (**Figure** 2).

The reaction path and its theoretical "superstructures": The reaction path concept

The potential energy surface concept is based on the Born-Oppenheimer approximation in quantum chemistry. This approximation decouples the electronic motion from that of the nuclei, thus allowing the determination of the electronic energy for any atomic configuration. The difference between the energy of a particular configuration and the energy of any other configuration represents the potential energy of interaction of the species considered. This concept creates the "reaction valleys" in the potential energy "mountains" and generates the TST approach because the idea of structure is allowed and the dimensionality problem of the potential energy surfaces is sufficiently reduced (systems with more than 4 atoms require an overwhelming number of point calculations). Seen in this manner, it is also clear that reaction theory always needs chemical intuition and a strong interaction with the experiment. However, the proper entity of the TST is the Gibbs energy ("free enthalpy"), so we have to consider reaction coordinates of free enthalpy profiles. In conventional TST [3,25], the Gibbs energy is calculated for the transition state and the reactants only. This theory provides an extremely useful conceptual framework which makes it possible to classify and characterize chemical reactions. In quantum chemistry, the quantitative aspects of conventional TST are frequently used to characterize potential energy barriers or the corresponding zero point energy differences, $\Delta^{\ddagger}E_o$, rather than the rate constant, k. The reasons are well-known (cf.[4a]). The most important reason is simply the error in calculating these energy differences for larger systems. The "uncertainties" in ΔE_o are exponentially enlarged by the e-function in Eyring's equation. In this manner, the rate constants k may be wrong by a factor of 10 if the error in the determination of the energy barriers is only 5,5 kJ/mol (at room temperature). Hence, a considerable uncertainty remains at least for larger systems or systems with elements of higher periods.

It is well-known that the occurrence of large thermal energies of the reactants relative to the energy of the transition state energy, challenges TST and related theories, because in this case, the trajectories do not only "feel" - if at all - the reaction valley alone, but they may also "feel" large parts of the potential energy surface (PES). This effect is also related to the situation of pumping energy into a certain normal mode of reacting species, and, in this manner, exceeding the transition state energy to a considerable extent. This may lead to the "vibrationally induced corner-cutting" in the reaction valley of the PES [26] in addition to the corner-cutting through tunneling. The characteristics of the ridge between the reaction and product valleys through which the reaction may tunnel, provide information for estimating tunnel probabilities.

In conventional TST, the geometric parameters of the PES saddle are presupposed to represent the geometry of the transition state. This is, in general, not true. Thus, for certain reactions we should analyse the reaction coordinate with respect to the free enthalpy surface at least near the saddle point. Then, a different position of the transition structure, more than one transition structure, or, in the case of RPs without a barrier, a particular (free enthalpy) transition structure may be found. These occurrences are the domain of a more developed form of TST, the VTST (Variational TST [27]) which follows one of the lines discussed by Laidler in his book *Theories of Chemical Reaction Rates* more than 20 years ago (pp 78).

Until now, the steepest descent path has been utilized to realize VTST and it has also been used to realize another approach, the method of the RP Hamiltonian. Here, information about the RP, together with information about the transverse vibration that defines the width of the valley, allow us to describe a RP valley potential [28]. This approach can be traced back to ideas of Hofacker, Marcus, and others [29] of how to use a coordinate related to the progress of the reaction and how to use the degrees of freedom perpendicular to it. The VTST aims at a calculation of the free-enthalpy-of-activation profile $\Delta G^\circ(T)$ along the RP in order to locate the true "transition state"(s), i.e., the true bottleneck(s) of the reaction, and, finally, to determine better reaction rates.

Miller et al. (1980) described the Hamiltonian for a reacting molecular system by using a reaction valley description of the energetics and dynamics in polyatomic systems. The whole valley floor with its inherent slope and widths of the valley is utilized to give us explanations for a variety of features of reaction dynamics. This includes specifically the dynamic coupling between the motion along the RP and the transverse vibration, i.e. the description of the energy flow from translation to vibration and vice versa.

When the reaction path is required over the whole distance from reactants to the transition structure, elementary problems may occur when using steepest descent. The main reason for this is the fact that the steepest descent path reaches the minimum from the direction given by the eigenvectors of the smallest eigenvalue of the reactant(s) [31]. This is, in general, not identical with the onset of the chemical reaction that we are interested in! The question remains: Will other definitions of RPs provide better results when using them within the framework of reaction theories mentioned above, and/or is it possible to find new approaches of reaction theories based on new or further developed definitions of RP?

When working in a larger coordinate space (i.e. treating larger systems), the question of defining a suitable and unique RP is a rather complex problem:

i. The definition has to meet mathematical as well as physical and chemical requirements. Steepest descent paths, gradient extremals, or, for certain cases, coordinate-driving procedures may coexist, and describe points of interest, depending on the reaction studied, as well as on the aims of the study. However, we know that the potential that is possibly hidden in the VF-GE algorithms is not yet worked out. Whether or not it is suitable as RP definition, GEs appear to be indispensable when, for instance, one tries to find bifurcation points, or if we are interested in other properties of a multiple valley structure.

ii. The question of a RP definition independent from the coordinate system used, or, in other words, the RP as a unique reaction feature, has to be worked out for each definition: the problem is solvable and the result is described for steepest descent [10], and in this book for GEs.

Gratitude is expressed to Prof.P.G. Mezey and Prof.D.G.Truhlar for stimulating discussions.

References

1. J.Phys.Chem. **87**, number 15, special issue dedicated to H.Eyring.
2. P.G.Mezey, *Potential Energy Hypersurfaces*, Elsevier, Amsterdam, 1987.
3. E.B.Wilson, J.C.Decius, and P.C.Cross, *Molecular Vibrations*, McGraw-Hill, New York, 1955.
4. For books and reviews published more recently see for instance: a) D.Heidrich, W.Kliesch, W.Quapp, *Properties of Chemically Interesting Potential Energy Hypersurfaces* (Lecture Notes in Chemistry, Vol.56), Springer, Berlin, 1991.
   b) Z.Havlas and R.Zahradnik, Int.J.Quantum Chem. **26**, 607 (1984). c) E.Kraka and T.H.Dunning, Jr., in Advances in Molecular Structure Theory, Vol.1 (pages 129-173), JAI Press Inc., 1990. d) M.L.McKee and M.Page, *Computing Reaction Pathways on Molecular Potential Energy Surfaces*, in K.B.Lipkowitz, D.B.Boyd (eds.) Reviews in Computational Chemistry, VCH, New York, 1993.
5. S.Glasstone, K.J.Laidler, and H.Eyring, *The Theory of Rate Processes* (International Chemical Series, L.P.Hammett, ed.), McGraw-Hill, New York, London, 1941.
6. K.Laidler, *Theory of Reaction Rates*, McGraw-Hill, New York, 1969.
7. For a short survey, cf. for instance: D.G.Truhlar and M.S.Gordon, Science **249**, 491 (1990).
8. W.J.Hehre, L.Radom, P.v.R.Schleyer, and J.A.Pople, *Ab initio Molecular Orbital Theory*, Wiley, New York 1986.
9. D.Heidrich and W.Quapp, Theor.Chim.Acta (Berl.) **70**, 89 (1986).
10. W.Quapp and D.Heidrich, Theor.Chim.Acta **66**, 245 (1984).
11. R.M.Minyaev, Int.J.Quantum Chem. **49**,105 (1994).
12. Cf. for instance a) M.J.Rothman, L.L.Lohr, Jr., C.S.Ewig, and J.R.Van Wazer,in D.G.Truhlar (ed) *Potential Energy Surfaces and Dynamics Calculations*, Plenum, New York, 1981. b) K.Müller, Angew.Chemie (Int. Ed.) **19**,1 (1980).
   c) M.J.Rothman and L.L.Lohr, Jr., Chem.Phys.Letters **70**, 405 (1980).
   d) R.Steckler and D.G.Truhlar, J.Chem.Phys. **93**, 6570 (1990).
13. K.Fukui, in R.Daudel and B.Pullman (eds.) *The World of Quantum Chemistry*,

Reidel, Dordrecht, 1974, p.113.
14. a) I.Shavitt, *The Tunnel Effect Corrections to the Rate of Reactions with Parabolic and Eckhart Barrier*, Report WIS-AEC-23, Theoret.Chem.Lab., University of Wisconsin, Madison, W.I., 1959. b) R.E.Weston, Jr., J.Chem.Phys. **31**, 892 (1959). c) I.Shavitt, ibid. **49**, 4048 (1968). d) R.A.Marcus, J.Chem.Phys. **45**, 4493 (1966), ibid. **49**, 2610, 2617 (1968). e) D.G.Truhlar and A.J.Kuppermann, J.Am.Chem.Soc. **93**, 1840 (1971). f) H.F.Schaefer III, Chem.Brit. **11**, 227 (1975).
15. K.Fukui, J.Phys.Chem. **74**, 4161 (1970).
16. a) A.Tachibana and K.Fukui, Theor.Chim.Acta **49**, 321 (1978); b) ibid. **57**, 81 (1980); c) A.Tachibana and K.Yamashita, Int.J.Quantum Chem.Symp. **15**, 621 (1981); d) K.Fukui, Int.J.Quant.Chem.Symp. **15**, 633 (1981).
17. **GAUSSIAN 92**, Revision C, M.J.Frisch, G.W.Trucks, M.Head-Gordon, P.M.W.Gill, M.W.Wong, J.B.Foresman, B.G.Johnson, H.B.Schlegel, M.A.Robb, E.S.Replogle, R.Gomberts, J.L.Andres, K.Raghavachari, J.S.Binkley, C.Gonzales, R.L.Martin, D.J.Fox, D.J.Defrees, J.Baker, J.J.P.Stewart, and J.A.Pople, Gaussian, Inc., Pittsburgh, PA, 1992.
18. C.Gonzales and H.B.Schlegel, J.Phys.Chem. **90**, 215 (1989).
19. J.Pancir̆, Collect.Czech.Chem.Commun. **40**, 1112 (1975).
20. D.K.Hofman, R.S.Nord, and K.Ruedenberg, Theor.Chim.Acta **69**,265 (1986); P.Jørgensen, H.J.A.Jensen, and T.Helgaker, Theor.Chim.Acta **73**, 55 (1988); N.Shida, J.E.Almlöf, and P.F.Barbara, Theor.Chim.Acta **76**, 7 (1989); W.Quapp, Theor.Chim.Acta **75**, 447 (1989).
21. C.J.Cerjan and W.H.Miller, J.Chem.Phys. **75**, 2800 (1981); J.Simmons, P.Jørgensen, H.Taylor, and J.Ozment, J.Phys.Chem. **87**, 2745 (1983). D.O'Neal, H.Taylor, and J.Simmons J.Phys.Chem. **88**, 1510 (1984); A.Banerjee, N.Adams, J.Simmons, R.Shepard, J.Phys.Chem. **89**, 52 (1985).
22. Cf. for instance D.J.Wales, J.Am.Chem.Soc. **115**, 11180 (1993).
23. K.Müller and L.D.Brown, Theor.Chim.Acta **53,** 75 (1979).
24. A.Tachibana and K.Fukui, Theor.Chim.Acta **51**, 189 (1979); A.R.Ruf and W.H.Miller, J.Chem.Soc. (Farad.Trans.2) **84**, 1523 (1988); P.Pechukas, J.Chem. Phys. **64**, 1516 (1976).
25. D.G.Truhlar, W.L.Hase, and J.T.Hynes, J.Phys.Chem. **87**, 2664 (1983); ibid., p. 5223(E); M.M.Kreevoy, and D.G.Truhlar, in C.Bernasconi (ed.): *Investigation of Rates and Mechanisms of Reactions*, Part 1, Wiley, New York, 1986.
26. B.Hartke and J.Manz, J.Am.Chem.Soc. **110**, 3063 (1988).
27. D.G.Truhlar, Annu.Rev.Phys.Chem. **35** 159 (1984).
28. W.H.Miller, N.C.Handy, and J.E.Adams, J.Chem.Phys. **72**, 99 (1980); for a survey see: Ref. [7].
29. For literature cf. [4c].
30. For references cf. [4a], p.14.

FROM REACTION PATH TO REACTION MECHANISM: FUNDAMENTAL GROUPS AND SYMMETRY RULES

PAUL G. MEZEY
Mathematical Chemistry Research Unit
Department of Chemistry and
Department of Mathematics and Statistics
University of Saskatchewan
Saskatoon, Canada, S7N 0W0

1. Introduction

A chemical reaction or a conformational change of a molecule can be regarded as a change of the nuclear arrangement that is often accompanied by changes in the electronic state. A continuous change of the nuclear arrangement of a molecule corresponds to a formal path in an abstract space defined by the internal coordinates of the molecules describing the mutual positions of the nuclei. This space, called the nuclear configuration space, has some unusual, and often counterintuitive properties. Some of these properties have important implications concerning the outcomes of reactions, and the patterns and interrelations of formal reaction paths and reaction mechanisms.

One of the main advantages of such models is based on macroscopic analogies: various paths in a nuclear configuration space are often used to represent chemical reactions. The earliest precise formulation of this approach is due to Fukui [1], who has introduced the concept of intrinsic reaction coordinate (IRC), and has given a definition in terms of the geometrical features of the potential energy surface associated with the given electronic state. Subsequent developments by Tachibana and Fukui have lead to generalizations, including the meta-IRC [2-4]. The IRC approach and related methods for the analysis of isoenergy contours and energy gradients [5-11] along potential energy hypersurfaces have formed the basis for a generalization of the concept of reaction path and for the topological description of reaction mechanisms [12-27].

The early topological analyses of potential energy hypersurfaces have been based on isoenergy contours [8,15] defined indirectly by the energy gradients using an

11

D. Heidrich (ed.), The Reaction Path in Chemistry:
Current Approaches and Perspectives, 11–38.
© 1995 Kluwer Academic Publishers. Printed in the Netherlands.

orthogonality condition, and on the distribution of domains defined by local curvature properties of the hypersurface along directions orthogonal to the gradient [12]. A subfamily of these domains, the "reactive domains", fulfill a stability condition for the formal reaction paths [12]. In later developments, topologies based on the catchment regions have been introduced [13,27], relying on the differential geometric model of reaction paths [1-4] and on tools and concepts originally described for geographical terrains [28,29]. Energy-dependent algebraic-topological representations of reaction mechanisms have been defined in terms of the homology groups and homotopy groups [23-25,27] of potential energy hypersurfaces.

The number of points of a potential energy hypersurface $E(K)$ where the energy gradient vanishes (the *critical points* of E) and the relative location of these points are of special importance in the study of reaction paths and reaction mechanisms. The Morse inequalities of algebraic topology provide lower bounds for the number of critical points of various types, e.g., minima, saddle points of one or more negative canonical curvatures [30-32].

Alternative relations to the Morse inequalities have been proposed in ref. [14] for both lower and upper bounds on the number of critical points along potential energy hypersurfaces. These bounds for the number of critical points of various types are important for the analysis of catchment regions [13], in the spirit of the mathematical model of "watersheds", originally derived for geographical terrains [28,29], and the differential geometric model [1-4].

The *catchment regions* of a nuclear configuration space represent chemical species, including stable conformers of molecules, transition structures, as well as the stable relative arrangements of two or more molecules with a specified overall stoichiometry [13,27]. A topological analysis of the pattern of various curvature domains of a potential energy hypersurface gives information on the reactive domains of the nuclear configuration space [12].

The theory of topological and differentiable manifolds is applicable to the study of topological patterns and features in domains of the potential energy hypersurface and the interrelations among local coordinate systems in the corresponding domains of the metric nuclear configuration space M [16,21,22]. Such a metric space M (that is not a vector space, but has a proper distance function defined in it) can be specified for each stoichiometric family of nuclei. A stoichiometric family contains every molecule, collection of molecules, ionic species and clusters composed precisely from the given set of nuclei. Any number of electrons in any electronic state associated with the given set of nuclei can be represented within the corresponding metric space M.

The concept of *reaction path* and the more general concept of *reaction mechanism* are of special importance in the interpretation of molecular processes. These are essentially classical concepts, serving as models to visualize a chemical reaction. Conventionally, a chemical reaction mechanism is specified if the essential, broad features of the assumed geometric rearrangement of the nuclei or larger molecular fragments (as well as possible changes in the electronic state) are given. If the reaction is confined to a single electronic state (that is, to one potential energy hypersurface), then a reaction mechanism may be *represented* by an assumed reaction path, usually, by a minimum energy path.

The classical mechanical model of a reaction path, however, is only a crude approximation, since any given path may serve, at most, as a rough guideline to the interpretation of the actual chemical process. Molecules are quantum mechanical objects subject to the Heisenberg uncertainty relation; they have wave-packet properties, consequently, their rearrangements cannot follow any specific path in the classical sense. Reactions do not occur along specific reaction paths, and the concept of *reaction channel* is a better model of reality. Since all formal, classical reaction paths following roughly the same general route from reactant to product can be thought of as representations of the same reaction mechanism, the notion of reaction channel can be generalized and represented topologically by a family of reaction paths that are equivalent in some sense. The corresponding equivalence classes of classical reaction paths may be taken to represent reaction mechanisms in a topological model. One approach [23-25,27] relies on the following equivalence: reaction paths that are deformable into one another below some energy bound are regarded as equivalent. The corresponding equivalence classes lead to the concept of *energy-dependent reaction mechanisms*.

The topological models of potential energy hypersurfaces represented as functions defined over such metric spaces M have been discussed in detail in refs. [17,18,23-27], with special emphasis on the pattern of reaction mechanisms. The *fundamental group of reaction mechanisms* [23-25,27] is an energy-dependent group-theoretical structure on each potential energy hypersurface, defined as the one-dimensional homotopy group of level sets of the potential energy hypersurface (hypersurface truncated at some energy bound A).

The elements of a fundamental group of reaction mechanisms are loop mechanisms (called fundamental reaction mechanisms, FRM, at the given energy bound A), which contain, as sub-mechanisms, all reaction mechanisms possible at the energy bound A. A change of the energy bound A (in a crude sense, a change of the temperature) does not necessarily change the pattern of reaction mechanisms, and each FRM group is an invariant of energy between two critical energy levels (between the energies of two critical points nearest in energy). However, the group can change at critical levels. The FRM groups themselves form an algebraic structure: they are elements of a lower semilattice [25,27].

The two rather different algebraic structures involved in the mechanisms of chemical reactions, the

   (i)   the point symmetry groups of the nuclear arrangements, and

   (ii) the fundamental groups (one-dimensional homotopy groups) of reaction mechanisms,

are interrelated. In this contribution some of the relations between these two families of groups will be discussed.

2. General comments on the interrelations between the distribution of point symmetry groups and critical points along potential energy hypersurfaces.

First we shall describe some relevant results concerning the first family of groups. Then some of the essential features of the second type of groups will be discussed, followed by the description of a framework where the interrelations among these two families of groups can be studied.

The point symmetry groups describe the static nuclear arrangement of a formal nuclear configuration, or a possible rearrangement where the point symmetry of the nuclear configuration remains invariant. These groups are independent of the electronic state, that is, of the actual potential energy hypersurface considered; nevertheless, they give constraints on the shapes on all the possible energy hypersurfaces defined over the nuclear configuration space M [33].

We shall follow the method of their analysis described in ref. [33]; in particular, we shall study the distribution of various point symmetry groups within the nuclear configuration space M.

Some fundamental features of a global point symmetry analysis of nuclear configurations along each potential energy hypersurface can be based on two theorems [33]. These theorems provide direct information on the *behavior of the potential energy surfaces* (the presence of a critical point within some region), *based on a property of the configuration space* M (symmetry conditions on boundary B and a test point K). *Symmetry* has a strong influence on *energy* in a rather general way. The application of these theorems requires only an inventory of the symmetry elements for a family of nuclear configurations, and conclusions on potential energy surfaces are obtained *without any quantum chemical calculation*.

The first of these theorems, the *catchment region point symmetry theorem* states the following:

> *Within each catchment region the highest point symmetry occurs at the critical point* [33].

Note that each catchment region contains precisely one critical point. Within this context, a higher point symmetry is interpreted by group - subgroup relations: a point symmetry group is of higher symmetry than another if it contains the other group as a subgroup. Here we regard each group as one of its own subgroups. Consequently, the theorem covers the case where all nuclear configurations of a catchment region have the same point symmetry (this is the case, for example, where all points of the catchment region have trivial symmetry).

In general, two point symmetry groups, chosen arbitrarily, need not be comparable by the group - subgroup relation. Each of the two groups may contain point symmetry operators as elements which are not occurring in the other group; in this case neither group is a subgroup of the other. One interesting aspect of the catchment region point symmetry theorem is that the possibility described above cannot occur for the point symmetry group of the critical point and any other group from any given catchment region: a catchment region never contains a nuclear configuration with a point symmetry group that is not comparable by the subgroup relation to the point symmetry group of the critical point configuration.

The second theorem, the *vertical point symmetry theorem of nuclear configuration spaces* interrelates the location and the point symmetries of nuclear configurations of critical points found on different potential surfaces corresponding to different electronic states [33]. Choose any (hyper)surface B that divides the nuclear configuration space M into two parts, M_1 and M_2, where the set M_1 of configurations contains surface B as its boundary (M_1 is a closed set). Some symmetry elements R'_i are present for each nuclear configuration K' along the boundary B. Let $\boldsymbol{R'}$ denote a family of such symmetry elements, $R'_1, R'_2, \ldots R'_p$,

$$\boldsymbol{R'} = \{ R'_1, R'_2, \ldots R'_p \} . \tag{1}$$

Select any nuclear configuration K from set M_1, and let $\boldsymbol{R}$ denote a family of symmetry elements $R_1, R_2, \ldots R_q$ that are present at point K :

$$\boldsymbol{R} = \{ R_1, R_2, \ldots R_q \} . \tag{2}$$

The *vertical point symmetry theorem of nuclear configuration spaces* states the following:

(i) if no configuration along B *possesses the family* $\boldsymbol{R}$ *of symmetry elements, or,*

(ii) if configuration K *does not have all the symmetry elements of family* $\boldsymbol{R'}$,

then the family M_1 *of configurations must contain at least one critical point for the potential energy surface of each electronic state (of each possible overall electronic charge).*

The proof of this theorem, some related theorems, and several examples can be found in ref. [33].

One should notice that if either one of conditions (i) and (ii) is fulfilled, then K must be an interior point of M_1, that is, K cannot fall on the boundary B.

There is considerable freedom in the number of symmetry elements to be considered. If one extends family $\boldsymbol{R}$ to *all* symmetry elements of configuration K, then the corresponding symmetry operators form the point symmetry group of K. Furthermore, if one takes *all common symmetry elements* along the boundary B then the family of symmetry operators corresponding to the family $\boldsymbol{R'}$ of symmetry elements is also a group. This elementary result can be verified easily as follows. If two symmetry elements R'_1 and R'_2 are present at each configuration along B then the symmetry element $R'_1 \times R'_2$ corresponding to the product symmetry operator within each local point symmetry group must also be present everywhere along B, hence closure follows. By a similar argument, associativity of product is also inherited from each local point symmetry group along B. The trivial symmetry element corresponding to the unit symmetry operator is present at each configuration, and if symmetry element R'_1 is present everywhere along B, then both the corresponding symmetry operator and its inverse operator are associated with each configuration along B. Consequently, if *all* common symmetry elements along the boundary B are taken, then, indeed, one obtains a group, the *maximal common point group of* B.

The important observation is that the theorem does not require the choice of this maximal common point group of B and one doesn't have to include all common symmetry elements in family R' ; the *vertical point symmetry theorem of nuclear configuration spaces* is valid even if one includes only *some* of the eligible symmetry elements in families R' and R .

The theorem does not specify the exact location, the type, and the number of the critical points, but the information obtained is useful. If for some arbitrarily small part M_1 of the configuration space M the condition of the theorem is fulfilled, then within M_1 there must exist some critical point for each electronic state and net charge. By subdividing M_1 further, the theorem can be applied to smaller and smaller subsets of M, and the uncertainty in the location of the critical point can be reduced. Since the theorem does not specify the exact number of critical points within M_1, some critical points may remain undetected by the theorem. For different electronic states the critical points may be of different types (e.g. minima, or saddle points) and may also have different locations within M_1. The chosen test point K itself does not have to be a critical point for any one of the potential surfaces.

These theorems on the influence of symmetry on general properties of potential energy hypersurfaces can be viewed as constraints inherent in the nuclear configuration space M, affecting all energy functions defined over M. In fact, M has more structure than a simple metric space, and M should never be considered without this additional structure.

One natural approach involves a partitioning of M into domains according to the 3D point symmetries of nuclear configurations of the corresponding stoichiometric family [33]. We denote by G_k the collection of all nuclear configurations of M which have a given point symmetry group g_k. It is possible that two nuclear configurations K_1 and K_2 of the same point symmetry g_k cannot be connected by a path along which all configurations K have the same symmetry g_k, consequently, a set G_k may be disconnected. A maximum connected component of set G_k is denoted by G_{kj}. Similarly to the catchment regions $C(\lambda,i)$ of each potential energy hypersurface $E(K)$, the configuration space symmetry domains G_{kj} generate a rigorous partitioning of the entire nuclear configuration space M into subsets.

There is a one-to-one correspondence between critical points $K(\lambda,i)$ and catchment regions $C(\lambda,i)$, however, there is no such simple correspondence between the symmetry domains G_{kj} and critical points, although important relations are implied by the theorems mentioned [33].

3. The fundamental groupoid of paths in a potential surface domain: a prelude to the fundamental group of reaction mechanisms

The second type of groups are the *fundamental groups of reaction mechanisms* (FRM groups) [23-25,27]. By contrast to the point symmetry groups associated with a nuclear configuration space M of the given stoichiometry, these groups are not determined by specifying M. The FRM groups are also dependent on the electronic state, that is, on the actual potential energy hypersurface E(K), as well as on an energy parameter A. A somewhat less elegant, but intuitively simpler algebraic structure, the *fundamental groupoid of paths* serves as the basis for the FRM groups. In this section we shall review the concepts and tools leading to the fundamental groupoid of paths, following the methods described in the original publications [23-25,27].

For a given potential energy hypersurface E(K) one may choose an energy parameter A, taken as an upper bound that defines a *level set* F(A) of the nuclear configuration space M:

$$F(A) = \{ \ K : \ E(K) < A \ \} . \tag{3}$$

Note that, in this contribution we shall consider *open level sets* which do not include points K where the "elevation" of the potential energy hypersurface E(K) equals the parameter value A. One may visualize a level set F(A) as that part of configuration space M where the value of the energy function of the given electronic state is less than the energy parameter (energy bound) A. A level set may be disconnected, as well as multiply connected.

A slightly different model is obtained if the strict inequality "<" is replaced by the inequality " ≤ " in the definition (3) of level sets, that is, if closed level sets are considered. This modified model leads to the same chemical conclusions and will not be discussed here. Note, however, that in the study of critical levels of potential energy hypersurfaces (the energy values of critical points) the closed level sets simplify the analysis.

A loop path within a level set F(A) is a path starting and ending at the same nuclear configuration K. Some loop paths of F(A) can be deformed continuously into one another within the level set F(A), while some others cannot. Such continuous deformations are called homotopies. Those loop paths which can be deformed into one another are said to be homotopically equivalent and to belong to the same homotopy class.

If one is interested in the general features of reaction mechanisms, it makes sense to consider homotopy classes of paths instead of individual paths. In a strict, quantum mechanical sense, individual paths having zero width are not compatible with the Heisenberg uncertainty relation: precise location orthogonal to the path implies infinite uncertainty in momentum for displacements of the path. Consequently, in the topological model of reaction mechanisms (Reaction Topology) a homotopy class of loop paths of a level set F(A) is regarded to represent a formal *loop reaction mechanism at the given energy bound* A.

Since each path can be regarded as a segment of a loop path, each formal reaction mechanism can also be regarded as a segment of a loop mechanism.

An algebraic relation (formal product) can be defined for some reaction paths, and by generalizing this product for reaction mechanisms, an algebraic structure (a group) of reaction mechanisms can be generated. In a formal sense, these groups are the one-dimensional homotopy groups of level sets of the potential energy hypersurface. In the following we shall briefly review the essential steps leading to the *fundamental group of reaction mechanisms* [23-25,27]. The notations and derivations follow those introduced earlier, for more detail the reader may consult ref [27].

Homotopy theory is the theory of continuous deformations. The topological representation of reaction mechanisms by homotopy equivalence classes of reaction paths is based on the actual chemical equivalence of all those reaction paths that lead from some fixed reactant to some fixed product, and are "not too different" from one another. This "chemical" condition, combined with an energy constraint, corresponds to a precise topological condition: two reaction paths are regarded as "not too different" *if they can be continuously deformed into each other below some fixed energy value* A, *that is, within a level set* F(A). Such paths are *homotopically equivalent* at energy bound A. This leads to a classification of all reaction paths: a collection of all paths that are deformable into one another within F(A) is a *homotopy equivalence class* of paths at energy bound A.

These homotopy classes are energy dependent. In the low energy range, an energy increase is usually associated with gradually reaching the hight of more and more reaction barriers. Consequently, a greater variety of reaction paths can be found within a level set F(A), and the homotopy classification of these paths leads to a greater number of classes. At higher energy bounds, however, another effect becomes dominant: as the energy bound A is further increased, more paths become deformable into one another, hence some classes may join to form a single class. Some equivalence classes that are distinct at lower energies may merge into a single homotopy class at a higher energy bound, if this bound allows the two collections of paths to become deformable into one another. In a high energy range, a further increase of the energy bound leads to fewer distinct classes. These trends are reflected in the actual interrelations between distinguishable reaction mechanisms at various temperatures. In the low energy range there are only a few reaction mechanisms. By increasing the energy bound, more reaction mechanisms become viable. By contrast, in the high energy range, a further increase of the energy allows a wider range of chemical transformations, fewer reaction mechanisms are distinguishable from one another. At very high energies (approaching the plasma state) no chemical species and no reaction mechanisms are distinguishable.

Within the above model the energy bound A may be taken as a parameter for the characterization of the *energy dependence* of reaction mechanisms. By an appropriate choice of the energy bound A, the detailed analysis of reaction mechanisms can be restricted to the *chemically most important low energy regions of the energy hypersurface*.

If there is no restriction on the value of the energy parameter A, then the above model is suitable for a classification of all reaction paths of the entire energy hypersurface into various homotopy equivalence classses. Instead of dealing with the very large set of all possible reaction paths, the analysis is reduced to the much smaller family of their equivalence classes.

It is useful to restrict our investigations to reaction paths and reaction mechanisms that fall below some energy bound A, that is, we shall study processes within a level set F(A). With respect to such energy bound A, a formal reaction path p is regarded as a mapping of the unit interval I,

$$I = [0, 1] \tag{4}$$

into level set F(A),

$$p: I \rightarrow F(A). \tag{5}$$

For the purposes of topological analysis it is useful to introduce a parametrization: the path p, as a mapping, assigns parameter values u from the unit interval I to points p(u) of the level set F(A). A path p is an assignment of parameter values u, $0 \leq u \leq 1$, to points p(u) of F(A) in a continuous manner. In particular, a path p is *not* a curve in F(A), that is, p is regarded not as a collection of points along some line, but as a continuous *function* defined by the above assignment. We must keep in mind that the resulting curve is merely the *image* of p. In particular, a given curve (as a point set) may be the common image obtained by several different assignments of u values to the same set of points of F(A), and all these different assignments are regarded as different paths.

The point p(0) of the image is the origin and the point p(1) is the extremity of the path. These points p(0) and p(1) are also called the beginning and the end of path p, respectively. In some cases the terms endpoints and extremities are used for these points.

We may consider all possible paths; a general path p is not constrained by any classical mechanical restriction. In particular, a path p is not necessarily a classical trajectory. The freedom in the selection of paths is an important aspect of the model: in a quantum mechanical sense none of the sharply defined lines in a nuclear configuration space M has direct physical meaning, just as sharply defined points as nuclear configurations or sharply defined electronic "positions" have no direct physical meaning. Instead of individual paths, the topological model uses *families* of paths to represent reaction processes. These families are defined by restrictions placed *collectively* on the members of these families, and these restrictions are topological rather than classical mechanical. In order to describe these restrictions, we need to discuss some additional concepts.

A *closed path* or a *loop* is a path p with origin and extremity at the same point:

$$p(0) = p(1) = K \in F(A) \tag{6}$$

Such paths are of special importance in the topological description of reaction mechanisms.

A *constant path* is another special case of paths where the image is just a single point K of the level set $F(A)$:

$$p: I \rightarrow K \in F(A). \tag{7}$$

The *inverse path* p^{-1} of path p is defined as the path parametrized in the sense opposite to p:

$$p^{-1}(u) = p(1-u). \tag{8}$$

The inverse path involves the same image, the same curve in $F(A)$ as the original path p, but the direction of the inverse path is reversed. This implies that the roles of origin and extremity are interchanged.

It is possible to provide a family of paths with a useful algebraic structure, by introducing the concept of *product path*. The product of two paths is defined as the first path continued by the second one, whenever this is possible. Of course, two randomly selected paths seldom follow one another, and the product path of two paths exists if and only if the endpoint of the first path coincides with the beginning of the second path.

If one reaction path p_2 is the continuation of another path p_1 then by joining them a new path is obtained. The the resulting overall path p_3 is called the *product path* of paths p_1 and p_2:

$$p_3 = p_1 \, p_2 \, , \tag{9}$$

where the parameter assignment is chosen as

$$p_3(u) = p_1(2u) \qquad \text{if } 0 \leq u \leq 1/2, \tag{10}$$

and

$$p_3(u) = p_2(2u-1) \quad \text{if } 1/2 \leq u \leq 1 . \tag{11}$$

A product $p_1 \, p_2$ exists if and only if the extremity of the first path p_1 coincides with the origin of the second path p_2:

$$p_1(1) = p_2(0). \tag{12}$$

The concepts of homotopy, homotopical equivalence, and homotopy equivalence classes are the main topological tools for the construction of quantum chemical reaction mechanisms within the potential energy hypersurface model.

Consider two paths, q_1 and q_2, within a level set $F(A)$. These paths q_1 and q_2 are *homotopic relative to their endpoints,* (or, using a simpler but somewhat imprecise terminology, *homotopic*), if they have common origin as well as common extremity and if they *can be continuously deformed into one another within* $F(A)$ while keeping their endpoints fixed:

$$q_1(0) = q_2(0), \tag{13}$$

$$q_1(1) = q_2(1), \tag{14}$$

$$q_1 \text{ is continuously deformable into } q_2. \tag{15}$$

The concise notation

$$q_1 \sim q_2 \tag{16}$$

is used to state that path q_1 is homotopic to q_2.

Consider four paths, $q_1, q_2, q_3,$ and q_4, within a level set $F(A)$. For simplicity, we may take a two-dimensional example, and regard $F(A)$ as a set of the plane. Let us assume that paths q_1 and q_2 are homotopically equivalent, $q_1 \sim q_2$. It is possible that a path q_3 has the same endpoints as q_1 and q_2, yet q_3 is not homotopic to q_1, and q_2. If there is a high-energy domain above energy A that is missing from the interior of level set $F(A)$, and if path q_3 leads along a side of the missing domain opposite to paths q_1 and q_2, then q_3 cannot be deformed continuously into either of q_1 and q_2, as long as these deformations are confined to the level set $F(A)$. Hence, there is no homotopical equivalence between path q_3 and either of paths q_1 and q_2 of the level set $F(A)$. (Note, however, that q_3 and q_1 may become equivalent if the energy bound A is raised above the missing mountaintop that separates them now.) Homotopical nonequivalence is most commonly due to a mismatch of extremities. Of course, if either the origin or the endpoint of a path q_4 is different from the corresponding extremity of paths q_1 and q_2, then q_4 does not fulfill the endpoint requirements, consequently, q_4 cannot be homotopic to paths q_1 and q_2. We say that the pair q_1 and q_2, the path q_3, and the path q_4 belong to three different homotopy classes.

In general, a family of paths which are homotopic to one another form a *homotopy equivalence class,* or in short, a *homotopy class.* (In fact, a precise statement requires a reference to the preservation of endpoints: these classes are homotopy classes relative to endpoints.) The homotopy class of all paths homotopic to some path p is denoted by $[p]$,

$$[p] = \{p': \ p' \sim p\}, \tag{17}$$

where it is understood that all these paths are confined to the actual level set $F(A)$.

Following a topological correspondence principle [27], the homotopy equivalence classes of formal reaction paths can be used to represent reaction mechanisms. The approach just described involves some geometrical restrictions: the fixed endpoints of these paths. This geometrical restriction also constrains the homotopy classes. As we shall see later, this restriction can be released, leading to a fully topological model of reaction mechanisms [27]. The topological model has some advantages: homotopy classes of formal paths of reactions are more compatible with quantum mechanics than the concept of individual reaction paths. In addition, homotopy classes also have much stronger algebraic properties than reaction paths, leading to the fundamental group of reaction mechanisms.

The next concept required as part of the background information for the description of the fundamental group of reaction mechanisms is the *fundamental groupoid of paths* within a level set $F(A)$.

Here we shall use the same conventions as in ref [27]: lower case notation p is used for reaction paths if we wish to allow the possibility of $p(0) \neq p(1)$, as opposed to the case of closed paths (also called loop paths or cycles) P for which $P(0) = P(1)$. The set of all paths within level set $F(A)$ of the potential energy hypersurface $E(K)$ is denoted by $\mathbf{P}$. In particular, each constant path P for which $P(I) = K \in F(A)$, that is, each path with image a single point K in level set $F(A)$, is also an element of $\mathbf{P}$.

For each path $p \in \mathbf{P}$ two mappings, L^* and R^*, are defined as

$$L^*: \mathbf{P} \to \mathbf{P}, \qquad L^*(p) = q \in \mathbf{P}, \tag{18}$$

where for the constant path q

$$q(I) = p(0) \in F(A), \tag{19}$$

and

$$R^*: \mathbf{P} \to \mathbf{P}, \qquad R^*(p) = q' \in \mathbf{P}, \tag{20}$$

where for the constant path q'

$$q'(I) = p(1) \in F(A). \tag{21}$$

Mapping $L^*(p)$ assigns the constant path q at the origin $p(0)$ of p to the path p, whereas mapping $R^*(p)$ assigns the constant path q' at the extremity $p(1)$ of p to the path p. Mapping L^* and mapping R^* are called the *left* and *right unit paths* of reaction path $p \in \mathbf{P}$, respectively.

If P is a closed path (a loop), then the equality $P(0) = P(1)$ implies that

$$L^*(P) = R^*(P). \tag{22}$$

The condition $p_1(1) = p_2(0)$ for the existence of the product path p_1p_2 generated by paths $p_1, p_2 \in P$, can be written in terms of mappings $L*$ and $R*$ as

$$R*(p_1) = L*(p_2). \tag{23}$$

The definition of homotopy equivalence classes $[p] = \{p': p' \sim p, p', p \in P\}$ relative to endpoints implies that

$$L*(p) = L*(p'), \tag{24}$$

and

$$R*(p) = R*(p') \tag{25}$$

for any two paths from the same homotopy class, $p, p' \in [p]$.

The family of all such homotopy classes is denote by $\Pi(F(A))$:

$$\Pi(F(A)) = \{ [p_\alpha]: p_\alpha \in P\}. \tag{26}$$

Although this set Π is simpler than the set P of all paths, it is helpful to introduce further simplifications, using an algebraic structure on set Π. The first step is to introduce two special transformations for the homotopy classes, defined as the two mappings L and R on set Π:

$$L: \Pi \to \Pi, \qquad L([p_\alpha]) = [L*(p_\alpha)] \in \Pi, \tag{27}$$

and

$$R: \Pi \to \Pi, \qquad R([p_\alpha]) = [R*(p_\alpha)] \in \Pi. \tag{28}$$

Classes $L([p])$ and $R([p])$ are called the *left and right units* of homotopy class $[p]$, respectively.

In order to introduce the algebraic structure, a *product* $[p_1][p_2]$ of the *equivalence classes* $[p_1]$ and $[p_2]$ of paths is defined as

$$[p_1][p_2] = [p_1p_2] \in \Pi. \tag{29}$$

The product $[p_1p_2]$ is the homotopy class which contains the products of paths from the homotopy classes $[p_1]$ and $[p_2]$.

This product $[p_1p_2]$ exists if and only if the condition

$$R([p_1]) = L([p_2]) \tag{30}$$

is fulfilled for the right unit $R([p_1])$ of the homotopy class $[p_1]$ and the left unit

$L([p_2])$ of the homotopy class $[p_2]$.

Homotopical equivalence within each homotopy class implies that this product, if it exists, is unique and does not depend on the choice of reaction paths $p_1, p_2 \in \mathbf{P}$, representing equivalence classes $[p_1], [p_2] \in \Pi$.

We shall assume that the family Π of all homotopy equivalence classes of the complete set $\mathbf{P}$ of all paths within level set $F(A)$ is equipped with the above product, a product that is not necessarily applicable for all pairs of equivalence classes.

Family Π, together with mappings L and R, fulfill the following conditions:

(i)   For mappings L and R

$$L \circ L = L = R \circ L, \tag{31}$$

and

$$L \circ R = R = R \circ R \tag{32}$$

where the symbol $\circ$ denotes the composition of mappings (one mapping followed by the other, in the order from right to left).

(ii) For any class $[p] \in \Pi$ the products $L([p])[p]$ and $[p]R([p])$ exist and

$$L([p])[p] = [p] = [p]R([p]) \in \Pi. \tag{33}$$

(iii) The products $L([p])L([p])$ and $R([p])R([p])$ exist for each $[p] \in \Pi$ and

$$L([p])L([p]) = L([p]) \in \Pi, \tag{34}$$

$$R([p])R([p]) = R([p]) \in \Pi, \tag{35}$$

that is, both the left unit $L([p])$ and the right unit $R([p])$ are *idempotent*.

(iv) For any two classes $[p_1], [p_2] \in \Pi$, fulfilling condition (30), we have

$$L([p_1][p_2]) = L([p_1p_2]) = L([p_1]), \tag{36}$$

and

$$R([p_1][p_2]) = R([p_1p_2]) = R([p_2]). \tag{37}$$

If in addition to (30), the condition

$$L([p_3]) = R([p_2]) \tag{38}$$

also holds for some homotopy class $[p_3] \in \Pi$, then the following products also exist:

$$([p_1] [p_2]) [p_3] \in \Pi, \tag{39}$$

and

$$[p_1] ([p_2] [p_3]) \in \Pi. \tag{40}$$

(v) If the products (39) and (40) exist, then they are the same, that is, the product of homotopy classes of paths is *associative*:

$$([p_1] [p_2]) [p_3] = [p_1] ([p_2] [p_3]), \tag{41}$$

and one may omit the parentheses and simply write $[p_1] [p_2] [p_3]$.

(vi) The existence of a unique inverse path p^{-1} for every path $p \in \mathbf{P}$ implies the existence of a *unique inverse* class $[p]^{-1}$ for every homotopy class $[p] \in \Pi$, defined as

$$[p]^{-1} = [p^{-1}] \in \Pi, \tag{42}$$

where for the pair $[p]$, $[p]^{-1}$ the following relations hold:

$$L([p]) = R([p]^{-1}) \tag{43}$$

and

$$R([p]) = L([p]^{-1}). \tag{44}$$

With respect to property (v) one should note that for the products $(p_1p_2)p_3$ and $p_1(p_2p_3)$ of paths p_1, p_2, and p_3 from the homotopy classes $[p_1]$, $[p_2]$, and $[p_3]$, respectively, associativity is not generally assured, since

$$(p_1p_2)p_3 \neq p_1(p_2p_3) \tag{45}$$

as long as p_1, p_2, and p_3 are not constant paths. The difference between these products is in the parametrization. In the product path $(p_1p_2)p_3$ the path segments p_1, p_2, and p_3 are assigned to the following intervals of parameter u : $[0,0.25]$, $[0.25,0.5]$, and $[0.5,1]$, respectively. By contrast, in the product path $p_1(p_2p_3)$, the path segments p_1, p_2, and p_3 are assigned to the following intervals of parameter u : $[0,0.5]$, $[0.5,0.75]$, and $[0.75,1]$, respectively. Consequently, these two paths, as mappings, are different. However, the point set images of the two paths are the same, and by a continuous reassignment of the parametrization, path $(p_1p_2)p_3$ can be converted into path $p_1(p_2p_3)$ using a continuous transformation. That is, paths $(p_1p_2)p_3$ and $p_1(p_2p_3)$ are homotopically equivalent,

$$(p_1p_2)p_3 \sim p_1(p_2p_3), \tag{46}$$

and the path product, although not associative in the ordinary sense, is *homotopically associative*. One of the important gains obtained by the introduction of homotopy classes is the fact that their product is associative in the ordinary sense.

The term *groupoid* is used in two different contexts. In some texts on algebra any set G with a product defined for some ordered pairs of its elements is called a groupoid. We follow an alternative, somewhat more demanding terminology: a set G is a groupoid if a product is defined for some ordered pairs of the elements of G, this product is associative, and for each element of G there exist left and right units.

Properties (i)-(vi) of set Π of homotopy classes of the complete set **P** of all reaction paths on the level set F(A), equipped with the product (29) and with mappings L and R, fulfill the conditions for a groupoid. The set Π of homotopy classes has an algebraic structure, Π is called the *fundamental groupoid* of reaction paths of level set F(A).

4. The fundamental group of reaction mechanisms (FRM groups)

Homotopy equivalence classes of reaction paths, restricted to a level set F(A) of some energy bound, form the fundamental groupoid Π of paths. This groupoid is an associative algebraic structure, with unit elements and with inverse. This algebraic structure has some disadvantages: there are many unit elements, and there is no closure property, since the product cannot be applied for all pairs of elements from the groupoid $\Pi(F(A))$. The main disadvantage is the lack of the closure property: if an algebraic operation can be carried out only for selected pairs of the elements of an algebraic structure, then this structure is of little practical use.

Fortunately, it is possible to consider an appropriate constraint for the selection of a more useful subset $\Pi'_1(K_0)$ of the groupoid $\Pi(F(A))$. This subset $\Pi'_1(K_0)$ has an algebraic structure with closure property, and with some additional properties which greatly enhance its chemical relevance. If one considers only *closed paths* P, with a common fixed endpoint K_0, then any such path can be continued by any other such path. Consequently, the product exists for any pairwise combination of these paths, hence the product (29) also exists for any pairwise combination of the homotopy classes [P] of these paths. In other words, for these constrained homotopy classes the product defined by eq. (29) has the closure property.

One can formulate this approach in precise topological terms. Choose an arbitrary point $K_0 \in F(A)$ and consider the following subset Π'_1 of groupoid Π:

$$\Pi'_1(K_0) = \{[P]: P(0) = P(1) = K_0, \ P \in [P], \ [P] \in \Pi \} . \tag{47}$$

The elements of this set $\Pi'_1(K_0)$ are the homotopy classes of all the closed paths (loops) P with the common endpoint $K_0 \in F(A)$. All left and right units, all inverses, and all possible pairwise products of these homotopy classes exist and are also elements of the same set $\Pi'_1(K_0)$:

$$L([P]), R([P]) \in \Pi'_1(K_0), \tag{48}$$

$$[P_1]^{-1} = [P_1^{-1}] \in \Pi'_1(K_0), \tag{49}$$

and

$$[P_1] [P_2] \in \Pi'_1(K_0), \tag{50}$$

for every $[P], [P_1], [P_2] \in \Pi'_1(K_0)$.

If a subset π of a groupoid Π has properties (48) - (50), then π is called a *stable subset,* and if the the mappings left unit L and right unit R are restricted to this subset, then π is called a *subgroupoid* of the original groupoid Π.

There are further advantages of $\Pi'_1(K_0)$; the definition (47) implies that, if the equivalence class $[P_0]$ contains the element constant path P_0 at point K_0,

$$P_0 \in [P_0], \tag{51}$$

where

$$P_0(I) = K_0 , \tag{52}$$

then

$$L([P]) = R([P]) = [P_0] \tag{53}$$

for *every* homotopy class $[P] \in \Pi'_1(K_0)$.

Consequently, the left unit mapping L and the right unit mapping R, when restricted to the subgroupoid $\Pi'_1(K_0)$, are *constant maps.* That is, there is a unique unit element $[P_0]$ of subgroupoid $\Pi'_1(K_0)$. Both the existence of a unique *unit element* and the *closure* property are special features of subgroupoid $\Pi'_1(K_0)$, whereas the existence of *inverse* and *associativity* are properties inherited from groupoid Π. The presence of these four criteria implies that the subgroupoid Π'_1 is a *group*, a *subgroup* of the original groupoid Π.

If at the energy bound A the level set F(A) is arcwise connected then this group $\Pi_1(K_0)$ is called the *fundamental group* of level set F(A). If explicit reference to the level set is required, then the notation

$$\Pi_1(F(A), K_0) \tag{54}$$

can be used. Note that the arcwise connectedness is often energy dependent: at low energies a level set may separate into several, disjoint subsets.

If the level set F(A) is arcwise connected, then the algebraic structure of the fundamental group, taken as an abstract group, is not affected by the choice of point

K_0. Within any arcwise connected set $F(A)$ the point K_0 can be interconnected with any other point K_1 by some path R. Consequently, any closed path P with both extremities at point K_0 can be extended into a homotopically equivalent closed path PRR^{-1}, by simply attaching a "detour" from K_0 to K_1 and back before concluding the path. That is, each closed path containing point K_0 is homotopically equivalent to a closed path containing K_1. Each loop P of reference point K_0 can be assigned to a loop containing K_1 as reference point. This implies that the choice of point K_0 is of little consequence, and the two *concrete* groups, $\Pi_1(F(A), K_0)$ and $\Pi_1(F(A), K_1)$ are *isomorphic:*

$$\Pi_1(F(A), K_0) \approx \Pi_1(F(A), K_1). \tag{55}$$

This group isomorphism has important topological consequences, since it implies that the specification of reference point K_0 can be omitted. This reference to a point was the last geometrical constraint of the model, all other aspects of the paths in the homotopy classes had no fixed, geometrical restriction. By eliminating this last geometrical restriction, the abstract *fundamental group* $\Pi_1(F(A))$ as an algebraic structure is fully topological. This fact facilitates the quantum chemical interpretation of this group.

Our goal is the representation of reaction mechanisms. An individual reaction path has many incidental features, some not compatible with quantum mechanics. Instead of a reaction path, a reaction mechanism can be better represented by a formal *reaction itinerary*, where the main, invariant features of the "journey" are relevant. Such a reaction itinerary can be represented by a whole family of similar paths, and it is natural to model a reaction itinerary by a homotopy equivalence class of paths.

For a comparison of the intuitive, chemical requirements placed on a model of reaction mechanisms and the tools offered by algebraic topology, we shall discuss various levels of the topological model.

On the first level, the homotopy equivalence classes [p] of the rather weak algebraic structure, the groupoid $\Pi(F(A))$, can be thought of as classical mechanical representations of configuration-to-configuration reaction mechanisms. The [p] reaction mechanisms are constrained by an upper bound A of energy, as implied by the given level set $F(A)$. Any two different homotopy equivalence classes [p_1] and [p_2] represent two fundamentally different chemical processes, that is, *two different reaction mechanisms* at energy bound A, since any two formal paths p_1 and p_2 from the two homotopy classes [p_1] and [p_2], respectively, cannot be deformed into each other below energy A. This model suggests an energy-dependent concept of classical, configuration-to-configuration reaction mechanisms; evidently, the homotopy classes do depend on the available energy A. This model is not fully topological. Groupoid $\Pi(F(A))$ involves geometrical, classical mechanical constraints, since the endpoints $p_1(0)$ and $p_1(1)$ of paths in each homotopy class [p_1] correspond to precisely defined nuclear configurations. These formal nuclear configurations are geometrical, classical and not quantum mechanical.

The next level of the topological model involves a concrete realization $\Pi_1(F(A), K_0)$

of the fundamental group with respect to a specific reference point K_0. Some of the classical mechanical, geometrical constraints of the groupoid model are released in this group $\Pi_1(F(A), K_0)$. The definition of group $\Pi_1(F(A), K_0)$ does not require the specification of an infinite family of fixed nuclear configurations as endpoints of configuration-to-configuration reaction mechanisms. However, the definition still involves a reference to a single, fixed, classical nuclear configuration K_0. The elements of this group $\Pi_1(F(A), K_0)$, the homotopy classes of closed paths, can be thought of as closed (loop or circular) reaction mechanisms. These mechanisms are constrained by the energy bound A of the level set F(A), and are also subject to the geometrical, classical mechanical restriction that they all involve a fixed nuclear configuration K_0.

On the highest level of the model, we find the abstract fundamental group $\Pi_1(F(A))$ of the level set F(A), that arises as a consequence of the invariance of any given realization $\Pi_1(F(A), K_0)$ of the fundamental group to the choice of point K_0 within any arcwise connected level set F(A). The fundamental group $\Pi_1(F(A))$ represents the algebraic structure of all closed reaction mechanism below energy A, with no classical mechanical reference to formal, fixed nuclear configurations.

Although the fundamental group $\Pi_1(F(A))$ involves directly only closed (loop) reaction mechanisms, this is not a serious limitation. In fact, any reaction mechanism within level set F(A) can be extended into a closed reaction mechanism; that is, any reaction mechanism that requires activation energy less than A may be viewed as a segment of some loop mechanism of level set F(A). In this sense, the family of all closed mechanisms represents all reaction mechanisms within F(A). Closed reaction mechanisms are of special importance, and the fundamental group $\Pi_1(F(A))$ represents the most essential interrelations among all reaction mechanism within level set F(A).

This model of reactions does not violate the Heisenberg uncertainty principle. Reaction mechanisms, defined as these homotopy equivalence classes, are fully quantum chemical within the context of any potential energy surface model.

It is possible to describe a fundamental group $\Pi_1(F(A))$ in terms of a set of generators. For any fixed energy bound A, the fundamental group $\Pi_1(F(A))$ of reaction mechanisms on level set F(A) is a *finitely generated free group* [27] with a finite family of *generator reaction mechanisms*

$$[P_1], [P_2], \ldots, [P_m] \tag{56}$$

as free generators.

This finite family of generator reaction mechanisms represents the chemically most important fundamental mechanisms within level set F(A). Any fundamental reaction mechanism $[P] \in \Pi_1$ can be expressed as some product of the generator mechanisms:

$$[P] = [P_{i_1}]^{\alpha_{i1}}[P_{i_2}]^{\alpha_{i2}} \cdots [P_{i_k}]^{\alpha_{ik}}, \tag{57}$$

where $\alpha_{ik} = \pm 1$ and repetitions are allowed.

A *reduced product* [27] is a product of form (57), where

$$[P_{ij}] = [P_{ij+1}] \tag{58}$$

implies that

$$\alpha_{ij} \neq - \alpha_{ij+1} , \tag{59}$$

that is, where no neighbors in sequence (57) are inverses of each other. If such consecutive inverse pairs occur in a product expression for a fundamental reaction mechanism, then this pair can always be replaced with the trivial mechanism [1], and also omitted from the product, unless the mechanism itself is $[P] = [1]$. Although by allowing various repetitions the same fundamental mechanism $[P]$ can be expressed in many (infinitely many) different ways, this reaction mechanism $[P]$ can be expressed in one and only one way as a *reduced product*.

Using this unique form in terms of the generator reaction mechanisms, it is convenient to express each fundamental reaction mechanism within each level set $F(A)$ of the nuclear configuration space M as a reduced product. The *length of a fundamental reaction mechanism* $[P]$ with respect to a generator set $\{[P_i]\}_{i=1,...,m}$ of group $\Pi_1(F(A))$ is the number k in expression (57) if $[P]$ is given as a reduced product.

The fundamental groups of reaction mechanisms are not in general commutative. The fundamental group Π_1 of level set $F(A)$ is non-Abelian whenever $m > 1$, since the reaction mechanisms $[P_1][P_2]$ and $[P_2][P_1]$ expressed in terms of two generator mechanisms $[P_1]$ and $[P_2]$ are two different reduced products in a free group:

$$[P_1] [P_2] \neq [P_2] [P_1]. \tag{60}$$

Consequently, the products $[P_1][P_2]$ and $[P_2][P_1]$ represent two different reaction mechanisms, hence the group is not commutative. However, every nontrivial fundamental group Π_1 always contains a special Abelian subgroup, the infinite cyclic group. Of course, in the case of $m=0$ the group is also commutative, since then

$$\Pi_1 = \{[1]\}, \tag{61}$$

the trivial group.

5. The energy dependence of the fundamental groups of reaction mechanisms

One important question is the following: How do fundamental groups $\Pi_1(F(A))$ of reaction mechanisms depend on the choice and on the changes of energy bound A? The level set $F(A)$ depends on the energy bound A. Consequently, the fundamental group of reaction mechanisms $\Pi_1(F(A))$ is also dependent on the energy bound A, however, $\Pi_1(F(A))$ is a discrete algebraic structure and the dependence is not continuous. Some energy changes do not effect the fundamental group, whereas some others do, leading to an energy dependent description of reaction mechanisms.

The fundamental group $\Pi_1(F(A))$ of reaction mechanisms is a topological invariant of the level set $F(A)$. Consider a change of the energy bound A to a new value A', leading to a new level set $F(A')$. If the new level set $F(A')$ is topologically equivalent to the old one,

$$F(A') \sim F(A), \tag{62}$$

then the two fundamental groups $\Pi_1(F_1(A))$ and $\Pi_1(F_2(B))$ of reaction mechanisms are isomorphic, and as abstract groups, are equal:

$$\Pi_1(F(A')) = \Pi_1(F(A)). \tag{63}$$

If, however, the change $A \rightarrow A'$ in the energy bound leads to a topologically significant change in the level set $F(A)$, then there is an accompanying change in the fundamental group $\Pi_1(F(A))$, as well, and eq. (63) no longer holds.

In order to characterize these changes, we need the concept of *critical levels:* a critical level is the energy of a critical point $K(\lambda, i)$ of the energy hypersurface $E(K)$. It is important to realize that essential, topological changes in level sets occur only at critical levels (where the usual topology of the metric nuclear configurtion space is taken [27]). Also note that the topology of a maximum connected component of a level set need not change at each critical level; for example, the topology of a lake does not have to change if a flooding spills over into a dry lakebed, as long as no new islands are created. Consequently, a critical level is only a necessary but not a sufficient condition for an energy induced change in the topology of the level set $F(A)$, and in the fundamental group $\Pi_1(F(A))$.

Consider two level sets $F_1(A)$ and $F_2(B)$ with reference to the same energy hypersurface $E(K)$, where the energy thresholds A and B are different,

$$A < B. \tag{64}$$

For simplicity, we assume that both level sets are simply or multiply connected. If during an $A \rightarrow B$ change of the energy threshold no critical level of the potential energy surface $E(K)$ is encountered, then the connectedness properties of the level sets must remain invariant [27]. This implies that there exists a homeomorphism between the two level sets. Consequently, the two fundamental groups $\Pi_1(F_1(A))$ and $\Pi_1(F_2(B))$ of reaction mechanisms in the two level sets $F_1(A)$ and $F_2(B)$, respectively, are isomorphic .

A necessary condition for a change in connectedness of level sets during an $A \rightarrow B$ change of the energy threshold is the existence of a critical level within the $[A, B)$ closed-open energy interval. The occurrence of a critical level, however, is not a sufficient condition for a change in connectedness, as our flooding lake example has indicated. Note that the definition of level set $F(A)$ as the set of nuclear configurations of energy *below* the threshold A implies the following: if a change in connectedness of a maximum connected component of a level set occurs, then the fundamental group does not remain invariant, that is, as concrete groups, $\Pi_1(F_1(A))$ and $\Pi_1(F_2(B))$ are no longer isomorphic. We still may have a homeomorphism between one level set and a subset of the other level set; in this case a homomorphism exists between the two fundamental groups of reaction mechanisms, $\Pi_1(F_1(A))$ and $\Pi_1(F_2(B))$.

The main results concerning the effects of a $A \rightarrow B$ change of the energy threshold can be summarized as follows:

(i)   A sufficient condition for topological invariance of level sets is the lack of critical level in the $[A, B)$ closed-open energy interval.

(ii)   If there is no change in connectedness during an energy change $A \rightarrow B$ then the fundamental groups of reaction mechanisms are isomorphic within $[A, B)$.

(iii)   If the net connectedness of level sets changes then at most a homomorphism can be guaranteed between the two groups.

(iv)   If in the interval $[A, B)$ the connectedness of level sets changes monotonically, then the fundamental group of reaction mechanisms in the level set of lower connectedness is a subgroup of the other group.

(v)   If A and B are two, subsequent critical levels, then the fundamental group $\Pi_1(F(A))$ of reaction mechanisms is invariant in the $(A,B]$ open-closed interval.

A crucial aspect of the family of feasible reaction mechanisms is the level of connectedness of the potential energy hypersurface domain at the given energy bound. Considering arcwise connectedness, an ordinary potato is simply connected, and a donut is multiply connected. If the analysis extends to all possible energy threshold values A within the potential energy hypersurface domain considered and to all the induced changes in the connectedness of the corresponding level sets $F(A)$ of the potential energy surface, then the family of the fundamental groups of reaction mechanisms obtained has an algebraic structure of its own. If each critical energy level C is nondegenerate within each connected level set component $F(C)$, then the fundamental groups form a *lower semilattice*. In the following paragraphs we shall review some of the concepts leading to the lower semilattice structure of fundamental groups.

A simple, two-dimensional example is useful to illustrate some of the relevant concepts. In a two-dimensional potential energy surface domain, a critical level A' of a saddle point is required for any new generator mechanism to enter the group Π_1. When the energy threshold reaches the level of a maximum, then one generator mechanism is eliminated. If during an increase of the energy threshold any given generator mechanism is eliminated from the group, then this generator mechanism never returns at higher energies.

Within any open-closed energy interval $(A,B]$, where A and B are two, subsequent critical levels, each group $\Pi_1(A,B]$ is invariant to changes of the level set value A', but these groups change at the limits of their respective energy intervals. The symbol $\Pi_1(A,B]$ stands for the fundamental group obtained in a level set $F(A')$ of energy bound A' falling within the open-closed energy interval $(A,B]$, $A' \in (A,B]$.

Group $\Pi_1(E_{max},\infty)$ is the trivial group $\{[1]\}$, where E_{max} is the maximum energy value within the potential energy hypersurface domain considered. This group $\Pi_1(E_{max},\infty)$ is exceptional, since the corresponding energy interval (E_{max},∞) is not bound. Each nontrivial group can be specified in terms of its family of generator mechanisms $[P_i]$.

A level set is the empty set, $F(A') = \varnothing$, for any threshold value A' that is lower than a lower bound on energy, for example, if $A' \le E_{min}$, where E_{min} is the minumum energy value within the potential energy hypersurface domain considered. For any such $A' \le E_{min}$ threshold value, no fundamental group of reaction mechanisms exists.

If the connectedness of a level set $F(A)$ changes *monotonically* during a change $A \to B$ of the energy threshold (where A and B are arbitrary energy thresholds), then the fundamental group Π_1 of the level set of lower connectedness is a subgroup of the group of the level set of higher connectedness. For example, if the connectedness increases monotonically during the change $A \to B$, implying that the degree of connectedness of $F(B)$ is greater than that of $F(A)$, then the abstract group $\Pi_1(F(A))$ is a subgroup of the abstract group $\Pi_1(F(B))$:

$$\Pi_1(F(A)) \subset \Pi_1(F(B)) \tag{65}$$

The trivial group of reaction mechanisms $\{[1]\}$ is a subgroup of any other group.

In each family of fundamental groups the subgroup relation is an ordering principle, that turns the family of groups into a *partially ordered set*. Here we consider the actual, concrete groups and their actual, concrete subgroups as specified by their generator mechanisms, which are, in turn, regarded as invariant between subsequent critical levels. In particular, a group is not necessarily regarded as a subgroup of another if it is merely isomorphic to a subgroup of the other group.

The binary lattice operations "meet" and "join" can be defined for pairs of fundamental groups [27] as having a largest common subgroup or a smallest common supergroup, respectively. (The smallest common supergroup of two groups is the smallest group containing both groups as subgroups.) There are, however, some restrictions. The lattice operation "meet" can be performed for any two elements of the family of fundamental groups, since the trivial group $\{[1]\}$ is a subgroup of any two groups. However, the "join" does not always exist for every pair of groups, since it is possible that there is no fundamental group of reaction mechanisms that contains both of these groups as subgroups. Consequently, the energy dependent fundamental groups of reaction mechanisms, associated with the family of level sets within a given domain of a potential energy hypersurface form a *lower semilattice,* that is a lattice only in special cases, where the join happens to exist for all pairs of groups of reaction mechanisms in the family.

In the extreme case, one may consider a complete energy hypersurface. Each potential energy hypersurface defined over a nuclear configuration space M is associated with a family of fundamental groups which form a lower semilattice.

The rank ℓ of each fundamental group can be associated with the formal level within the lower semilattice; the higher the rank, the higher the level, since the group-subgroup relations are related to an increase of the number of free generators. The effects of changes in the energy threshold A can be monitored within such a semilattice and, in particular, the energy threshold changes are related to changes in the semilattice level ℓ .

The higher the level ℓ , the larger the group of reaction mechanisms (with more free generators), which implies a richer family of reaction mechanisms. Level ℓ is a simple measure of the complexity of the family of reaction mechanisms. At low energies, for example below and at A_{min}, no fundamental group exists; there are no chemical species, hence no reactions occur. Immediately above the value A_{min} that necessarily belongs to an energy minimum within the potential energy hypersurface domain considered, the fundamental group is the trivial group $\{[1]\}$. This group represents a single energy basin, that is, a single chemical species, with no actual reactions at this energy threshold. When the energy threshold is further increased, initially one may expect to obtain more free generators, that corresponds to moving to higher levels ℓ in the semilattice.

The increase in the level ℓ with an increasing energy threshold is not necessarily monotonic even at low energies, nevertheless, the general trend is an initial increase. Eventually, however, this trend is reversed; beyond some energy threshold the dominant trend is a decrease in the level ℓ of complexity of the family of reaction mechanisms as the energy threshold is further increased.

Finally, above an energy threshold A_{max}, the fundamental group of reaction mechanisms becomes again the trivial group $\{[1]\}$. At such high energy thresholds all reaction mechanisms become topologically equivalent, hence essentially indistinguishable within the given subset of the potential energy surface. At such high energies, this subset of the surface behaves as a single, simple basin, where no detailed features are topologically significant.

6. Interrelations of point symmetry groups of nuclear arrangements, the catchment regions of chemical species, and the fundamental groups of reaction mechanisms

There exist several links between two important topological models: the model of catchment regions as topological open sets of a nuclear configuration space, and the model of quantum chemical reaction mechanisms, based on homotopy equivalence classes of formal reaction paths. The connections between these two models are of some importance in computer based quantum chemical synthesis planning and molecular design.

One of these links is in fact geometrical: the point symmetry of nuclear configurations. If the energy threshold A of a level set F(A) is changed continuously, then the fundamental group of reaction mechanisms can change only at critical levels. On the other hand, within any local neighborhood of a critical point, the nuclear configuration of the critical point has the highest point symmetry. The change of a fundamental group is the consequence ot the inclusion of a new generator mechanism, that is, a new homotopy class of formal reaction paths. At an energy threshold only infinitesimally above the critical point, all these formal reaction paths must encounter nuclear configurations that are infinitesimally close to having the symmetry of the critical point nuclear configuration.

This statement can be made precise using the concepts of symmetry deficiency measures [34]. The derivation of this and additional rules is facilitated if we adopt a different family of level sets. As mentioned in section 3, a *closed level set* F(A) of the nuclear configuration space M is defined by:

$$F'(A) = \{ K : E(K) \le A \} . \tag{66}$$

In addition to the points of level sets F(A), these *closed level sets* F'(A) include nuclear configurations K where the "elevation" of the potential energy hypersurface E(K) equals the parameter value A.

The following general rule, the *symmetry rule of the fundamental group of reaction mechanisms* holds:

> *At an energy threshold A where the fundamental group changes, all paths in all the newly accessible homotopy classes encounter a local maximum point of the path where the nuclear configuration has a point symmetry that is locally maximal along the reaction coordinate.*

Note that if the rule is applied to the special case of a minimum energy path, then the resulting symmetry condition is a well-known property of minimum energy paths, however, the rule is more general, as it applies to all paths of the new homotopy classes of paths.

Additional rules can be deduced if one considers the energy values at critical points $K(\lambda,i)$ corresponding to the maximum energy in closed level sets F'(A), where

$$A = E(K(\lambda,i)) \tag{67}$$

Consider all closed critical level sets of a potential energy hypersurface E(K), that is, all closed level sets F'(A) such that the potential energy hypersurface E(K) has a critical point $K(\lambda,i)$ of energy value A. These closed level sets F'(A) generate a topology, called the critical level topology of potential energy hypersurfaces [15].

Here we shall be concerned with a subfamily of special critical levels: those where the point symmetry $g(K(\lambda,i))$ of the nuclear configuration of the critical point $K(\lambda,i)$ is unique within a small open neighborhood $N(K(\lambda,i))$ of the critical point within its catchment region $C(\lambda,i)$:

$$g(K(\lambda,i)) \neq g(K), \tag{68}$$

if

$$K \in N(K(\lambda,i)), \qquad K \neq K(\lambda,i), \tag{69}$$

where

$$N(K(\lambda,i)) \subset C(\lambda,i), \quad N(K(\lambda,i)) \text{ open in } C(\lambda,i). \tag{70}$$

This condition already implies that *the point symmetry* $g(K(\lambda,i))$ *at the critical point* $K(\lambda,i)$ *is unique within the entire catchment region* $C(\lambda,i)$ of the critical point, that is,

$$g(K(\lambda,i)) \neq g(K), \tag{71}$$

if

$$K \in C(\lambda,i)), \qquad K \neq K(\lambda,i). \tag{72}$$

The proof of this statement follows from the fact that if the point symmetry $g(K)$ of a nuclear configuration $K \in C(\lambda,i))$, different from the critical point, $K \neq K(\lambda,i)$, is the same as the point symmetry $g(K(\lambda,i))$ at the critical point $K(\lambda,i)$ of the same catchment region $C(\lambda,i)$, then this symmetry $g(K(\lambda,i))$ must be present along the entire steepest descent path from the point K to the critical point $g(K(\lambda,i))$. Since the steepest descent path must enter the open neighborhood $N(K(\lambda,i))$ in order to reach the critical point $K(\lambda,i)$, the open neighborhood $N(K(\lambda,i))$ must also contain a continuum of points with the same symmetry $g(K(\lambda,i))$. Consequently, if $N(K(\lambda,i))$ does not have points $K \neq K(\lambda,i)$ with symmetry $g(K(\lambda,i))$, then the catchment region $C(\lambda,i)$ cannot have such points either, and the point symmetry $g(K(\lambda,i))$ is unique within the entire catchment region $C(\lambda,i)$.

As a consequence of the above result, a formal reaction path p can pass through the catchment region $C(\lambda,i)$ without ever encountering a nuclear configuration of point symmetry $g(K(\lambda,i))$. For the segment

$$p_{C(\lambda,i)} = p \cap C(\lambda,i) \tag{73}$$

of such paths, the general condition on point symmetries,

$$g(K) \subset g(K(\lambda,i)) \qquad \text{for } K \in p_{C(\lambda,i)} \tag{74}$$

can be sharpened,

$$g(K) \subset g(K(\lambda,i)), \quad g(K) \neq g(K(\lambda,i)) , \quad \text{for} \quad K \in p_{C(\lambda,i)}. \tag{75}$$

Of course, if the critical point of the catchment region is an element of the level set, then within the homotopy classes of paths passing through the catchment region the maximal local symmetry is $g(K(\lambda,i))$. Nevertheless, within the family of homotopic paths, the paths with this maximal local symmetry form a subset of measure zero.

7. Summary

The distribution of point symmetry groups of nuclear arrangements of molecules has a strong influence on the distribution of critical points along potential energy hypersurfaces. These interrelations of geometrical symmetry conditions and the essentially topological critical point relations can be used to study the role of symmetry in the generalization of reaction paths into formal reaction mechanisms. After a review of the topological model of the fundamental groupoid and the fundamental group of reaction mechanisms, the energy dependence of the fundamental groups, their relations to the point symmetry groups of nuclear arrangements, and and to the catchment region distribution of formal chemical species are discussed.

References

[1]   K. Fukui, J. Phys. Chem., **74**, 4161 (1970).
[2]   A. Tachibana and K. Fukui, Theor. Chim. Acta, **49**, 321 (1978).
[3]   A. Tachibana and K. Fukui, Theor. Chim. Acta, **51**, 189 (1979).
[4]   A. Tachibana and K. Fukui, Theor. Chim. Acta, **51**, 275 (1979).
[5]   P. Pulay, Mol. Phys., **17**, 197 (1969).
[6]   P. Pulay, Mol. Phys., **18**, 473 (1970).
[7]   P. Pulay, Direct use of gradients for investigating Molecular Energy Surfaces, in Applications of Electronic Structure Theory, H. F. Schaefer, Ed. Plenum, New York, 1977.
[8]   P.G. Mezey, Progr. Theor. Org. Chem., **2**, 127 (1977).
[9]   M.V. Basilevsky, Chem. Phys., **24**, 81 (1977).
[10]   B.C. Garrett, D.G. Truhlar, and R.S. Grev, in Potential Energy Surfaces and Dynamic Calculations, D.G. Truhlar, Ed., Plenum, New York, 1981.
[11]   W. Quapp and D. Heidrich, Theor. Chim. Acta, **66**, 245 (1984).
[12]   P.G. Mezey, Theor. Chim. Acta, **54**, 95 (1980).
[13]   P.G. Mezey, Theor. Chim. Acta, **58**, 309 (1981).
[14]   P.G. Mezey, Chem. Phys. Letters, **82**, 100 (1981), **86**, 562 (1982).
[15]   P.G. Mezey, Theor. Chim. Acta, **60**, 97 (1981).
[16]   P.G. Mezey, Internat. J. Quantum Chem., Quant. Biol. Symp., **8**, 185-196 (1981).
[17]   P.G. Mezey, Theor. Chim. Acta, **62**, 133 (1982).
[18]   P.G. Mezey, Theor. Chim. Acta, **63**, 9 (1983).
[19]   P.G. Mezey, J. Chem. Phys., **78**, 6182 (1983).

[20] P.G. Mezey, Can. J. Chem., **61**, 956 (1983). (Volume dedicated to Prof. H. Gunning).

[21] P.G. Mezey, The Topological Model of Non-Rigid Molecules and Reaction Mechanisms, in "Symmetries and Properties of Non-Rigid Molecules: A Comprehensive Survey",Eds. J. Maruani and J. Serre, Elsevier Sci. Publ. Co., Amsterdam, 1983, pp. 335-353.

[22] P.G. Mezey, Int.J. Quant. Chem., **26**, 983 (1984).

[23] P.G. Mezey, Int.J. Quant. Chem. Symp., **18**, 77 (1984).

[24] P.G. Mezey,Theor. Chim. Acta, **67**, 43 (1985).

[25] P.G. Mezey, Theor. Chim. Acta, **67**, 91 (1985).

[26] P.G. Mezey, Theor. Chim. Acta, **67**, 115 (1985).

[27] P.G. Mezey, Potential Energy Hypersurfaces, Elsevier, Amsterdam, 1987.

[28] C.A. Cayley, Philos. Mag., 18, 264 (1859).

[29] J.C. Maxwell, Philos. Mag., 40, 233 (1870).

[30] M. Morse, Calculus of Variation in the Large, Am. Math. Soc. Colloq. Publ., Vol. **18**, 1934.

[31] M. Morse and S.S. Cairns, Critical Point Theory in Global Analysis and Differential Topology, Academic Press, New York, 1969.

[32] J. Milnor, Morse Theory, Annals of Math. Studies, Vol. **51**, Princeton Univ. Press, Princeton, 1973.

[33] P.G. Mezey, J. Amer. Chem. Soc., **112**, 3791 (1990).

[34] P.G. Mezey, Shape in Chemistry: An Introduction to Molecular Shape and Topology, VCH Publishers, New York, 1993.

LOOSE DEFINITIONS OF REACTION PATHS

XAVIER CHAPUISAT

Laboratoire de Chimie Théorique
CNRS, URA 506
Bâtiment 490, Faculté des Sciences
Université de Paris-Sud
91405 - ORSAY Cédex, France

1. Introduction

A long time ago in the history of chemistry, the concept of "reaction energy profile" has been associated with the graph of a one-variable continuous function, sketchily representing the amount of free energy or enthalpy to be brought to (or delivered by) the reacting system, in order to allow it to pass from the reagents' to the products' conformations, continuously through a variety of nuclear arrangements. In this sketchy representation, the abcissa, the so-called "reaction coordinate", is normally not carefully defined from the mathematical viewpoint, just qualitatively intended to feature a continuous configurational sequence (for a review on energy profiles, see Ref.[1]). It should be emphasized that, since they explicitly refer to the notion of molecular conformation identified with the geometrical arrangement of the nuclei (whose motion is supposed to be slow, the electrons permanently and almost instantly adjusting themselves to it), the concepts of reaction profile, reaction coordinate and (see below) reaction path are, from the phenomenological point of view, typically chemical. Indeed, the physical view of a molecule is rather that of an aggregate of nuclei and electrons, to be described, according to standard quantum mechanics, in terms of states : then the shape and size of the molecule would rather be associated to mean values of quantities represented by operators and normally not to the fixed arrangement of the only nuclei, just because of the uncertainty principles. Nevertheless, under the tutelar protection of the Born-

39

D. Heidrich (ed.), The Reaction Path in Chemistry:
Current Approaches and Perspectives, 39–75.
© 1995 Kluwer Academic Publishers. Printed in the Netherlands.

Oppenheimer approximation, chemists like to think of a molecule as an object endowed with a shape and a size that are, in addition, continuously variable.

The concepts of reaction profile, reaction coordinate and so on, turned out to be fairly successful : so, later on in the history of chemistry, there were many attempts to make them more quantitative. Within the framework of theoretical chemistry, the decisive additional concept is that of Potential Energy Surface (PES), a particularly practical mean of representing the values of the molecular electronic energy V for an infinity of nuclear configurations, as the graph of the function $V = V(q^1, q^2, \ldots q^n)$, where $\underline{q} = \{q^1, q^2, \ldots q^n\}$ is a set of generalized internal coordinates allowing one to uniquely describe all molecular (i.e. nuclear) arrangements of interest (for a recent review, see Ref.[2]). There are otherwise three coordinates (generally the Cartesian components of the total center of mass vector position $\underline{r}$) accounting for the overall translation and three coordinates (generally the Eulerian angles φ, θ and χ, collectively noted $\underline{\Theta}$) accounting for the overall rotation of the molecular system. For being well suited as internal coordinates, the q^i must be translation and rotation-invariant, which implies that the analytical expressions $\{q^i = {}^\circ q^i(\underline{x}), i=1,\ldots n\}$ must be independent from the frame in which the atomic Cartesian coordinates $\underline{x} = \{x_0^1, y_0^1, z_0^1, x_0^2, \ldots z_0^N\}$ are measured. It should be emphasized that the catalogue of the internal coordinates is, as such, very limited, actually consisting of (there are quite few exceptions, e.g. hyperspherical coordinates) :

(i) distances :
$$r_{12} = [(X_2 - X_1)^2 + (Y_2 - Y_1)^2 + (Z_2 - Z_1)^2]^{1/2}, \tag{1}$$

(ii) planar angles :

$$\theta_{213} = \cos^{-1} \left\{ \frac{(X_2 - X_1)(X_3 - X_1) + (Y_2 - Y_1)(Y_3 - Y_1) + (Z_2 - Z_1)(Z_3 - Z_1)}{r_{12} \, r_{13}} \right\} \tag{2}$$

and (iii) dihedral angles :

$$\phi_{3124} = \cos^{-1} \left\{ \cot\theta_{213} \cot\theta_{124} + \frac{(X_3 - X_1)(X_4 - X_2) + (Y_3 - Y_1)(Y_4 - Y_2) + (Z_3 - Z_1)(Z_4 - Z_2)}{r_{13} \, r_{24} \, \sin\theta_{213} \, \sin\theta_{124}} \right\}$$
$$\tag{3}$$

(ϕ_{3124} is the angle between vector $\vec{12}\times\vec{13}$ of norm $r_{12}\,r_{13}\sin\theta_{213}$ and vector $\vec{12}\times\vec{24}$ of

norm $r_{12}\,r_{24}\sin\theta_{124}$, see Figure 1). $X_k = \dfrac{\sum_j m_j x_0^j}{\sum_j m_j}$ (k=1, ... 4) is the x-th Cartesian

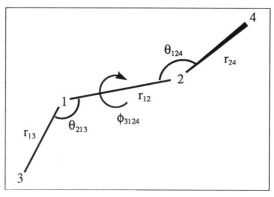

Figure 1

component of the center of mass of an atomic group (the k-th one, possibly reduced to one atom) made up with atoms of current indices j and Cartesian coordinates $\{x_0^j, y_0^j, z_0^j\}$.

For an N-atom system, there is $n \le 3N-6$: if $n = 3N-6$, the molecular system is formally free, i.e. subjected to no inner constraints, whereas if $n < 3N-6$ the system is considered (from the theoretical viewpoint) as being subjected to (3N-6-n) model constraints. Model constraints are very often resorted to for describing molecules, in particular for large N : indeed, whatever its power, no human intelligenge is able to conceive precise things in spaces of dimensionality as large as 3N-6 for large N. Therefore chemists, whatever the size of the molecule in which the reaction studied actually takes place, describe the configurational changes in the molecule by means of a reduced number (n) of geometrical parameters. We shall see below, in sections 2.3 and 3.2 , that the very concept of model constraint still needs to be clarified. In order to provide a simplified (but satisfactory from the phenomenological point of view) description of an event as complicated as a chemical reaction is, the last (but not least !) concept is that of Reaction Path (RP) on the PES

2. The concept of Reaction Path. Mathematical framework.

2.1.   MINIMUM ENERGY PATH

In 1984, Quapp and Heidrich [3] presented a thorough mathematical discussion of the concept of minimum energy path on the PES of a chemically reacting system (other important contributions are by Schlegel and Gonzales [4-6]; for good surveys, see also Refs.[7,8]). After pointing out some misleading statements about this question, Quapp and Heidrich (following Fukui [9a] and McIver and Komornicki [9b]) proposed a *suitable definition of a reaction path as a continuous line connecting, in any configuration space, reagents and products through all intermediate stationary points (transition-state saddle points and reaction-intermediate minima).* In analyzing the requirement that a path of steepest descent, regarded as a RP, must be coordinate invariant, Quapp and Heidrich in fact adopt the point of view of Pechukas [10], according to whom the RP is nothing but a curve (that can fit in with a variety of mathematical definitions), a branch of which goes from a higher to a lower potential energy region, no dynamical relevance being given to such a path of steep but not necessarily steepest descent. Moreover, they manage to get rid of the statement of Tachibana and Fukui [11-14], that the steepest descent path started from a saddle point and calculated as a gradient vector field line in the mass-weighted Atomic Cartesian Coordinate Configuration Space (ACCCS), the so-called Intrinsic Reaction Path (IRP), should be preferred to any other path (other worthwhile contributions about the IRP can be found in Refs.[4,5,7,9,15-21]). The IRP reaches any minimum along the local lowest-frequency normal mode [7,9,12,15].

2.2.   STEEPEST DESCENT PATHS FOR UNCONSTRAINED SYSTEMS [3,18,22]
        (for ascent paths, see Refs.[23-26])

2.2.1. *Genuine or mass-weighted reaction path ? SF or BF frame ?*
        Atomic Cartesian or curvilinear internal coordinates ? [3-6,8,12,27-31]

For an N-atom molecular system, a 3N-dimensional ACCCS being given [from now on, we shall note the atomic Cartesian coordinates, collectively $\underline{x} = \{x^1, \ldots x^{3N}\}$, so that $x_0^j = x^{3j-2}$, $y_0^j = x^{3j-1}$, $z_0^j = x^{3j}$ $(j=1,\ldots N)$], the branch of a path of steepest descent for the potential energy function $u(\underline{x})$ is defined by the gradient system of ordinary differential equations :

$$\frac{dx^k}{dt} = -\frac{\partial u[\underline{x}(t)]}{\partial x^k} \qquad (k=1, \dots 3N) \qquad (4)$$

where the parameter t has a dimension that is not a length (see section 2.2.2 for a comprehensive view and more details on the notation). Quapp and Heidrich [3] have demonstrated that, after introducing (as it is usual for unconstrained systems) 3N-6 internal curvilinear coordinates adequately describing all molecular deformations of interest :

$$\underline{q} = {}^\circ\underline{q}(\underline{x}) = \{{}^\circ q^i(x^1, \dots x^{3N}); i=1, \dots 3N-6\} \qquad (5)$$

Eq.(4) is transformed into (insofar as the internal deformations are only concerned) :

$$\frac{dq^i}{dt} = -\sum_{j=1}^{3N-6} {}^\circ g^{ij}[\underline{x}(\underline{q})] \frac{\partial V(\underline{q})}{\partial q^j} \qquad (i=1, \dots 3N-6) \qquad (6)$$

where t is the (scaled) curvilinear displacement coordinate along the RP considered a line in the ACCCS $(\underline{x})$, $V(\underline{q})$ is the expression of the potential in internal coordinates :

$$V[{}^\circ\underline{q}(\underline{x})] = u(\underline{x}) \qquad (7)$$

and :
$$^\circ g^{ij}(\underline{x}) = \sum_{k=1}^{3N} \frac{\partial {}^\circ q^i}{\partial x^k} \frac{\partial {}^\circ q^j}{\partial x^k} \qquad (i,j=1, \dots 3N-6) \qquad (8)$$

is the current contravariant component of the metric tensor of transformation (5).

Two points are worth mentioning :

(a) if $\underline{x}$ denotes the genuine ACCCS, Eqs.(4) and (6) are the defining systems of ordinary differential equations for the invariant path of steepest descent on the genuine potential energy surface of $V(\underline{q})=u(\underline{x})$. In particular, the RP so defined is mass-independent (see also Eqs.(5) and (8));

(b) if, on the other hand, $\underline{X}$ denotes the mass-weighted ACCCS :

$$\underline{X} = \{X^{a+3(k-1)} = m_k^{1/2} x^{a+3(k-1)}; k=1,\dots N; a=1,2,3\} \qquad (9)$$

then the defining system of ordinary differential equations for the descending gradient vector field line starting from a saddle point (the so-called Intrinsic Reaction Path, IRP) is written as :

$$\frac{dX^{a+3(k-1)}}{dT} = - \frac{\partial U(\underline{X})}{\partial X^{a+3(k-1)}} \qquad (k=1,\dots N; \ a=1,2,3) \qquad (10)$$

where $V[^{\circ}\underline{Q}(\underline{X})] = U(\underline{X})$ and $\underline{q} = ^{\circ}\underline{Q}(\underline{X}) = \{^{\circ}Q^i(\underline{X}) = ^{\circ}q^i(m_1^{-1/2}X^1, \dots m_N^{-1/2}X^{3N}); i=1, \dots$
$3N-6\}$. In Eq.(10), T is the (scaled, see section 2.2.2.) curvilinear displacement coordinate along the IRP considered a line in the ACCCS $(\underline{X})$. For rescaled mass-weighted coordinates, see Ref.[30].

The sequence of molecular configurations along the IRP also define a line in the genuine ACCCS $(\underline{x})$, for which the defining system of ordinary differential equations is [3] :

$$m_k \frac{dx^{a+3(k-1)}}{dT} = - \frac{\partial u(\underline{x})}{\partial x^{a+3(k-1)}} \qquad (k=1,\dots N; \ a=1,2,3) \qquad (11)$$

since $\quad \dfrac{\partial}{\partial X^{a+3(k-1)}} = m_k^{-1/2} \dfrac{\partial}{\partial x^{a+3(k-1)}} \quad (k=1,\dots N; \ a=1,2,3)$ and $U(\underline{X}) = u(\underline{x})$. The latter path is clearly coordinate-independent, but mass-dependent and therefore different from the path obtained from Eqs.(4) and (6). After introducing the same set of 3N-6 curvilinear internal coordinates as previously, see Eq.(5), and as far as, once again, the internal coordinates only are concerned, Eqs.(10) or equivalently (11) are transformed into :

$$\frac{dq^i}{dT} = - \sum_{j=1}^{3N-6} {}^{\circ}G^{ij}[\underline{x}(\underline{q})] \frac{\partial V(\underline{q})}{\partial q^j} \qquad (i=1, \dots 3N-6) \qquad (12)$$

where : $\qquad {}^{\circ}G^{ij}(\underline{x}) = \sum_{k=1}^{N} m_k^{-1} \sum_{a=1,2,3} \frac{\partial {}^{\circ}q^i}{\partial x^{a+3(k-1)}} \frac{\partial {}^{\circ}q^j}{\partial x^{a+3(k-1)}} \qquad (i,j=1, \dots 3N-6) \qquad (13)$

The main advantage of Eqs.(11-13) is that the position of the centre of mass of the molecule is preserved along the IRP : constant $= \displaystyle\sum_{k=1}^{N} m_k \, x^{a+3(k-1)} \quad (a=1,2,3) \qquad (14)$

which is not the case of the reaction path defined by Eqs.(4) and (6). The latter actually preserve the three quantities $\displaystyle\sum_{k=1}^{N} x^{a+3(k-1)} \quad (a=1,2,3)$.

It should be noticed that (i) $°G^{ij}(\underline{x}) = °\mathfrak{g}^{ij}(\underline{X}) = \sum_{k=1}^{3N} \dfrac{\partial °Q^i}{\partial X^k}\dfrac{\partial °Q^j}{\partial X^k}$ (i,j=1, ... 3N-6)

and (ii) $°G^{ij}(\underline{x})$ is nothing but the current element of the $°\Sigma$-matrix, defined in Ref.[32], Eq.(47) as :

$$°\Sigma^{ij}(\underline{x}) = \sum_{k=1}^{N}\dfrac{1}{m_k}\left[\dfrac{\partial °q^i}{\partial x^k}\dfrac{\partial °q^j}{\partial x^k} + \dfrac{\partial °q^i}{\partial y^k}\dfrac{\partial °q^j}{\partial y^k} + \dfrac{\partial °q^i}{\partial z^k}\dfrac{\partial °q^j}{\partial z^k}\right] \quad (i,j=1, ... 3N-6) \quad (15)$$

The elements of this matrix, transcribed in terms of $\underline{q}$ only [i.e. in the form $°\underline{\underline{\Sigma}}[\underline{x}(\underline{q})]= \underline{\underline{\Sigma}}(\underline{q})$, which is unrestrictrdly possible because of the translation and rotation invariance of the expressions $°\underline{q}(\underline{x})$], are to be used in the exact expression of the quantum kinetic energy operator, obtained in following the so-called $\underline{q}(\underline{x})$-approach, in the general form of Eq.(3), Ref.[32]. They play a role in the part of the Hamiltonian operator that accounts for pure internal deformations, which can be entirely depicted with the help of the 3N-6 internal coordinates. Indeed, the overall exact quantum-mechanical kinetic energy operator can be written as [32] :

$$\hat{\mathfrak{T}} = -\dfrac{\hbar^2}{2}\left[\Sigma^{kl}\dfrac{\partial}{\partial q^k} + \dfrac{\partial \Sigma^{kl}}{\partial q^k} - \dfrac{\Sigma^{kl}}{2}\tilde{g}_k\right]\dfrac{\partial}{\partial q^l} - \dfrac{i\hbar}{2}\left[2\Gamma^{ak}\dfrac{\partial}{\partial q^k} + \dfrac{\partial \Gamma^{ak}}{\partial q^k} - \dfrac{\Gamma^{ak}}{2}\tilde{g}_k\right]\hat{J}_\alpha + \dfrac{\mu^{\alpha\beta}}{2}\hat{J}_\alpha\hat{J}_\beta$$

$$(\alpha,\beta = x,y,z\ ;\ k,l = 1, ... 3N-6) \quad (16)$$

where $\tilde{g}_k(\underline{q}) = \tilde{g}^{-1}\dfrac{\partial \tilde{g}}{\partial q^k}$ is the logarithmic derivative of $\tilde{g}(\underline{q})$, the determinant of the 3N-3-

dimensional matrix $\begin{pmatrix} \underline{\underline{\Sigma}}(\underline{q}) & \underline{\underline{\Gamma}}^T(\underline{q}) \\ \underline{\underline{\Gamma}}(\underline{q}) & \underline{\underline{\mu}}(\underline{q}) \end{pmatrix}$. $\tilde{g}(\underline{q})$ also appears in the expression of the volume

element used for calculating the matrix elements, $dv = [\tilde{g}(\underline{q})]^{-1/2}dq^1...dq^{3N-6}\sin\theta d\theta\, d\varphi\, d\chi$. In Eq.(16), $\hat{J}_\alpha$ (α = x,y,z) denotes the α-th Body-Fixed (BF) component of the total angular momentum vector operator; in addition, $\underline{\underline{\Gamma}}(\underline{q})$ [3x(3N-6)] and $\underline{\underline{\mu}}(\underline{q})$ [3x3] are, respectively, the Coriolis matrix and the inverse effective inertia tensor, whose definitions and current components are given in Ref.[32], Eqs.(48-51) and (52,53). See also section 3.1. below.

2.2.2. *Overall view and connections between the various approaches.*

For a given potential, there are several possible definitions of the steepest descent path, depending on which ACCCS is referred to. Pointing out that there are no *a priori* kinetic considerations in the definition of the RP, Pechukas [10] initially suggested to define a branch of the RP as the steepest descent line from a saddle point (or possibly an infinitely removed dissociation area) towards a minimum of the potential (that can occasionally be also infinitely removed), in the genuine ACCCS ($\underline{x}$), according to the following set of differential equations :

$$\frac{d\underline{x}}{ds} = - \frac{\nabla_x u(\underline{x})}{\|\nabla_x u(\underline{x})\|} \quad \text{(normalized vector)} \qquad (17)$$

On the contrary, because of its dynamical relevance (possibility of fixing the centre of mass and proportionality of the total classical kinetic energy $\mho$ to the square of the time derivative of the displacement S, $2\mho = (dS/d\tau)^2$), Miller et al [33] on the one hand, and Nauts and Chapuisat [34] on the other hand, defined the reaction path as the path of steepest descent in the mass-weighted ACCCS ($\underline{X}$), the IRP, as :

$$\frac{d\underline{X}}{dS} = - \frac{\nabla_X U(\underline{X})}{\|\nabla_X U(\underline{X})\|} \quad \text{(normalized vector)} \qquad (18)$$

Previously, Fukui et al [11,27,35], also resorting to the definition of the IRP, suggested to use either of the two following sets of differential equations :

$$m_k \frac{dx^{a+3(k-1)}}{dT} = - \frac{\partial u(\underline{x})}{\partial x^k} \qquad (k=1,...N; \; a=1,2,3) \qquad (19')$$

$$\frac{dX^k}{dT} = - \frac{\partial U(\underline{X})}{\partial X^k} \qquad (k=1,3N) \qquad (19'')$$

depending on whether pure cartesian ($\underline{x}$) or mass-weighted atomic cartesian ($\underline{X}$) coordinates are used. However, they instead recommend integrating the differential equations in terms of the 3N-6 curvilinear internal coordinates only, in the form [11] :

$$\frac{dq^1}{G^{1j}(\underline{q})\partial_j V(\underline{q})} = \cdots = \frac{dq^{3N-6}}{G^{3N-6\,j}(\underline{q})\partial_j V(\underline{q})} \qquad (20)$$

where $G^{ij}(\underline{q}) = {}^\circ G^{ij}[\underline{x}(\underline{q})] = {}^\circ g^{ij}[\underline{X}(\underline{q})]$ is the current contravariant component of the appropriate metric tensor (see below, section 3.1.).

The connections between all these approaches are outlined in the following :

(a)   The Pechukas - Fukui *et al* - Quapp and Heidrich connection :

In the genuine ACCCS ($\underline{x}$) :   *(Pechukas)* $\dfrac{d\underline{x}}{ds} = -\dfrac{\nabla_x u(\underline{x})}{\|\nabla_x u(\underline{x})\|}$, cf. Eq.(17)

$$\Rightarrow \quad \frac{dx^k}{ds} = -\frac{\partial u(\underline{x})/\partial x^k}{\|\nabla_x u(\underline{x})\|} \quad (k=1,3N)$$

$\Rightarrow$ passage to the local chart ($\underline{q}$) : $\dfrac{dq^i}{ds} = -\dfrac{g^{ij}(\underline{q})\partial_j V}{\sqrt{g^{ij}(\underline{q})\partial_i V \partial_j V}}$ (i=1,3N-6)

($g^{ij}(\underline{q}) = {}^{\circ}g^{ij}[\underline{x}(\underline{q})]$, $\partial_i V$ stands for $\dfrac{\partial V(\underline{q})}{\partial q^i}$ and Einstein's implicit summation convention is

used)

$$\Rightarrow \quad -dt = \frac{-ds}{\sqrt{g^{ij}(\underline{q})\partial_i V \partial_j V}} = \frac{dq^1}{g^{1j}(\underline{q})\partial_j V} = \ldots = \frac{dq^{3N-6}}{g^{3N-6\,j}(\underline{q})\partial_j V} \quad (\textit{Fukui et al})$$

$$\Rightarrow \quad \frac{dq^i}{dt} = -{}^{\circ}g^{ij}[\underline{x}(\underline{q})]\,\partial_j V \quad (i=1,3N-6)$$

$$\Rightarrow \quad \text{come back to ACCCS } (\underline{x}) : \quad \frac{dx^k}{dt} = -\frac{\partial u(\underline{x})}{\partial x^k} \quad (k=1,3N)$$

$$\Rightarrow \qquad \boxed{\frac{d\underline{x}}{dt} = -\nabla_x u(\underline{x})} \quad \textit{(Quapp and Heidrich)}, \text{ cf. Eq.(4)}$$

(b)   The Miller *et al* - Fukui *et al* - Quapp and Heidrich connection :

In the mass-weighted ACCCS ($\underline{X}$) :   *(Miller et al)* $\dfrac{d\underline{X}}{dS} = -\dfrac{\nabla_X U(\underline{X})}{\|\nabla_X U(\underline{X})\|}$, cf. Eq.(18)

$$\Rightarrow \quad \frac{dX^k}{dS} = -\frac{\partial U(\underline{X})/\partial X^k}{\|\nabla_X U(\underline{X})\|} \quad (k=1,3N)$$

$\Rightarrow$ passage to the local chart ($\underline{q}$) : $\dfrac{dq^i}{dS} = -\dfrac{G^{ij}(\underline{q})\partial_j V}{\sqrt{G^{ij}(\underline{q})\partial_i V \partial_j V}}$ (i=1,3N-6)

$$\Rightarrow -dT = \frac{-dS}{\sqrt{G^{ij}(q)\partial_i V \partial_j V}} = \frac{dq^1}{G^{1j}(q)\partial_j V} = \dots = \frac{dq^{3N-6}}{G^{3N-6\,j}(q)\partial_j V} \quad \text{(Fukui et al)}$$

$$\Rightarrow \frac{dq^i}{dT} = -\,^\circ\mathfrak{g}^{ij}[\underline{X}(q)]\,\partial_j V \quad (i=1,3N-6) \tag{21}$$

$$\Rightarrow \text{ come back to ACCCS } (\underline{X}): \frac{dX^k}{dT} = -\frac{\partial U(\underline{X})}{\partial X^k} \quad (k=1,3N)$$

$$\Rightarrow \qquad \boxed{\frac{d\underline{X}}{dT} = -\,\underline{\nabla}_X U(\underline{X})} \quad \text{cf. Eq.(19'')}$$

Owing to the fact that the mass-weighted and pure Cartesian atomic coordinates are related by :

$$\underline{X} = \underline{\underline{M}}^{1/2}.\underline{x} \tag{22}$$

(this relation is frame-independent), where :

$$\underline{\underline{M}} = \begin{pmatrix} m_1 & & & & & & & \\ & m_1 & & & & & 0 & \\ & & m_1 & & & & & \\ & & & m_2 & & & & \\ & & & & \ddots & & & \\ & & & & & \ddots & & \\ & & & & & & m_{N-1} & \\ & 0 & & & & & m_N & \\ & & & & & & & m_N & \\ & & & & & & & & m_N \end{pmatrix} \tag{23}$$

it is possible to transcribe any scalar function of $\underline{x}$ in terms of $\underline{X}$, and conversely, e.g. :

$$q = {}^\circ q(\underline{x}) = {}^\circ q(\underline{\underline{M}}^{-1/2}.\underline{X}) = {}^\circ Q(\underline{X})$$

$$u(\underline{x}) = V[{}^\circ q(\underline{x})] = V[{}^\circ Q(\underline{X})] = U(\underline{X})$$

A similar situation exists for the metric tensors (see below) :

$${}^\circ G^{ij}(\underline{x}) = {}^\circ G^{ij}(\underline{\underline{M}}^{-1/2}.\underline{X}) = {}^\circ\mathfrak{g}^{ij}(\underline{X}) = \Sigma^{ij}(q)$$

In view of this last expression, Eq.(21) can be rewritten as :

$$\frac{dq^i}{dT} = - {}^{\circ}\Sigma^{ij}[\underline{x}(\underline{q})]\partial_j V \quad (i=1,3N-6) \qquad (24)$$

from where it is possible to come back from the local chart ($\underline{q}$) to the genuine ACCCS ($\underline{x}$) [instead of the mass-weighted ACCCS ($\underline{X}$), as previously] :

$$m_k \frac{dx^k}{dT} = - \frac{\partial u(\underline{x})}{\partial x^k} \quad (k=1,3N)$$

$$\Rightarrow \quad \boxed{\underline{\underline{M}} \frac{d\underline{x}}{dT} = - \underline{\nabla}_x u(\underline{x})} \quad \textit{(Quapp+Heidrich)}, \text{ cf. Eq.(19')}$$

It clearly appears [3] that the two definitions of the RP as a steepest descent path (mass-weighted IRP or genuine RP) give rise to two different curves (a double-lined and a single-lined border is respectively put round the corresponding differential equations above), when viewed in the same ACCCS [above, the genuine one ($\underline{x}$)].

As far as the metric tensors are concerned, by analogy with the definition given in Eq.(8) for the ACCCS ($\underline{x}$), ${}^{\circ}g^{ij}(\underline{x}) = \sum_{k=1}^{3N} \frac{\partial {}^{\circ}q^i}{\partial x^k} \frac{\partial {}^{\circ}q^j}{\partial x^k}$ (i,j=1, ... 3N-6), the definition

appropriate for the ACCCS ($\underline{X}$) is : ${}^{\circ}g^{ij}(\underline{X}) = \sum_{k=1}^{3N} \frac{\partial {}^{\circ}Q^i}{\partial X^k} \frac{\partial {}^{\circ}Q^j}{\partial X^k}$ (i,j=1, ... 3N-6)

and since $\frac{\partial {}^{\circ}Q^i}{\partial X^k} = \frac{\partial {}^{\circ}q^i}{\partial x^{k'}} \frac{\partial x^{k'}}{\partial X^k} = \frac{\partial {}^{\circ}q^i}{\partial x^{k'}} \frac{\delta^{kk'}}{\sqrt{m_k}}$, it comes out :

$${}^{\circ}g^{ij}(\underline{\underline{M}}^{1/2} \cdot \underline{x}) = \sum_{k=1}^{3N} \frac{1}{m_k} \frac{\partial {}^{\circ}q^i}{\partial x^k} \frac{\partial {}^{\circ}q^j}{\partial x^k} = {}^{\circ}G^{ij}(\underline{x}) \quad (i,j=1, ... 3N-6),$$

a result similar with those in Eqs.(13) and (15).

All this holds, whatever the frame of reference in which the atomic cartesian components are measured, e.g. Space-Fixed (SF) or BF. Moreover, s, S, t and T are just parameters. Mathematically , they are all on the same footing. Physical considerations only lead to distinguishing them from one another, since they do not have the same physical dimension : s is [L], S is $[M^{1/2}L]$, t is $[M^{-1}T^2]$ and T is $[T^2]$.

2.3.   LOOSE DEFINITIONS OF REACTION PATHS [10]

Being a line that goes, in an ACCCS, from the reagents' region to the products' region, through all stationary points in between (transition states and reaction intermediates), any steepest-descent path is definitely a minimum-energy path and, as such, can be considered as a RP. However, since (i) it is non-negligibly heavy to calculate a steepest-descent line by integration of the adequate set of coupled differential equations [4,5,18-21] and (ii) there are different lines (at least two, the genuine one and the mass-weighted one, IRP) that can be termed steepest-descent paths and further consideration only allows one to decide in favour of one or the other (e.g. dynamical considerations are in favour of the IRP), it is by no means shocking to think of alternative, simpler definitions for a RP. The only point which is a matter of universal agreement, is that *a RP must be a minimum-energy path joining up all stationary points of the PES*. But in between the stationary points, choosing one particular line is, to a large extent, of secondary importance, as long as is not specified the use of this line that will be made later on.

Moreover, even in view of dynamical calculations, the IRP has a rather severe drawback (a similar situation prevails for the genuine steepest-descent path) : according to the Reaction Path Hamiltonian philosophy, initially developed by Miller, Handy and Adams [33] and extended by others [15,18,31,36,37], the curvilinear displacement along the IRP (i.e. the parameters t or T above) should be taken as one of the internal coordinates describing the system[1]. This is undoubtedly a clever idea, since the value of that very coordinate would, throughout the dynamical calculations, be a measure of the advancement of the reactive process. The complications stem from the fact that : (i) it can be fairly complicated to define the complementary 3N-7 internal coordinates (except for few isomerizations in which they can be considered small-amplitude vibrational coordinates) and (ii) in many cases, multivaluation difficulties cannot be avoided, especially when large-amplitude motions drive the representative point of the system in

---

[1]The RPH approach is somewhat connected with the Transition State Theory (TST) of bimolecular reactions [48-53].

regions of the ACCCS ($\underline{X}$) far from the IRP, this last being locally intrinsically curved [37-39] : then it is no trivial matter to refer one distant point to one given point along the IRP, in such a way that the value of T (or t) for this point would uniquely be also attributable to the distant point. Several attempts have been made for preserving the physically clear and useful character of the Reaction Path Hamiltonian approach, in using reference lines that are close to but different from the IRP, so that the displacement coordinate along that line and the complementary coordinates do not suffer from the above mentioned drawbacks [15,40-47]. But this is an illustration of the IRP or any other steepest descent line being no sacred definition of a RP.

In this context, there is a real advantage in using geometrically defined coordinates, one of which (the reaction coordinate) can be viewed as a fair approximation to some steepest-descent curvilinear coordinate. It should be emphasized that, for polyatomic systems, there are always several sets of internal coordinates that uniquely describe the shape and size of the system. For instance, for the simple case of a triatomic system, once again excluding the hyperspherical parametrization, there are at least three sets of coordinates describing unambiguously the system :

(1) the Jacobi coordinates $\{R,r,\alpha\}$ (note that there are in fact three sets of Jacobi coordinates, depending on which atom is 1 in Figure 2a)

(2) the valence coordinates $\{R_1,R_3,\theta\}$, see Figure 2b (same remark as for the Jacobi coordinates)

(3) the lengths of the sides of the triangle $\{r_1,r_2,r_3\}$, see Figure 2c.

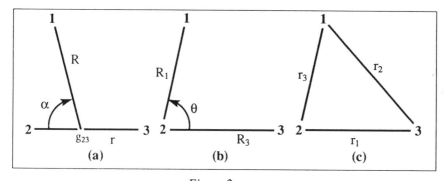

Figure 2

For tetratomic and larger systems, the situation is even more diversified. Selecting a set of internal coordinates among which one can be treated as an appropriate reaction coordinate is, by itself, a physical question. For example, if the dissociation of a triatomic molecule according to $123 \rightarrow 1 + 23$ is considered, the Jacobi coordinate set of Figure 2a is advisable if the **23** fragment is likely rotating strongly (and then the reaction coordinate is R), whereas if the fragment is known not to rotate much, the valence coordinate set of Figure 2b can be preferable (and the reaction coordinate then is R_1).

Otherwise, Jacobi coordinates can also be used to account for isomerization reactions, e.g. HCN $\rightleftarrows$ CNH and $HO_2 \rightleftarrows O_2H$, in which case the reaction coordinate is rather the bending angle α in Figure 2a. In addition, it is clear from Figure 3 that the connection between α and S, the curvilinear displacement along the IRP is nearly linear, for both systems, whatever the PES in the case of O_2H and everywhere throughout the reaction.

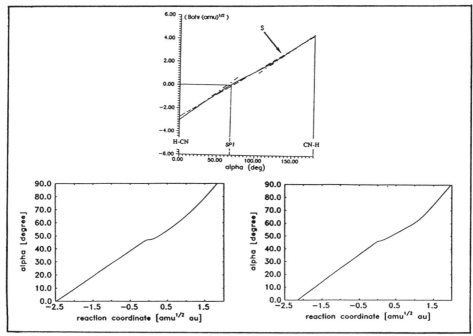

Figure 3. The relation between S (the curvilinear displacement coordinate along the IRP) and the Jacobi bending angle α, is almost linear, for HCN $\rightleftarrows$ CNH and the Murrell et al PES [54] (a) as well as for $HO_2 \rightleftarrows O_2H$ and two different PES by Varandas et al [55,56] (b and c)

Once a geometrical coordinate has been identified as a suitable reaction coordinate, the question of the corresponding RP and reaction profile can be posed as follows : for a fixed value of that geometrical reaction coordinate, there is at least one molecular configuration for which the electronic (potential) energy is minimal. Moreover, to one configuration correspond points in any ACCCS or in any local chart; thus the locus of all that points for all values of the reaction coordinate, meets the conditions for being a minimum energy path (for mathematical details, see below section 3.). Since (i) a set of points may correspond to one single configuration, unaffected by overall rotation and translation, and (ii) more than one minimum may exist at a time, a continuity requirement must be carefully checked to be fulfilled by the locus before definitely deciding for its status of RP. Such definitions of RP's we term "loose" (in contrast with the more stringent ones, of the steepest-descent type). The plot of the values of the local potential energy minima as a function of the geometrical reaction coordinate is called the reaction energy profile.

A last point needs to be carefully appreciated. In actual practice, for any reactive event involving a large molecular system (let us say more than pentatomic), they are not 3N-6 coordinates that can actually be considered by the chemist for visualizing or describing the reaction, but a much smaller number n (< or <<3N-6): this means that many coordinates (actually 3N-6-n) are viewed as being either frozen throughout the reaction or adiabatically (i.e. stepwisely) adjusted to the variations of the n coordinates retained. For instance, almost all reactions (unimolecular as well as bimolecular) basically consisting in the migration of an H atom have couterpart reactions in which alkyl groups substitute for H, e.g. HCN $\rightleftarrows$ CNH and CH_3CN $\rightleftarrows$ $CNCH_3$. For this last hexatomic isomerizing species (3N-6=12), most likely, the detailed internal geometry of the methyl group plays a secondary role when the reaction is in progress and the 3x4-6=6 coordinates (whatever they are) allowing its description, can be treated as mentioned above. Of the six remaining coordinates, two are obviously important (R and α in Figure 4 next page, α being still possibly considered as the reaction coordinate) and deciding whether the last four coordinates (r, the CN triple bond length and three Eulerian angles accounting for the methyl group orientation, among which the torsional angle ϕ around R, see Figure 4) are to be considered or not, is just a matter of sophistication of the theoretical description invoked, in particular depending on (i) the cost of the calculations one is ready to pay and (ii) the size of the calculations one is able to implement.

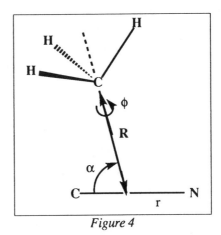

Figure 4

In this context, the idea (introduced by Miller) of using one atomic Cartesian coordinate of the migrating atom or ion, straightforwardly as a reaction coordinate ([15,40,41]), appears to be a particular case of the proposal in the two preceding paragraphs. Similarly, in few cases, the 'active' coordinates accounting for a reactive event in a polyatomic system (see below section 3.2.1), can all be taken Cartesian.

As an extreme case, let an N-atom molecular system be such that, for physical reasons, it is established that what is essential in the evolution of the molecule configuration in the course of a reaction, is adequately described by one single curvilinear internal coordinate (that will obviously be also the reaction coordinate), whatever the other $3N-7$ internal coordinates. For example, the cis-trans isomerization process in ethylene is generally considered to be well accounted for by one internal rotation dihedral angle, which can be mathematically defined as the angle between the inter-H segments in both CH_2 groups, when looking at the molecule along the direction of the CC bond (see Figure 5, next page). Then, substituting the torsional angle ϕ for the reaction coordinate and defining a reaction path as the sequence of all those configurations that correspond to potential energy minima at constant values of ϕ, is reasonable.

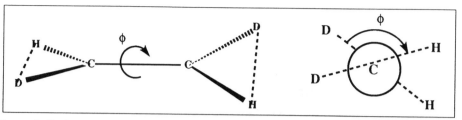

Figure 5. Definition of the torsional angle φ accounting for the intramolecular cis-trans isomerization process in a substituted ethylene molecule. Lateral view of the molecule and schematic illustration of φ (on the left) and Newman projection side view along the CC bond (on the right).

3. The concept of Reaction Path and the dimensionality reduction requirement.

3.1. UNCONSTRAINED SYSTEMS.

3.1.1. *Transformations of coordinates.*

For the sake of clarity, we summarize in this section the definitions and notations that have already been used from place to place or that are going to be used below, in a systematic way.

Basically, three generic sets of coordinates can be used for describing the nuclear arrangement in space, of an N-atom molecule :
1. atomic cartesian coordinates : $\underline{x} = \{x^1, \dots x^{3N}\}$
2. mass-weighted atomic cartesian coordinates : $\underline{X} = \{X^1, \dots X^{3N}\}$ $\Rightarrow$ ACCCS.

The first of these two sets, $\underline{x} = \{x_0^1, y_0^1, z_0^1, x_0^2, \dots z_0^N\}$, is adequate for describing the positions of the atoms, in the BF frame as well as in the SF frame, whereas the second one is well suited for dynamical calculations;

3. internal curvilinear coordinates : $\underline{q} = \{q^1, \dots q^{3N-6}\} \Rightarrow$ local chart.

This last set uniquely describes the size and shape of the molecule, but neither its orientation nor its location in the SF frame. The molecular center of mass being considered fixed at the origin of the BF frame, the orientation of the BF frame with respect to the SF frame is measured by means of three Euler angles : $\underline{\Theta} = \{\theta, \varphi, \chi\}$, and the coordinates of the center of mass are, in the SF frame : $\underline{r} = \{r^1, r^2, r^3\}$.

For the three sets of 3N coordinates, $\underline{x}$, $\underline{X}$ and $\{\underline{q} \oplus \underline{\Theta} \oplus \underline{r}\}$, to be equivalent, the transformations between them must be bijective, i.e. non-singular. In other words, the Jacobians of the applications in the triangular scheme below, must be non-zero, except, possibly, in sets of zero measure. As far as the various applications are concerned, those that are actually not obtainable straightforwardly are indicated by question marks in the following scheme :

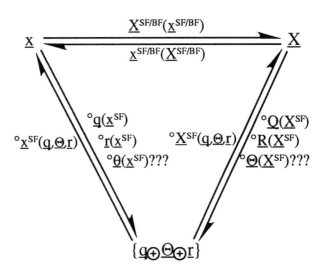

The various applications mentioned are as follows :

1. $\underline{X}(\underline{x}) = \underline{\underline{M}}^{1/2}.\underline{x}$ and $\underline{x}(\underline{X}) = \underline{\underline{M}}^{-1/2}.\underline{X}$, where $\underline{\underline{M}}$ is the diagonal matrix of the atomic masses given in section 2.2.2., whatever SF or BF is the frame in which the x's or X's are measured;

2. $^{\circ}x^{SF, a+3(j-1)}(\underline{q}, \underline{\Theta}, \underline{r}) = r^a + \sum_{a'=1,2,3} \mathcal{E}^a_{a'}(\underline{\Theta}) \, {}^{\circ\circ}x^{BF, a'+3(j-1)}(\underline{q})$ (25')

and : $\quad {}^{\circ}X^{SF, a+3(j-1)}(\underline{q}, \underline{\Theta}, \underline{r}) = \sqrt{m_j} \, {}^{\circ}x^{SF, a+3(j-1)}(\underline{q}, \underline{\Theta}, \underline{r})$ (j=1,... N; a=1,2,3) (25") where $\mathcal{E}^a_{a'}(\Theta)$ is an Euler rotation matrix element and $^ox^{BF,a'+3(j-1)}(q)$ is the expression of an atomic BF Cartesian components in terms of q only (which is always easily established, with the help of some geometry and trigonometry). Notice that (i) the three conditions for the centre of mass at the origin of the BF frame, $\sum\limits_{j=1}^{N} m_j {}^ox^{BF,a+3(j-1)}(q)=0$ (a=1,2,3), must hold and (ii) the expressions $x^{BF,a'+3(j-1)}(q)$ implicitly depend on the way the BF frame is tied to certain atoms in the molecule;

3. $^oq^i(\underline{x})$ (i=1, ... 3N-6) is one of those expressions belonging to the (very limited !) catalogue of internal coordinates evoked in the introduction of this chapter, whatever $\underline{x}$, either $\underline{x}^{BF}$ or $\underline{x}^{SF}$, and therefore $^oQ^i(\underline{X}) = {}^oq^i(\underline{M}^{-1/2}.\underline{X})$. In addition, there is obviously :

$$^or^a(\underline{x}^{SF}) = m_{tot}^{-1} \sum_{j=1}^{N} m_j x^{SF,a+3(j-1)} \qquad (a=1,2,3) \qquad (26')$$

where $m_{tot} = \sum\limits_{j=1}^{N} m_j$ is the total mass of the molecule, and :

$$^oR^a(\underline{X}^{SF}) = {}^or^a(\underline{M}^{-1/2}.\underline{X}^{SF}) = m_{tot}^{-1} \sum_{j=1}^{N} \sqrt{m_j} X^{SF,a+3(j-1)} \qquad (a=1,2,3). \qquad (26'')$$

The only real difficulty regarding the determination of the inverse applications of both applications $\underline{x}^{SF} = {}^o\underline{x}^{SF}(q,\Theta,r)$ and $\underline{X}^{SF} = {}^o\underline{X}^{SF}(q,\Theta,r)$, concerns the relations $\Theta = {}^o\Theta(\underline{x}^{SF})$ [or equivalently $\Theta = {}^o\Theta(\underline{X}^{SF})$]. Actually and most unfortunately, these last relations are not straightforwardly obtainable in closed form (see the Appendix for a summary of Sutcliffe's proposal concerning this question [57,58]). This is the reason why it is usual, in molecular physics, to prescribe the orientation of the BF axes in an explicit way, e.g. (i) the principal axes as BF axes, coincide permanently with the inertial principal axes, or (ii) the z-axis is parallel with the line joining two atoms and the xz-plane is parallel to the plane of the same two atoms plus a third one whose x-component is taken positive, or still (iii) the z-axis passes through a given atom, or is parallel with the bisector of the angle of three atoms, or still is perpendicular to the plane of three atoms, etc... [32,59]. Then, inverting the relations $\underline{x}^{BF} = {}^o\underline{x}^{BF}(q)$ between the 3N atomic BF Cartesian coordinates and the 3N-6 internal curvilinear coordinates, is an easy task. Indeed, for given curvilinear internal coordinates, $q = {}^oq(\underline{x}^{BF})$, since these do not depend on the frame, the orientation of the BF frame must be explicitly specified by three additional relations, the so-called 'axial constraints', denoted $0 = {}^\circ C^a(\underline{x}^{BF})$ (a=1,2,3). Similarly, the three constraints concerning the centre of mass lying at the origin of the BF frame, $0 = \sum_{j=1}^{N} m_j x^{BF,a+3(j-1)}$ (a=1,2,3), must be fulfilled.

Although the application $\underline{x}^{SF} = {}^\circ\underline{x}^{SF}(\underline{q},\underline{\Theta},\underline{r})$ can hardly be inverted, the fact that the alternative application, $\underline{x}^{BF} = {}^\circ\underline{x}^{BF}(\underline{q})$, actually can be inverted in the usual form :

$$\begin{cases} \underline{q} = {}^\circ\underline{q}(\underline{x}^{BF}) & : \textit{internal coordinates} \\ \underline{0} = {}^\circ\underline{C}(\underline{x}^{BF}) & : \textit{axial constraints} \\ 0 = \sum_{j=1}^{N} m_j x^{BF,a+3(j-1)} & (a=1,2,3) \end{cases} \qquad (27)$$

allows the derivation of the metric tensor for the inverse application of $\underline{x}^{SF} = {}^\circ\underline{x}^{SF}(\underline{q},\underline{\Theta},\underline{r})$:

$$\begin{cases} \underline{q} = {}^\circ\underline{q}(\underline{x}^{SF}) & : \textit{internal coordinates} \\ \underline{\Theta} = {}^\circ\underline{\theta}(\underline{x}^{SF}) \ ? : \textit{Euler angles} \\ \underline{r}^a = m_{tot}^{-1} \sum_{j=1}^{N} m_j x^{SF,a+3(j-1)} & (a=1,2,3) \end{cases} \qquad (28)$$

We shall not give the expression of this tensor here, but rather that of the metric tensor for the inverse application of the mass-weighted atomic Cartesian coordinates, i.e. $\underline{X}^{SF} = {}^\circ\underline{X}^{SF}(\underline{q},\underline{\Theta},\underline{r})$:

$$\begin{cases} \underline{q} = {}^\circ\underline{Q}(\underline{X}^{SF}) \\ \underline{\Theta} = {}^\circ\underline{\Theta}(\underline{X}^{SF}) \\ \underline{r}^a = m_{tot}^{-1} \sum_{j=1}^{N} \sqrt{m_j} \, X^{SF,a+3(j-1)} & (a=1,2,3) \end{cases} \qquad (29)$$

in the guise of matrix $\underline{\underline{G}}(\underline{q},\underline{\Theta})$ which has been already invoked, cf. ${}^\circ\underline{\underline{g}}[\underline{X}^{SF}(\underline{q},\underline{\Theta},\underline{r})]$, and the general expression of which (that comes out rather tediously in Refs.[32,59,60]) can be written in the form of Eq.(30) below (the dimensions of the blocks are explicitly indicated).

$$
\underline{\underline{G}}(q,\Theta) = \begin{pmatrix} \underline{\underline{\Sigma}}(q) & [\underline{\underline{\Gamma}}(q)]^T \cdot [\underline{\underline{\Omega}}(\Theta)]^{-1} & \underline{\underline{0}} \\[2em] [\underline{\underline{\Omega}}(\Theta)]^{-1T} \cdot \underline{\underline{\Gamma}}(q) & [\underline{\underline{\Omega}}(\Theta)]^{-1T} \cdot \underline{\underline{\mu}}(q) \cdot [\underline{\underline{\Omega}}(\Theta)]^{-1} & \underline{\underline{0}} \\[2em] \underline{\underline{0}} & \underline{\underline{0}} & \underline{\underline{m}}_{tot}^{-1} \end{pmatrix} \begin{matrix} 3N\text{-}6 \\[2em] 3 \\[2em] 3 \end{matrix} \quad (30)
$$

$$
\begin{matrix} 3N\text{-}6 & \qquad 3 & \qquad 3 \end{matrix}
$$

The various matrices appearing in Eq.(30) are as follows :

(i) $\underline{\underline{\Sigma}}(q)$, in fact $^\circ\underline{\underline{\Sigma}}[x^{SF/BF}(q)]$, is a $(3N\text{-}6)$-dimensional square matrix defined in Eq.(15) above. It is independent of the particular choice of the BF frame. In other words, *it depends exclusively on the choice of the internal coordinates*;

(ii) $\underline{\underline{\Gamma}}(q)$, or rather $\underline{\underline{\Gamma}}[x^{BF}(q)]$, is a $3\times(3N\text{-}6)$ matrix defined in Eqs.(48-50) of Ref.[7]. *It depends on both choices (BF frame and internal coordinates).* Its elements can be expressed in terms of the internal coordinates q only;

(iii) $\underline{\underline{\Omega}}^{-1T}(\Theta) = \begin{pmatrix} \sin\chi & \cos\chi & 0 \\ \dfrac{-\cos\chi}{\sin\theta} & \dfrac{\sin\chi}{\sin\theta} & 0 \\ \cot\theta\cos\chi & -\cot\theta\sin\chi & 1 \end{pmatrix}$ is a 3-dimensional non-

singular matrix [except at the poles, subset of zero measure, cf. $\det\underline{\underline{\Omega}}^{-1} = \dfrac{1}{\sin\theta}$];

(iv) $\underline{\mu}[\underline{x}^{BF}(q)]$ is a 3x3 square matrix defined in Eqs.(49,50,52,53) of Ref.[32] and often called the inverse "effective inertia tensor". *It depends exclusively on the choice of the BF frame.* However, its elements can be transcribed in terms of q only;

(v) $\underline{m}_{tot}^{-1}$ is a diagonal 3x3 square matrix with diagonal elements equal to m_{tot}^{-1} .

The Jacobian of transformation (29) is, by construction, equal to the square root of the determinant of matrix $\underline{\underline{G}}(q,\underline{\Theta})$, i.e. $J(q,\underline{\Theta}) = \dfrac{\sqrt{\tilde{g}(q)}\ m_{tot}^{-3/2}}{\sin\theta}$, since $\underline{\underline{G}}$ can be

factorized as $\begin{pmatrix} \underline{\underline{1}} & \underline{\underline{0}} & \underline{0} \\ \underline{\underline{0}} & \underline{\underline{\Omega}}^{-1T}(\underline{\Theta}) & \underline{0} \\ \underline{0} & \underline{0} & 1 \end{pmatrix} \times \begin{pmatrix} \underline{\underline{\Sigma}}(q) & \underline{\underline{\Gamma}}^{T}(q) & \underline{0} \\ \underline{\underline{\Gamma}}(q) & \underline{\underline{\mu}}(q) & \underline{0} \\ \underline{0} & \underline{0} & m_{tot}^{-1} \end{pmatrix} \times \begin{pmatrix} \underline{\underline{1}} & \underline{\underline{0}} & \underline{0} \\ \underline{\underline{0}} & \underline{\underline{\Omega}}^{-1}(\underline{\Theta}) & \underline{0} \\ \underline{0} & \underline{0} & 1 \end{pmatrix}$. $\tilde{g}(q)$ has been

defined in section 2.2.1 and appears in Eq.(16). For more details, see Refs.[32,59,60].

Insofar as the frame-independent expressions of the potential energy function are concerned, let us simply recall that :

$$V(q) \quad = \quad u(\underline{x}) \quad = \quad U(\underline{X}).$$
$$\| \qquad\qquad \|$$
$$V[{}^{\circ}q(\underline{x})] \quad V[{}^{\circ}\underline{Q}(\underline{X})] = V[{}^{\circ}q(\underline{\underline{M}}^{-1/2}.\underline{X})]$$

3.1.2. *Various reaction paths.*

As mentioned in section 2.2.1, various steepest-descent paths can be defined, depending on the ACCCS referred to, in particular :

(i) if it is the normal ACCCS, the differential equations to be integrated are (4) or (6), depending on whether Cartesian coordinates $\underline{x}$ or curvilinear coordinates q are used. Relations between $\underline{x}$ and q are always known, that are frame-independent in the form q(x) but are frame-dependent (particularly SF or BF) in the form x(q). Consequently, the metric tensor elements $g^{ij}(q) = {}^{\circ}g^{ij}[\underline{x}(q)]$, necessary for integrating (6), can always be derived from (8);

(ii) if it is the mass-weighted ACCCS that is referred to, the differential equations are (10), (11) or (12), depending on whether mass-weighted Cartesian coordinates $\underline{X}$, Cartesian coordinates $\underline{x}$ or curvilinear coordinates $\underline{q}$ are used.

It is no matter of two different definitions of the same path, but definitely of two different paths, e.g. compare Eqs.(4) and (11). The following properties should be noticed :

(i) both paths are coordinate-, i.e. $\underline{q}$-independent;

(ii) the genuine steepest-descent path is mass-independent and the molecular centre of mass moves along it;

(iii) the mass-weighted steepest-descent path (IRP) is mass-dependent, but the molecular centre of mass is fixed along it;

(iv) considered as continuous lines in their respective ACCCS, overall rotation takes place along both paths (see Ref.[61]);

(v) all stationary points of the PES of $V(\underline{q})$ belong to both paths.

However, as mentioned in section 2.3, it may be simpler or more profitable to substitute one *ad hoc* internal (geometrical) coordinate for the reaction coordinate. Next, for a fixed value of this reaction coordinate, the configuration that minimizes the potential energy is determined. Within the framework of what we call the *loose definition*, this configuration corresponds to the point of the RP for this value of the reaction coordinate. Actually, if q^1 denotes the supposed reaction coordinate, the *loosely defined RP* is defined as the parametric solution of the following set of 3N-7 equations of 3N-6 unknowns :

$$\begin{cases} \dfrac{\partial V(\underline{q})}{\partial q^2} = 0 \\ \quad \vdots \\ \quad \vdots \\ \dfrac{\partial V(\underline{q})}{\partial q^{3N-6}} = 0 \end{cases} \tag{31}$$

written in the form :

$$\begin{cases} q^2 = q^2_{AEP}(q^1) \\ \quad\vdots \\ q^{3N-6} = q^{3N-6}_{AEP}(q^1) \end{cases} \tag{32}$$

We call this RP the Adiabatic (i.e. q^1-adjusted) Equilibrium Path (AEP). It should be emphasized that, the defining conditions (31) being less restrictive than the condition $\underline{\nabla}_q V(\underline{q}) = \underline{0}$ that defines the stationary points (where $\underline{\nabla}_q$ denotes the 3N-6-dimensional gradient vector in terms of internal coordinates), all stationary points from reagents to products (saddle points corresponding to transition states as well as potential minima for equilibrium conformations) belong to the AEP, so that its status as a RP is confirmed, according to the loose definition. Clearly, up to now, the AEP has been defined regardless of any considerations concerning an ACCCS or the nature (SF or BF) of a reference frame. Moreover, the plot of $V_{AEP}(q^1) = V[q^1, q^2_{AEP}(q^1), \ldots q^{3N-6}_{AEP}(q^1)]$ as a function of q^1 actually constitutes the reaction energy profile. It should be emphasized that, *when chemists invoke reaction coordinates and reaction profiles, the concept that is implicitly referred to, is (in most cases) the AEP.*

3.2.   DIMENSION-REDUCED, i.e. CONSTRAINED SYSTEMS : REACTION PATHS AND CURVILINEAR COORDINATES.

3.2.1. *Active and inactive internal coordinates.*

For values of N larger or equal to say five, deriving (in 3N-6 dimensions) the PES of an N-atom molecular system undergoing large-amplitude deformations is just out of question; studying the overall dynamics of the system is similarly out of reach (here, we consider neither molecular dynamical studies in which there are many molecules of the same species, the mechanics resorted to being classical and the intermolecular potential being empirical, nor the quantal small-amplitude vibrational problem of polyatomic molecules). For such "large" systems, the chemist is usually only interested in few relevant deformations, which can be adequately described by means of a limited number ($n \ll 3N-6$) of appropriate internal coordinates. It should be emphasized that the qualitative nature of the problem posed has changed : the question is no longer "What may happen to the system in all generality, including the reaction ?", but rather "What should happen to the system in order for it to achieve a given reactive process ?". This is the qualitative basis for partitioning a complete set of 3N-6 internal coordinates $q = \{q^1, q^2, \ldots q^{3N-6}\}$ into two subsets, respectively the *active coordinates* :

$$q' = \{q^1, q^2, \ldots q^n\} \tag{33'}$$

and the *inactive coordinates* :

$$q'' = \{q^{n+1}, q^{n+2}, \ldots q^{3N-6}\} \tag{33''}$$

i.e. $q = \{q' \oplus q''\}$. It is worth noticing that, in addition to the multiplicity of choices for the internal coordinates q describing the shape of the unconstrained system, the above mentioned partition is *specific to one particular reactive event* concerning the system, so that, to a new reaction, does correspond a new partition. Moreover, the inactive coordinates q'' are normally unspecified : any set of coordinates complementing q' to q is the same adequate.

When a decision has been made concerning the active coordinates q' that are intended to describe adequately the molecular system all along the reaction, i.e. the potential energy function is expressed as a function of q' and the dynamics in terms of q' is fully considered, there is still another decision to be taken concerning the role, i.e. the dynamical status of the inactive coordinates q''. More explicitly, the question to be answered is "Are the inactive coordinates q'' rigidly frozen ($q'' = q_0''$) or gradually adjusted to the active coordinates $[q'' = {}^{\circ}q''(q')]$?".

3.2.2. *Model molecular systems with frozen inactive coordinates.*

If the inactive coordinates are frozen (e.g. a bond length is held constant, an entire substituent group is supposed to keep rigid throughout its displacement in the course of the reaction, etc...), we term the corresponding molecular model *rigidly constrained*. The system is viewed as divided into subsystems, some of which keep rigid and are articulated with other subsystems next to them by means of hinges or any other perfect joints. From the physical viewpoint, a model system clearly substitutes for the real system, whose full description is supposed to be either unnecessary or unattainable.

Obtaining a potential energy function $V_0(\underline{q}')$ for a rigidly constrained model system is standard job, and there is no conceptual difficulty for joining $V_0(\underline{q}')$ to the free system potential energy function $V(\underline{q})$:

$$\left.\begin{array}{c} q^{n+1} = q_0^{n+1} \\ \cdots \\ q^{3N-6} = q_0^{3N-6} \end{array}\right\} \Rightarrow V_0(\underline{q}') = V(q^1, q^2, \ldots q^n, q_0^{n+1}, \ldots q_0^{3N-6}) = [V(\underline{q})]_{q_0''}$$

Expressing $V_0(\underline{q}')$ in terms of Cartesian or mass-weighted Cartesian coordinates is also a simple task, since expressions for $\underline{q}'(\underline{x}) = \{{}^\circ q^i(\underline{x}); i=1,2,\ldots n\}$ are always at disposal :

$$V_0(\underline{q}') \quad = \quad u_0(\underline{x}) \quad = \quad U_0(\underline{X}).$$
$$\| \qquad\qquad \|$$
$$V_0[{}^\circ\underline{q}'(\underline{x})] \quad V_0[{}^\circ\underline{Q}'(\underline{X})] = V_0[{}^\circ\underline{q}'(\underline{M}^{-1/2}\cdot\underline{X})]$$

On this basis, defining a RP for the PES of $V_0(\underline{q}') = u_0(\underline{x})$ is a well posed problem, for which it has just to be specified which definition of the RP is retained :

(i) the genuine steepest-descent path of $V_0(\underline{q}')$ is derived by solution of either sets of coupled differential equations :

$$\frac{dx^k}{dt} = -\sum_{l=1}^{3N} \mathfrak{T}_0^{kl}(\underline{x}) \frac{\partial u_0(\underline{x})}{\partial x^l} \qquad (k=1, \ldots 3N) \qquad (34)$$

in Cartesian coordinates or, alternatively, in curvilinear coordinates :

$$\frac{dq^i}{dt} = -\sum_{j=1}^{n} {}^\circ g_0^{ij}[\underline{x}(\underline{q}')] \frac{\partial V_0(\underline{q}')}{\partial q^j} \qquad (i=1, \ldots n) \qquad (35)$$

In Eq.(34), $\mathfrak{T}_0^{kl}(\underline{x})$ is a contravariant component of a tensor, $\underline{\mathfrak{T}}_0(\underline{x})$, the role of which is to account for the fact that the integration is no longer in the full ACCCS but, because of the internal constraints, in a manifold imbedded into the ACCCS, such that $\underline{q}''(\underline{x})=\underline{q}_0''$.

For the same reason, in Eq.(35), ${}^\circ g_0^{ij}(\underline{x})$ is not equal to $\sum_{k=1}^{3N} \dfrac{\partial {}^\circ q^i}{\partial x^k} \dfrac{\partial {}^\circ q^j}{\partial x^k}$, but can

nevertheless be always obtained. This last statement stems from the fact that relationships such that $\underline{x}^{BF} = {}^{\circ}\underline{x}^{BF}(\underline{q}')$ are always at disposal, whatever the BF frame and the constraints;

(ii) similarly, the two sets of differential equations for calculating the mass-weighted IRP of $V_0(\underline{q}')$ are, respectively :

$$m_k \frac{dx^{a+3(k-1)}}{dT} = -\sum_{l=1}^{3N} x_0^{a+3(k-1),l}(\underline{x}) \frac{\partial u_0(\underline{x})}{\partial x^l} \quad (k=1,\dots N;\ a=1,2,3) \tag{36}$$

$$\frac{dq^i}{dT} = -\sum_{j=1}^{n} {}^{\circ}G_0^{ij}[\underline{x}(\underline{q}')] \frac{\partial V_0(\underline{q}')}{\partial q^j} \quad (i=1,\dots n) \tag{37}$$

Various methods for calculating ${}^{\circ}G_0^{ij}(\underline{x}) = \Sigma_0^{ij}(\underline{q}')$ $(i,j=1,\dots n)$ are presented in Refs. [62,63];

(iii) if q^1 still denotes the supposed reaction coordinate, the AEP for the PES of $V_0(\underline{q}')$ (loose definition) is obtained as the parametric solution of the following set of n-1 equations of n unknowns :

$$\begin{cases} \dfrac{\partial V_0(\underline{q}')}{\partial q^2} = 0 \\ \quad\vdots \\ \dfrac{\partial V_0(\underline{q}')}{\partial q^n} = 0 \end{cases} \tag{38}$$

which can be written as :

$$\begin{cases} q^2 = q_{ad}^2(q^1) \\ \quad\vdots \\ q^n = q_{ad}^n(q^1) \end{cases} \tag{39}$$

Although the problem of defining and deriving RP's for molecular systems with rigidly frozen degrees of freedom, i.e. for $V_0(\underline{q}')$, is mathematically well posed, it should be emphasized, from the physical point of view, that no link can be *a priori* established between these paths and the corresponding RP's that would be otherwise defined for the free system. *A lot of care must therefore be taken when using such RP's for physical purposes.*

But there is a way of bypassing this drawback.

3.2.3. Model systems with adiabatically adjusted inactive coordinates.

If, instead of being frozen ($q''=q_0''$), the inactive coordinates q'' are allowed to change by gradual adjustment to the variations of the active coordinates all along the reaction, which can be formulated mathematically by means of relationships of the type :

$$q''={}^\circ q''(q')$$

the physical nature of the connection between q'' and q' still must be specified. Such molecular models we term "adiabatically constrained" or q'-adjusted.

Because (i) those parts of the free system PES that are actually explored by either classical trajectories or quantal wave packets describing a chemical reaction, are preferentially the lowest ones and (ii) modern quantum-chemical methods for calculating PES include geometry optimizations, generally gradient optimizations (for gradient extremal methods, see Refs.[64-66]), it is natural to retain for q'', at given q', the values which minimize $V(q)$. Mathematically, this is expressed as :

$$\left(\frac{\partial V(q)}{\partial q''} \right)_{q'} = 0 \quad \text{i.e.} \quad \left. \begin{array}{c} \dfrac{\partial V(q)}{\partial q^{n+1}} = 0 \\[2mm] \vdots \\[2mm] \dfrac{\partial V(q)}{\partial q^{3N-6}} = 0 \end{array} \right\} \tag{40}$$

$$\Rightarrow \begin{cases} q^{n+1}=q_{eq}^{n+1}(q') \\ \vdots \\ q^{3N-6}=q_{eq}^{3N-6}(q') \end{cases} \quad \text{i.e.} \quad q'' = q_{eq}''(q') \tag{41}$$

This particular adiabatic dependence of q'' upon q', we call "partial-equilibrium representation". It is denoted by the subscript eq.

The potential energy function $V_{eq}(q')$ for this particular adiabatically constrained model system is :

$$V_{eq}(q') = V[q', q''_{eq}(q')]$$

$V_{eq}(q')$ can also be expressed in terms of Cartesian or mass-weighted Cartesian coordinates :

$$V_{eq}(q') \;=\; u_{eq}(x) \;=\; U_{eq}(X).$$

$$\begin{array}{cc} \| & \| \\ V_{eq}[^{\circ}q'(x)] & V_{eq}[^{\circ}Q'(X)] = V_{eq}[^{\circ}q'(\underline{\underline{M}}^{-1/2}\cdot X)] \end{array}$$

The defining conditions (40) are less restrictive than the condition $\underline{\nabla}_q V(q) = \underline{0}$ defining the stationary points. Therefore all stationary points from reagents to products (saddle points for transition states and potential minima for equilibrium conformations) belong to the n-dimensional PES of $V_{eq}(q')$, and the search for a RP on that PES is, according to the loose definition, well-founded. In contrast with the previous case, even if the steepest-descent paths so obtained are not exactly the same as their counterparts for the free systems (the AEP is actually the same), the physical link between the latter and the former is guaranteed, in virtue of the loose definition of RP's. But it should be emphasized that, since the partition of q into q' and q'' is, to some extent, arbitrary, the number of definitions for a RP is formally unlimited, in actual practice far from unique !

In this approach, the only additional point to fix is "Which definition of the RP is resorted to ?" :

(i) the genuine steepest-descent RP on the PES of $V_{eq}(q')$, can be derived from :

$$\frac{dx^k}{dt} = -\sum_{l=1}^{3N} \mathcal{X}^{kl}_{eq}(x)\frac{\partial u_{eq}(x)}{\partial x^l} \qquad (k=1, \dots 3N) \qquad (42)$$

in Cartesian coordinates or, alternatively, in curvilinear coordinates :

$$\frac{dq^i}{dt} = -\sum_{j=1}^{n} {}^{\circ}g^{ij}[x(q')]\frac{\partial V_{eq}(q')}{\partial q^j} \qquad (i=1, \dots n) \qquad (43)$$

Here again, the tensors $\underline{\underline{\mathcal{X}}}_{eq}(x)$ and ${}^{\circ}\underline{\underline{g}}_{eq}(x)$ account for the differential equations being modified by the internal constraints, i.e. the solution curve entirely lies in a manifold embedded in the ACCCS (x) and defined by $q''(x)=q''_{eq}[q'(x)]$;

(ii) similarly, the defining differential equations for the IRP of an adiabatically constrained system, supposedly well accounted for by $V_{eq}(q')$, are :

$$m_k \frac{dx^{a+3(k-1)}}{dT} = -\sum_{l=1}^{3N} \mathcal{X}_{eq}^{a+3(k-1),l}(\underline{x}) \frac{\partial u_{eq}(\underline{x})}{\partial x^l} \qquad (k=1,\ldots N; \; a=1,2,3) \qquad (44)$$

$$\frac{dq^i}{dT} = -\sum_{j=1}^{n} {}^{\circ}G_{eq}^{ij}[\underline{x}(\underline{q}')] \frac{\partial V_{eq}(\underline{q}')}{\partial q^j} \qquad (i=1,\ldots n) \qquad (45)$$

Methods for calculating ${}^{\circ}G_{eq}^{ij}[\underline{x}(\underline{q}')] = \Sigma_{eq}^{ij}(\underline{q}')$ are developped in Ref.[67];

(iii) as far as the AEP is concerned, q^1 still denoting the reaction coordinate, the same path is obtained in the adiabatically constrained, partial equilibrium case as in the unconstrained case, loose definition. Indeed, in the partial-equilibrium representation, the system of equations for the AEP :

$$\begin{cases} \dfrac{\partial V_{eq}(\underline{q}')}{\partial q^2} = 0 \\ \quad \vdots \\ \quad \vdots \\ \dfrac{\partial V_{eq}(\underline{q}')}{\partial q^n} = 0 \end{cases} \qquad (46)$$

turns out to be nothing but a subset of the system of the AEP defining equations (31), section 3.1.2, the complementary subset being given in (40) above. Therefore, the solution of Eqs.(46), written as :

$$\begin{cases} q^2 = q^2_{AEP}(q^1) \\ \quad \vdots \\ \quad \vdots \\ q^n = q^n_{AEP}(q^1) \end{cases} , \qquad (47)$$

along with Eqs.(41) above, in which (47) has been inserted :

$$\begin{cases} q^{n+1} = q_{eq}^{n+1}[q^1, q^2_{AEP}(q^1),\ldots q^n_{AEP}(q^1)] = q^{n+1}_{AEP}(q^1) \\ \quad \vdots \\ \quad \vdots \\ q^{3N-6} = q_{eq}^{3N-6}[q^1, q^2_{AEP}(q^1),\ldots q^n_{AEP}(q^1)] = q^{3N-6}_{AEP}(q^1) \end{cases} \qquad (48)$$

constitute the very definition of the AEP. Mathematically, this interesting property :

> *" The same AEP is obtained from the free system PES, V(q), as*
> *well as from the adiabatically constrained system PES, $V_{eq}(q')$ "*

stems from the fact that the loose definition that applies to the AEP, is expressed by means of explicit coupled equations and not by coupled differential equations, as is the case for the definitions of the RP's as steepest-descent paths.

4. Conclusion.

In this chapter, we have attempted to show that, in spite of the multiple definitions which have been proposed, during the last three decades, for the reaction path of a chemical reaction on a given PES, all the more so since there are multiple ways of expressing dimension-reduced potential energy functions for large polyatomic molecular systems, it is possible to derive, in all cases, defining equations, either explicit equations or differential equations. The mathematical and physical connections between these various RP's have also been investigated : certain definitions have been shown to be interesting from particular viewpoints, e.g. the IRP from the dynamical viewpoint, but are practically impossible to implement numerically, as is the case of the IRP for molecular systems as small as pentatomics. Moreover, definitely, no one definition is by itself better than another. Selecting one particular definition rather than another, is just a matter of convenience, generally in view of subsequent use [10].

In this context, defining the RP as an AEP has several advantages over other more sophisticated definitions :
(i) selecting (or designing) one particular geometrical coordinate as reaction coordinate, does correspond to usual processes in chemistry;
(ii) the AEP meets the criteria of a minimum energy path, i.e. the minimal requirement upon which there is universal agreement for a path on a PES to be accepted as a RP;
(iii) calculating an AEP comes down to the simple problem of solving a set of coupled (non-differential) equations;
(iv) the same AEP results from calculations making use of either the free system PES $V(q)$ or the adiabatically constrained, quasi-equilibrium system PES $V_{eq}(q')$;

(v) almost all modern quantum-chemistry programs propose geometry optimizations by gradient methods, and therefore allow direct calculation of quasi-equilibrium potential energy functions $V_{cq}(q')$;

(vi) it may allow preserving the clear analysis of a reactive event, based on the Reaction Path Hamiltonian approach, but without the severe drawbacks of the original version of this approach, based on the use of the IRP;

(vii) in many instances, for small free systems, it has been possible to design one geometrical reaction coordinate (q^1) such that the reaction energy profile $V_{AEP}(q^1)$ looks very similar to steepest descent reaction energy profiles.

For all these reasons, the loose definition of RP's is certainly interesting. But claiming that it is a cure-all, would be a real mistake : for certain purposes (in particular for small systems when either dynamical or topographical considerations are important), the steepest-descent definitions can well be more appropriate !

It should be kept in mind that the qualifier "steepest-desent" for a RP, or even IRP, is by itself, at least for large polyatomic systems, insufficient as long as the PES which is concerned has not been fully specified : is it the free system PES, or a constrained system PES and, in this last case, which degrees of freedom are considered, what about the status of the other degrees of freedom (are they frozen or adiabatically adjusted) etc... ? Indeed, the steepest descent RP's for $V(q)$, $V_0(q')$ and $V_{eq}(q')$ are definitely not the same, which can be easily demonstrated in representing the curves corresponding to the various paths in the same ACCCS, say $(\underline{x})$. Insofar as steepest-descent paths for free systems are concerned, Ref.[3] is an exhaustive account.

Choosing an adequate definition for the RP of a given chemical reaction according to (i) the information at disposal about the reaction (particularly the PES) and (ii) the physicals goals of the study, is a matter of personal experience. The present book aims at providing helpful instruments in that field. The book "Properties of Chemically Interesting Potential Energy Surfaces", by Heidrich, Kliesch and Quapp, is also a useful guide [2].

Appendix. $\underline{\Theta} = {}^{\circ}\underline{\theta}(x^{SF})$ [or $\underline{\Theta} = {}^{\circ}\underline{\Theta}(\underline{X}^{SF})$].

In section 3.1.1, it was stated that the inverse of the basic application :

$${}^{\circ}_{x}{}^{SF,\,a+3(j-1)}(\underline{q},\underline{\Theta},\underline{r}) = r^{a} + \sum_{a'=1,2,3} \mathcal{C}^{a}_{a'}(\underline{\Theta}) \, {}^{\circ}_{\circ}x^{BF,\,a'+3(j-1)}(\underline{q}) \qquad (a=1,2,3;\ j=1,\ ...N)$$

can hardly be put in tractable form ($\mathcal{C}^{a}_{a'}(\underline{\Theta})$ is an Euler rotation matrix element and ${}^{\circ}_{x}{}^{SF,\,a+3(j-1)}$ and ${}^{\circ}_{x}{}^{BF,\,a'+3(j-1)}$ are atomic Cartesian components), namely :

$$\begin{cases} \underline{q} = {}^{\circ}\underline{q}(x^{SF}) \ : internal\ coordinates \\ r^{a}=m_{tot}^{-1}\sum_{j=1}^{N} m_{j}x^{SF,\,a+3(j-1)} \quad (a=1,2,3)\ : SF\ coordinates\ of\ the\ centre\ of\ mass \\ \underline{\Theta} = {}^{\circ}\underline{\theta}(x^{SF})\ : Euler\ angles \end{cases}$$

In fact, the relations concerning the internal coordinates and the coordinates of the centre of mass, are just straightforward expressions and the only difficulty concerns the Euler angles, $\underline{\Theta} = {}^{\circ}\underline{\theta}(x^{SF})$. Let us examine this point more precisely.

An essential decision to be made about a molecular model is "How are the BF axes tied to the molecule ?" Let us note that it is possible to define these axes in terms of the positions of three points, either individual atoms or centres of mass of atomic groups, e. g. two axes in the plane, perhaps one parallel with the segment joining two points, and the third axis perpendicular to this plane.

Let $\underline{e}_{1}$, $\underline{e}_{2}$, and $\underline{e}_{3}$ represent three unit vectors along these BF axes; in the SF frame, they are expressible in terms of the atomic Cartesian coordinates :

$$\underline{e}_{i} = \underline{e}_{i}(x^{SF}) \quad (i=1,2,3) \tag{A1}$$

According to the usual theory of rotations, the BF axes are related to the SF axes by rotations through Euler angles and, in terms of these angles, $\underline{e}_{1}$, $\underline{e}_{2}$, and $\underline{e}_{3}$ are given by the columns of the folloxing Euler matrix :

$$\begin{pmatrix} \cos\varphi\cos\theta\cos\chi-\sin\varphi\sin\chi & -\cos\varphi\cos\theta\sin\chi-\sin\varphi\cos\chi & \cos\varphi\sin\theta \\ \sin\varphi\cos\theta\cos\chi+\cos\varphi\sin\chi & -\sin\varphi\cos\theta\sin\chi+\cos\varphi\cos\chi & \sin\varphi\sin\theta \\ -\sin\theta\cos\chi & \sin\theta\sin\chi & \cos\theta \end{pmatrix} \tag{A2}$$

From Eqs.(A1,A2), expressions are obtained for the Euler angles in terms of $\underline{x}^{SF}$, for instance :

$$\left. \begin{array}{l} \cos\theta = e_{3z}(\underline{x}^{SF}) \\ \sin\varphi\sin\theta = e_{3y}(\underline{x}^{SF}) \\ \sin\chi\sin\theta = e_{2z}(\underline{x}^{SF}) \end{array} \right\} \Rightarrow \left\{ \begin{array}{l} \theta = \cos^{-1}[e_{3z}(\underline{x}^{SF})] \\[2mm] \varphi = \sin^{-1}\dfrac{e_{3y}(\underline{x}^{SF})}{\{1-[e_{3z}(\underline{x}^{SF})]^2\}^{1/2}} \\[4mm] \varphi = \sin^{-1}\dfrac{e_{2z}(\underline{x}^{SF})}{\{1-[e_{3z}(\underline{x}^{SF})]^2\}^{1/2}} \end{array} \right. \quad \text{(A3)}$$

As an example, let us define the BF axes as follows :

(i) $\underline{e}_3$ is parallel with $\overrightarrow{21}$, the vector pointing from atom 2 to atom 1 :

$$\underline{e}_3 = \frac{\overrightarrow{21}}{|\overrightarrow{21}|} \Rightarrow \left\{ \begin{array}{l} e_{3y}(\underline{x}^{SF}) = \dfrac{y_0^{2SF} - y_0^{1SF}}{\sqrt{[x_0^{2SF}-x_0^{1SF}]^2+[y_0^{2SF}-y_0^{1SF}]^2+[z_0^{2SF}-z_0^{1SF}]^2}} \\[5mm] e_{3z}(\underline{x}^{SF}) = \dfrac{z_0^{2SF} - z_0^{1SF}}{\sqrt{[x_0^{2SF}-x_0^{1SF}]^2+[y_0^{2SF}-y_0^{1SF}]^2+[z_0^{2SF}-z_0^{1SF}]^2}} \end{array} \right. \quad \text{(A4)}$$

(remind that $x_0^j=x^{3j-2}$, $y_0^j=x^{3j-1}$ and $z_0^j=x^{3j}$ (j=1,...N)), so that :

$$\theta = \cos^{-1}\frac{z_0^{2SF} - z_0^{1SF}}{\sqrt{[x_0^{2SF}-x_0^{1SF}]^2+[y_0^{2SF}-y_0^{1SF}]^2+[z_0^{2SF}-z_0^{1SF}]^2}} \in [0,\pi] \Rightarrow \sin\theta > 0 \quad \text{(A5)}$$

and :

$$\varphi = \sin^{-1}\frac{y_0^{2SF} - y_0^{1SF}}{\sqrt{[x_0^{2SF}-x_0^{1SF}]^2+[y_0^{2SF}-y_0^{1SF}]^2}} \in [0,2\pi] \quad \text{(A6)}$$

(ii) $\underline{e}_2$ is perpendicular to the plane of the atoms 1, 2 and 3 (and therefore $\underline{e}_1$ lies in the plane, perpendicular to $\underline{e}_3$, so that the BF frame of unit vectors $\{\underline{e}_1,\underline{e}_2,\underline{e}_3\}$ is direct, as is also the SF frame) :

$$\underline{e}_2 = \frac{\overrightarrow{31}\mathbf{x}\overrightarrow{32}}{|\overrightarrow{31}\mathbf{x}\overrightarrow{32}|} \quad \text{(A7)}$$

Inverting (A7) is straightforward, but the result written in a form similar to (A5 or A6) would be heavy. Consequently, the expressions of the components of the metric tensors,

$$°g^{ij}(\underline{x}) = \sum_{k=1}^{3N} \frac{\partial °q^i}{\partial x^k} \frac{\partial °q^j}{\partial x^k} \text{ in Eq.(8) or } °G^{ij}(\underline{x}) = \sum_{k=1}^{N} m_k^{-1} \sum_{a=1,2,3} \frac{\partial °q^i}{\partial x^{a+3(k-1)}} \frac{\partial °q^j}{\partial x^{a+3(k-1)}} \text{ in Eq.(13)}$$

(for values of i and/or j beyond 3N-6, q^{3N-5}, q^{3N-4} and q^{3N-3} standing respectively for θ, φ, and χ), would be even heavier. This approach, initially suggested by Sutcliffe [57] and which can be implemented in computer assisted methods for deriving molecular Hamiltonians [58], seems to be less efficient than the one described in the text above, in particular for calculating $\underline{G}(\underline{q},\underline{\Theta})$ in Eq.(30), which is nothing but $°\underline{G}[\underline{x}^{SF}(\underline{q},\underline{\Theta},\underline{r})]$. Indeed, the matrices appearing in (30) are either pre-established, cf. $\underline{\underline{\Omega}}^{-1}(\underline{\Theta})$, or obtainable from the only expressions $\underline{x}^{BF}(\underline{q})$, cf. $\underline{\underline{\Gamma}}(\underline{q})$ and $\underline{\mu}(\underline{q})$ (remind that $\underline{\underline{\Sigma}}(\underline{q})$ is frame-invariant).

References

1. P.G. Mezey, Theor.Chim.Acta, **54**, 95 (1980)
2. D. Heidrich, W. Kliesch and W. Quapp, *Properties of chemically interesting potential energy surfaces*, Springer-Verlag, Berlin, 1991
3. W. Quapp and D. Heidrich, Theor.Chim.Acta, **66**, 245 (1984)
4. H.B. Schlegel, Adv.Chem.Phys., **68**, 253 (1987)
5. C. Gonzales and H.B. Schlegel, J.Chem.Phys., **90**, 2154 (1990)
6. C. Gonzales and H.B. Schlegel, J.Phys.Chem., **94**, 5523 (1990)
7. P.G. Mezey, *Potential energy hypersurfaces* (Studies in physical and theoretical chemistry 53), Elsevier, Amsterdam, 1987
8. B. Friedrich, Z. Herman, R. Zahradnik and Z. Havlas, in Advances in quantum chemistry, Vol. 19, Academic Press, 1988, p. 247
9a. K. Fukui, J.Phys.Chem., **74**, 4161 (1970)
9b. J.W. McIver and A. Komornicki, J.Am.Chem.Soc., **94**, 2625 (1972)
10. P. Pechukas, J.Chem.Phys., **64**, 1516 (1976)
11. A. Tachibana and K. Fukui, Theor.Chim.Acta, **49**, 321 (1978)
12. A. Tachibana and K. Fukui, Theor.Chim.Acta, **51**, 189, 275 (1979)
13. A. Tachibana and K. Fukui, Theor.Chim.Acta, **57**, 81 (1980)

14. A. Tachibana, Theor.Chim.Acta, **58**, 301 (1981)

15. B.A. Ruf and W.H. Miller, J.Chem.Soc., **84**, 1523 (1988)

16. A. Tachibana, I. Okazaki, M. Koizumi, K. Hori and T. Yamabe, J.Am.Chem.Soc., **107**, 1190 (1985)

17. A. Tachibana, H. Fueno and T. Yamabe, J.Am.Chem.Soc., **108**, 4346 (1986)

18. M. Page and J. McIver, J.Chem.Phys., **88**, 922 (1988)

19. M. Page, C. Doubleday and J. McIver, J.Chem.Phys., **93**, 5634 (1990)

20. N.R. Walet, A. Klein and G. Do Dang, J.Chem.Phys., **91**, 2848 (1989)

21. B.L. Garrett, M.J. Redmon, R. Steckler, D.G. Truhlar, K.K. Baldridge, D. Bartol, M.V. Schmidt and M.S. Gordon, J.Phys.Chem., **92**, 1476 (1988)

22. B.H. Schlegel, in K.P. Lawley (ed.) *Ab initio methods in quantum chemistry I*, Wiley, New York, 1987, p.249

23. A. Banerjee, N. Adams and J. Simons, J.Phys.Chem., **89**, 52 (1985)

24. J. Baker, J.Comput.Chem., **7**, 385 (1986)

25. W. Kliesch, K. Schenk, D. Heidrich and H. Dachsel, J.Comput.Chem., **9**, 810 (1988)

26. M.V. Basilevski and A.G. Shamov, Chem.Phys., **60**, 347 (1981)

27. K. Fukui, S Kato and S. Fujimoto, J.Am.Chem.Soc., **97**, 1 (1975)

28. D.G. Truhlar and A. Kupperman, J.Am.Chem.Soc., **93**, 1840 (1971)

29. D.G. Truhlar, F.B. Brown, R. Steckler and A.D. Isaacson, in D.C. Clary (ed.) *Theory of chemical reaction dynamics*, Reidel, Dordrecht, 1986, p.285

30. C.E. Dykstra, Acc.Chem.Res., **21**, 355 (1988)

31. W. Quapp, H. Dachsel and D. Heidrtich, J.Molec.Struct. (Theochem), **205**, 245 (1990)

32. X. Chapuisat, A. Nauts and J.P. Brunet, Molec.Phys., **72**, 1 (1991)

33. W.H. Miller, N.C. Handy and J.E. Adams, J.Chem.Phys., **72**, 99 (1980)

34. A. Nauts and X. Chapuisat, Chem.Phys.Lett., **85**, 212 (1982)

35. K. Fukui, in R. Daudel and B. Pullman (eds.) *The world of quantum chemistry*, Reidel, Dordrecht, 1974, p.113

36. A. Nauts et X. Chapuisat, Chem.Phys., **76**, 349 (1983)

37. B. Hartke and J. Manz, J.Am.Chem.Soc., **110**, 3063 (1988)

38. G.C. Schatz, Chem.Rev., **87**, 81 (1987)

39. C.A. Parr, J.C. Polanyi and W.H. Wong, J.Chem.Phys., **58**, 5 (1973)

40. W.H. Miller, B.A. Ruf and Y-T Chang, J.Chem.Phys., **89**, 6298 (1988)

41. N. Shida, P.F. Barbara and J.E. Almf, J.Phys.Chem., **99**, 4061 (1989)

42. R.A. Marcus, J.Chem.Phys., **49**, 2610 (1968)

43. K. Morokuma and S. Kato, in D.G. Truhlar (ed.) *Potential energy surfaces and dynamics calculations*, Plenum, New York , 1981

44. W.H. Miller, in D.C. Clary (ed.) *The theory of chemical reaction dynamics*, Reidel, Dordrecht, 1986, p.27

45. M.V. Basilevski, Chem.Phys., **24**, 81 (1977)

46. S.M. Colwell, Theor.Chim.Acta, **74**, 123 (1988)

47. D.G. Truhlar, R. Steckler and M.S. Gordon, Chem.Rev., **87**, 217 (1987)

48. H. Eyring, J.Chem.Phys., **3**, 107 (1935)

49. S. Glasstone, K.J. Laidler and H. Eyring, *The theory of rate processes*, McGraw-Hill, New York, 1941

50. K.J. Laidler, *Theories of chemical reaction rates, McGraw-Hill*, New York, 1969, p.78

51. Special issue, dedicated to H. Eyring, J.Phys.Chem., **87**, number 15 (1983)

52. K.J. Laidler and M.C. King, J.Phys.Chem., **87**, 2657 (1983)

53. M.M. Kreevoy and D.G. Truhlar, in A. Weissenberg (ed.) *Techniques of chemistry*, vol. 6, C.F. Bernasconi (ed.) Part 1, Wiley, New York, 1986, p.13

54. J.N. Murrell, S. Carter and L.O. Halonen, J.Molec.Spectrosc., **93**, 307 (1982)

55. A.J.C. Varandas, J. Brabdao and L.M. Quintales, J.Phys.Chem., **93**, 3732 (1988)

56. M.R. Pastrana, L.A.M. Quintales, J. Brandao and A.J.C. Varandas, J.Phys.Chem., **94**, 8073 (1990)

57. B.T. Sutcliffe, in R. Carbo (ed.) *Current aspects of quantum chemistry*, Elsevier, 1982

58. N.C. Handy, Molec. Phys., **61**, 207 (1987)

59. X. Chapuisat, Molec.Phys., **72**, 1233 (1991)

60. A. Nauts and X. Chapuisat, Molec.Phys., **55**, 1287 (1985)

61. X. Chapuisat and A. Nauts, Molec.Phys., **82**, 1131 (1994)

62. A. Nauts and X. Chapuisat, Chem.Phys.Lett., **136**, 164 (1987)

63. M. Menou and X. Chapuisat, J.Molec.Spectry., **159**, 300 (1993)

64. M.V. Basilevski, Chem.Phys., **67**, 337 (1982)

65. D.K. Hoffman, R.S. Nord and K. Ruedenberg, Theor.Chim.Acta, **69**, 265 (1986)

66. W. Quapp, Theor.Chim.Acta, **75**, 447 (1989)

67. X. Chapuisat, J.Molec.Spectry., to be published

ROLE AND TREATMENT OF ZERO EIGENVALUES OF ROTATION IN THE CARTESIAN FORCE CONSTANT MATRIX ALONG REACTION PATH

AKITOMO TACHIBANA[a] and TOSHIHIRO IWAI[b]

Division of Molecular Engineering[a]
and Department of Applied Mathematics and Physics,[b]
Faculty of Engineering, Kyoto University,
Kyoto 606-01, Japan

1 Introduction

Rotational vectors in chemistry appear as eigenvectors associated with three zero eigenvalues of the force constant matrix at an equilibrium configuration of an isolated molecule. Another set of three zero eigenvalues of the force constant matrix is related with translational motions of the center of mass. Accordingly, for N-nucleus molecular system, $f = 3N - 6$ degrees of freedom are necessary to describe the vibrational vectors at the equilibrium configuration. In GF theory of molecular vibrations, internal coordinate systems have been introduced for describing the vibrational motions of a molecule at a stable equilibrium point [1]. The internal coordinates consists of the f independent configurational parameters $q^1, q^2, \ldots, q^f$ such as bond lengths, valence angles, angles between a bond and the plane determined by three nuclei, and angles between two planes.

A chemical reaction system is thought of as a supermolecule, a gigantic molecule. Reactants (R) and products (P) are particular states of the supermolecule. The supermolecule is first arranged at a stable equilibrium position that corresponds to the R, and then a chemical reaction takes place along a large amplitude vibrational path to form the P. In describing reaction paths, the concept of "intrinsic reaction coordinate" (IRC) introduced by Fukui [2] is of great use, since the IRC is a path connecting R and P through a transition state (TS), and is concerned with internal coordinates only, being independent of translational and rotational degrees of freedom. The IRC is looked upon as a steepest desent path through a saddle point. In the string model of vibronic interaction, the moving frame associated with the IRC subject to the boundary condition at TS has been called the IRC frame [3]. The vibrational motion along the IRC plays an important role in the theory of reaction ergodography [4].

In this article, the rotational and vibrational motions are to be examined for the supermolecule in generic state, $i.e.$, without reference to equilibrium configurations.

D. Heidrich (ed.), The Reaction Path in Chemistry:
Current Approaches and Perspectives, 77–94.

Before going into the rotation and vibration theory in the center-of-mass system with the Cartesian coordinates, we are to make a brief review of the theory of reaction paths in internal coordinates.

2   A review of reaction path theory

2.1   The internal space

In order to define the internal space and internal coordinates, it is strict and of great use to exploit notions from differential geometry. Let X_0 be the space of all the ennuples $x = (x_1, \cdots, x_N)$ of the nucleus position vectors $x_\alpha \in \mathbf{R}^3, \alpha = 1, \ldots, N$. As is well-known, the translational degrees of freedom are gotten rid of from X_0 to give rise to the center-of-mass system

$$X = \{x = (x_1, \ldots, x_N) | \sum_{\alpha=1}^N m_\alpha x_\alpha = 0,\ x_\alpha \neq x_\beta \text{ if } \alpha \neq \beta\}. \tag{2.1}$$

The rotation group $SO(3)$ acts on X in a natural manner

$$gx := (gx_1, \ldots, gx_N),\quad g \in SO(3),\ x \in X. \tag{2.2}$$

By the action of g, the molecule is rotated rigidly about the center of mass. If the molecular configurations are not rectilinear, the group $SO(3)$ acts freely, *i.e.*, without fixed points, on X. This means that if $hx = x$ for $h \in SO(3)$ and for x whose configuration is not rectilinear then $h =$ identity. For x corresponding to a rectilinear configuration, the equation $hx = x$ determines a subgroup $\{h \in SO(3);\ hx = x\} \cong SO(2)$. In order to make the quotient space $X/SO(3)$ to be a manifold, we restrict X to the subset consisting of non-rectilinear molecular configurations. Then the X restricted is made into a principal fibre bundle over the quotient space [5]

$$\pi :\ X \longrightarrow M := X/SO(3), \tag{2.3}$$

where X is understood as restricted, and π is the natural projection which maps x to the equivalence class it belongs to. By definition of the quotient space, a point of M is an equivalence class of molecular configurations which are mutually connected by rotations, so that any molecular configuration of an equivalence class has the same form. The M is referred to as the internal space, whose local coordinates are called internal coordinates.

Since a potential function, originally defined on the center-of-mass system X, should be invariant under the $SO(3)$ action, it may be viewed as a function on M (see (3.32) for definition). Critical points of the potential function determine the form of the molecule at equilibrium. The M can be equipped with a natural metric

$$d\sigma^2 = \sum_{i,j=1}^f a_{ij} dq^i dq^j, \tag{2.4}$$

the definition of which will be given in Sec. 3 (see (3.20)). We here note that the metric tensor is defined on the whole internal space M, and not restricted to the vicinity of an equilibrium point in M. The inverse $(a^{ij}) = (a_{ij})^{-1}$ is identified with the Wilson's G matrix for nonrigid molecule [6].

2.2   Intrinsic reaction coordinates

For the potential function U, the IRC equation [7] is defined to be

$$\frac{dq^i}{\displaystyle\sum_{j=1}^{f} a^{ij}\frac{\partial U}{\partial q^j}} = d\tau, \quad i = 1, 2, \ldots, f, \tag{2.5}$$

which means that reaction paths run in the direction of the gradient of the potential U. The IRC is a particular solution to the IRC equation subject to the boundary condition at the transition state (TS); the displacement vector for nuclei at the TS should be in the unstable vibrational mode uniquely defined at the TS. Contrary to the usual way to treat gradient flows, we view the reaction path as starting with the reactant state for the transition state. The meta-IRC is a generic solution to the IRC equation without the boundary condition at TS, which is suitable for studying the excited state chemical reactions [7,8]. The geodesic variational principle and the related principle of least action are established in [9,10]. The meta-IRC converges to the normal coordinate with smallest force constant at the equilibrium point [8]. In Sec.5, IRC's and meta-IRC's are illustrated.

2.3   Normal vibrations at an equilibrium position

Let us denote the Hessian matrix of the potential U by H_{ij};

$$H_{ij} = \frac{\partial^2 U}{\partial q^i \partial q^j} - \sum_{k=1}^{f} \Gamma_{ij}^k \frac{\partial U}{\partial q^k}, \quad i, j = 1, 2, \ldots, f, \tag{2.6}$$

where Γ_{ij}^k's are the Christoffel symbols defined by

$$\Gamma_{jk}^i = \frac{1}{2}\sum_{\ell=1}^{f} a^{i\ell}\left(\frac{\partial a_{\ell k}}{\partial q^j} + \frac{\partial a_{j\ell}}{\partial q^k} - \frac{\partial a_{jk}}{\partial q^\ell}\right), \quad i, j, k = 1, 2, \ldots, f. \tag{2.7}$$

At equilibrium positions, the definition (2.6) becomes the usual one, owing to $\partial U/\partial q^i = 0$. In the theory of small vibrations, the vibrational coordinates at the equilibrium position are defined through the eigenvectors for the Hessian matrix at the equilibrium position. The eigenvalue equation for the Hessian matrix is described as

$$\sum_{j=1}^{f} H_{ij}v^j = \mu \sum_{j=1}^{f} a_{ij}v^j. \tag{2.8}$$

Let $v_n = (v_n^i)$, $n = 1, 2, \ldots, f$, be normalized eigenvectors associated with the eigenvalues μ_n, $n = 1, 2, \ldots, f$, along with the normalization conditions

$$\sum_{i,j=1}^{f} a_{ij}v_n^i v_m^j = \delta_{nm}, \quad n, m = 1, 2, \ldots, f. \tag{2.9}$$

Let $\Delta q = (\Delta q^i)$ denote a tangent vector at the equilibrium point in M. Then the normal vibrational coordinates (ΔQ^n), $n = 1, 2, \ldots, f$, at the equilibrium point are defined through

$$\Delta q = \sum_{n=1}^{f} v_n \Delta Q^n. \tag{2.10}$$

Note here that the coordinates (ΔQ^n) are not defined in a vicinity of the equilibrium point p in M, but defined originally in the tangent space to M at p, $T_p(M)$.

The conventional theory of small vibrations at the equilibrium position is set up in the tangent space $T_p(M)$. The kinetic energy T and the potential function U are approximately described as

$$T = \frac{1}{2} \sum_{i,j=1}^{f} a_{ij} \Delta q^i \Delta q^j, \quad U(\Delta q) - U_0 = \frac{1}{2} \sum_{i,j=1}^{f} H_{ij} \Delta q^i \Delta q^j, \tag{2.11}$$

respectively, where a_{ij} and H_{ij} are both evaluated at p. Hence the equation of motion results in

$$\sum_{j=1}^{f} a_{ij} \frac{d^2 \Delta q^j}{dt^2} = -\sum_{j=1}^{f} H_{ij} \Delta q^j, \quad i = 1, 2, \ldots, f. \tag{2.12}$$

In terms of the normal vibrational coordinates ΔQ^n, this equation is put in the form

$$\frac{d^2 \Delta Q^n}{dt^2} = -\mu_n \Delta Q^n. \tag{2.13}$$

2.4   Normal vibrations across the reaction path

The reaction path theory studies local normal vibrational motions across the reaction path, and thereby provides the understanding of the vibrational energy transfer along the reaction path. Since we are considering things at non-equilibrium points, the Taylor series expansion of the potential U contains first-order derivatives;

$$U(\Delta q) - U_0 = \sum_{i=1}^{f} \frac{\partial U}{\partial q^i} \Delta q^i + \frac{1}{2} \sum_{i,j=1}^{f} H_{ij} \Delta q^i \Delta q^j. \tag{2.14}$$

Hence, we have to consider the Hessian matrix along with the gradient vector. We wish to understand that the reaction proceeds in the direction of gradient vector and "vibrational modes" across the reaction path take place in the hyperplane orthogonal to the gradient vector. It is of great use accordingly to decompose the tangent space $T_p(M)$ into the direct sum of the one-dimensional subspace in the direction of the gradient vector and its orthogonal complement, where p is a point of the meta-IRC, i.e., a point of a steepest descent path. For this purpose, we introduce the extended Hessian matrix [11] by

$$\overline{H}_{ij} := H_{ij} - \frac{1}{|\nabla U|^2} S_{ij}, \quad i, j = 1, 2, \ldots, f, \tag{2.15}$$

with

$$|\nabla U|^2 = \sum_{i,j=1}^{f} a^{ij} \frac{\partial U}{\partial q^i} \frac{\partial U}{\partial q^j}, \qquad (2.16)$$

$$S_{ij} = \left(\sum_{k,\ell} H_{ik} a^{k\ell} \frac{\partial U}{\partial q^\ell} \right) \frac{\partial U}{\partial q^j} + \frac{\partial U}{\partial q^i} \left(\sum_{k,\ell} H_{kj} a^{k\ell} \frac{\partial U}{\partial q^\ell} \right) \qquad (2.17)$$

Let $\overline{v}_f = (\overline{v}_f^i)$ denote the normalized gradient vector for U,

$$\overline{v}_f^i = \frac{1}{|\nabla U|} \sum_{j=1}^{f} a^{ij} \frac{\partial U}{\partial q^j}. \qquad (2.18)$$

Then a straightforward calculation shows that $\overline{v}_f$ is an eigenvector of the extended Hessian matrix;

$$\sum_{j=1}^{f} \overline{H}_{ij} \overline{v}_f^j = -\overline{\mu}_f \sum_{j=1}^{f} a_{ij} \overline{v}_f^j, \qquad \overline{\mu}_f = \sum_{i,j=1}^{f} H_{ij} \overline{v}_f^i \overline{v}_f^j. \qquad (2.19)$$

Since $\overline{H}_{ij}$ is a symmetric matrix, we can find vectors orthogonal to $\overline{v}_f$ as eigenvectors of $\overline{H}_{ij}$ other than $\overline{v}_f$. This accounts for why we have introduced the extended Hessian matrix $\overline{H}_{ij}$. We denote by $v_n = (\overline{v}_n^i)$, $n = 1, 2, \ldots, f - 1$, the remaining $f - 1$ normalized eigenvectors associated with eigenvalues $\overline{\mu}_n$;

$$\sum_{j=1}^{f} \overline{H}_{ij} \overline{v}_n^j = \overline{\mu}_n \sum_{j=1}^{f} a_{ij} \overline{v}_n^j, \quad n = 1, 2, \ldots, f - 1. \qquad (2.20)$$

The $\overline{v}_m$, $m = 1, \ldots, f - 1, f$, all together form an orthonormal system of tangent vectors subject to the normalization conditions

$$\sum_{i,j=1}^{f} a_{ij} \overline{v}_n^i \overline{v}_m^j = \delta_{nm}, \quad n, m = 1, 2, \ldots, f. \qquad (2.21)$$

The first $f - 1$ vectors $\overline{v}_n$, $n = 1, \ldots, f - 1$, are concerned with "vibrational modes" across the meta-IRC and the last vector $\overline{v}_f$ indicates the diretion of the reaction.

The normal modes $\Delta \overline{Q}^n$, $n = 1, 2, \ldots, f$, are defined through

$$\Delta q^i = \sum_{n=1}^{f} \overline{v}_n^i \Delta \overline{Q}^n, \quad i = 1, 2, \ldots, f. \qquad (2.22)$$

Then the Taylor expansion (2.14) becomes expressed as

$$U(\Delta q) - U_0 = |\nabla U| \Delta \overline{Q}^f + \frac{1}{2} \sum_{n=1}^{f} \overline{\mu}_n (\Delta \overline{Q}^n)^2 + \text{coupling terms}, \qquad (2.23)$$

where the coupling terms are those among f normal modes $\Delta \overline{Q}^n$, $n = 1, 2, \ldots, f$. The right-hand side of (2.23) shows that the first term induces the driving force for the reaction and the remaining terms are concerned with vibrational modes.

3   Cartesian Hessian Matrix

3.1   Rotational and vibrational vectors

In this section, we do not restrict ourselves to small vibrations. Following our previous article [6], we start with necessary geometric setting up. Let e_a, $a = 1, 2, 3$, be a moving frame in $\mathbf{R}^3$, which is designated by the Euler angles ϕ^a, $a = 1, 2, 3$. Any molecular configuration can then be assigned by both local internal coordinates q^i, $i = 1, 2, \ldots, f$, relative to the moving frame and the Euler angles. In what follows, the metric ds^2 and the connection form ω, which are closely related with the kinetic energy and the angular momentum of the molecule, respectively, play a key role; the metric is defined to be

$$ds^2 = \sum_{\alpha=1}^{N} m_\alpha (dx_\alpha | dx_\alpha), \quad x = (x_1, \ldots, x_N) \in X, \tag{3.1}$$

and the connection form to be

$$\omega = I^{-1} \sum_{\alpha=1}^{N} m_\alpha x_\alpha \times dx_\alpha, \tag{3.2}$$

where I is the inertial tensor given by

$$I = \sum_{\alpha=1}^{N} m_\alpha [(x_\alpha | x_\alpha) - |x_\alpha)(x_\alpha|]. \tag{3.3}$$

Since we treat molecular configurations in generic state, we have to define rotational and vibrational vectors definitely. Rotational vectors are defined to be infinitesimal generators of the $SO(3)$ action, which turn out to be expressed as

$$(\psi \times x_1, \ldots, \psi \times x_N) \quad \text{for } \psi \in \mathbf{R}^3. \tag{3.4}$$

In terms of differential operators, the rotational vector takes the form

$$\sum_{\alpha=1}^{N} (\psi \times x_\alpha | \frac{\partial}{\partial x_\alpha}) = (\psi | \sum_{\alpha=1}^{N} x_\alpha \times \frac{\partial}{\partial x_\alpha}) =: (\psi | J). \tag{3.5}$$

A tangent vector $v = (v_1, \cdots, v_N)$ to X at x is called a vibrational vector [5], if it is orthogonal to any rotational vector at x with respect to the metric (3.1). Hence, the tangent vector $v = (v_1, \ldots, v_N)$ is shown to be a vibrational vector, if and only if

$$\sum_{\alpha=1}^{N} m_\alpha x_\alpha \times v_\alpha = 0. \tag{3.6}$$

This is also equivalent to

$$\omega(\sum_{\alpha=1}^{N} (v_\alpha | \frac{\partial}{\partial x_\alpha})) = 0, \tag{3.7}$$

as is easily seen from (3.2). Moreover, the ω satisfies, for any rotational vector (3.5),

$$\omega(\sum_{\alpha=1}^{N}(\psi \times x_\alpha|\frac{\partial}{\partial x_\alpha})) = \psi, \quad \psi \in \mathbf{R}^3. \tag{3.8}$$

We have here to notice that vibrational vectors defined above should be distinguished from vibrational modes used in the theory of small vibrations.

We proceed to describe rotational and vibrational vectors in local coordinates (ϕ^a, q^i). Let ω^a and J_a be the components of ω and J with respect to the moving frame e_a, respectively;

$$\omega = \sum_{a=1}^{3} e_a \omega^a, \quad J = \sum_{a=1}^{3} e_a J_a. \tag{3.9}$$

Then, from (3.8) it follows that

$$\omega(J_a) = e_a, \tag{3.10}$$

which implies that ω^a are dual to J_a. Since the forms ω^a and dq^i constitute a local basis of the space of one-forms on X, we can determine vibrational vectors ξ_i through

$$\omega^a(\xi_j) = 0, \quad dq^i(\xi_j) = \delta^i_j. \tag{3.11}$$

The ξ_i and J_a are put together to form a local basis of tangent vector fields on X, satisfying

$$\omega^a(J_b) = \delta^a_b, \quad dq^i(J_b) = 0. \tag{3.12}$$

We put ω^a in the form

$$\omega^a = \sum_{b=1}^{3} \theta^a_b d\phi^b + \sum_{i=1}^{f} \beta^a_i dq^i. \tag{3.13}$$

Then, from (3.11) and (3.12) it follows that

$$J_a = \sum_{b=1}^{3} (\theta^{-1})^b_a \frac{\partial}{\partial \phi^b}, \quad \xi_i = \frac{\partial}{\partial q^i} - \sum_{a=1}^{3} \beta^a_i J_a, \tag{3.14}$$

where θ^{-1} is the inverse matrix of $\theta = (\theta^a_b)$ given in (3.13). The rotational vectors J_a and the vibrational vectors ξ_i satisfy the following commutation relations

$$[J_a, J_b] = \sum_{c=1}^{3} \varepsilon_{abc} J_c,$$

$$[\xi_i, \xi_j] = -\sum_{c=1}^{3} F^c_{ij} J_c, \quad F^c_{ij} := \frac{\partial \beta^c_j}{\partial q^i} - \frac{\partial \beta^c_i}{\partial q^j} - \sum_{a,b=1}^{3} \varepsilon_{abc} \beta^a_i \beta^b_j, \tag{3.15}$$

$$[\xi_i, J_a] = \sum_{b,c=1}^{3} \beta^b_i \varepsilon_{abc} J_c,$$

where ε_{abc} is the antisymmetric symbol with $\varepsilon_{123} = 1$. The middle equation of (3.15) means that two independent vibrational vectors are coupled to give rise to an infinitesimal rotation. This fact implies that molecular vibrations cannot be separated from rotations [5].

3.2   Moving frames in the center-of-mass system

We proceed to study the infinitesimal displacement of $x \in X$. Since ω^a and dq^i is a local basis of the space of one-forms, $dx = (dx_1, \ldots, dx_N)$ is put in the form

$$dx = \sum_{a=1}^{3} B_a \omega^a + \sum_{i=1}^{f} B_i dq^i, \tag{3.16}$$

or

$$dx_\alpha = \sum_{a=1}^{3} B_a^\alpha \omega^a + \sum_{i=1}^{f} B_i^\alpha dq^i, \quad \alpha = 1, \ldots, N,$$

where B_a and B_i, taking the $3 \times N$ matrix form, are determined to be

$$\begin{aligned} B_a &= J_a x, \quad i.e., \quad B_a^\alpha = J_a x_\alpha, \\ B_i &= \xi_i x, \quad i.e., \quad B_i^\alpha = \xi_i x_\alpha. \end{aligned} \tag{3.17}$$

The (B_a, B_i) form a moving frame in the center-of-mass system X. From (3.17), one has

$$J_a = \sum_{\alpha=1}^{N} (B_a^\alpha | \frac{\partial}{\partial x_\alpha}), \quad \xi_i = \sum_{\alpha=1}^{N} (B_i^\alpha | \frac{\partial}{\partial x_\alpha}). \tag{3.18}$$

Further, the vectors (B_a^α, B_i^α) are shown to satisfy

$$\begin{aligned} \sum_{\alpha=1}^{N} m_\alpha (B_a^\alpha | B_b^\alpha) &= (e_a | I | e_b) =: I_{ab}, \\ \sum_{\alpha=1}^{N} m_\alpha (B_a^\alpha | B_i^\alpha) &= 0. \end{aligned} \tag{3.19}$$

Moreover, the B_i^α's define a metric tensor a_{ij} on the internal space M;

$$a_{ij} := ds^2(\xi_i, \xi_j) = \sum_{\alpha=1}^{N} m_\alpha (B_i^\alpha | B_j^\alpha). \tag{3.20}$$

Note that ds^2 is invariant under the $SO(3)$ action and that ξ_i and $\partial/\partial q^i$ are in one-to-one correspondence. Thus the metric ds^2 given by (3.1) comes to be expressed as

$$ds^2 = \sum_{a,b=1}^{3} I_{ab} \omega^a \omega^b + \sum_{i,j=1}^{f} a_{ij} dq^i dq^j. \tag{3.21}$$

The second term of the right-hand side of (3.21) is the metric $d\sigma^2$ given in (2.4).

We turn to the gradient vector $\partial/\partial x = (\partial/\partial x_1, \ldots, \partial/\partial x_N)$, which can be written as

$$\frac{\partial}{\partial x} = \sum_{a=1}^{3} s^a J_a + \sum_{i=1}^{f} s^i \xi_i, \tag{3.22}$$

or

$$\frac{\partial}{\partial x_\alpha} = \sum_{a=1}^{3} s_\alpha^a J_a + \sum_{i=1}^{f} s_\alpha^i \xi_i,$$

where $s^a = (s_\alpha^a)$ and $s^i = (s_\alpha^i)$ are dual to B_a and B_i;

$$\sum_{\alpha=1}^{N} (s_\alpha^b | B_a^\alpha) = \delta_a^b, \qquad \sum_{\alpha=1}^{N} (s_\alpha^j | B_a^\alpha) = 0,$$

$$\sum_{\alpha=1}^{N} (s_\alpha^a | B_i^\alpha) = 0, \qquad \sum_{\alpha=1}^{N} (s_\alpha^j | B_i^\alpha) = \delta_i^j,$$

$$(3.23)$$

and satisfy

$$\sum_{c=1}^{3} s_\alpha^c I_{ca} = m_\alpha B_a^\alpha, \qquad \sum_{k=1}^{f} s_\alpha^k a_{ki} = m_\alpha B_i^\alpha. \qquad (3.24)$$

These equations can be proved by using (3.18-22). Further, in the dual manner to (3.19) and (3.20), the s^a and s^i are shown to satisfy

$$\sum_{\alpha=1}^{N} \frac{1}{m_\alpha} (s_\alpha^a | s_\alpha^b) = (I^{-1})_{ab},$$

$$\sum_{\alpha=1}^{N} \frac{1}{m_\alpha} (s_\alpha^i | s_\alpha^a) = 0, \qquad (3.25)$$

$$\sum_{\alpha=1}^{N} \frac{1}{m_\alpha} (s_\alpha^i | s_\alpha^j) = a^{ij}.$$

We proceed to study the derivatives of the moving frames (B_i, B_a) and (s^i, s^a). Since B_i and B_a take the $3 \times N$ matrix form (the same matrix form as x takes), so do their derivatives. Thus the first-derivatives of (B_i, B_a) are expressed as linear combinations of the independent "vectors" (B_i, B_a) as follows:

$$J_a B_b = \sum_{c=1}^{3} G_{ab}^c B_c + \sum_{k=1}^{f} G_{ab}^k B_k,$$

$$J_a B_i = \sum_{c=1}^{3} G_{ai}^c B_c + \sum_{k=1}^{f} G_{ai}^k B_k,$$

$$\xi_i B_a = \sum_{c=1}^{3} G_{ia}^c B_c + \sum_{k=1}^{f} G_{ia}^k B_k,$$

$$\xi_i B_j = \sum_{c=1}^{3} G_{ij}^c B_c + \sum_{k=1}^{f} G_{ij}^k B_k,$$

$$(3.26)$$

where G_{ab}^c's are coefficients so defined. From (3.15), (3.17), and (3.26), the coefficients which appear in the right-hand side of (3.26) are shown to satisfy

$$G_{ab}^c - G_{ba}^c = \varepsilon_{abc}, \qquad G_{ab}^k = G_{ba}^k,$$

$$G_{ij}^a - G_{ji}^a = -F_{ij}^a, \qquad G_{ij}^k = G_{ji}^k,$$

$$G_{ia}^c - G_{ai}^c = \sum_{b=1}^3 \beta_i^b \varepsilon_{abc}, \qquad G_{ia}^k = G_{ai}^k. \tag{3.27}$$

Operating (3.19) and (3.20) with J_a and ξ_k, and using (3.26) along with the fact that I_{ab} and a_{ij} are invariant against rotation, we find that the coefficients satisfy the following equations;

$$J_d I_{ab} = \sum_{c=1}^3 G_{da}^c I_{cb} + \sum_{c=1}^3 G_{db}^c I_{ac} = 0,$$

$$\sum_{c=1}^3 G_{aj}^c I_{cb} + \sum_{k=1}^f G_{ab}^k a_{jk} = 0, \tag{3.28}$$

$$J_a a_{ij} = \sum_{k=1}^f G_{ai}^k a_{kj} + \sum_{k=1}^f G_{aj}^k a_{ik} = 0,$$

and

$$\xi_k I_{ab} = \frac{\partial I_{ab}}{\partial q^k} = \sum_{c=1}^3 G_{ka}^c I_{cb} + \sum_{c=1}^3 G_{kb}^c I_{ac},$$

$$\sum_{c=1}^3 G_{ij}^c I_{bc} + \sum_{k=1}^f G_{ib}^k a_{kj} = 0, \tag{3.29}$$

$$\xi_k a_{ij} = \frac{\partial a_{ij}}{\partial q^k} = \sum_{\ell=1}^f G_{ki}^\ell a_{\ell j} + \sum_{\ell=1}^f G_{kj}^\ell a_{i\ell}.$$

In particular, from the third equations of (3.29) together with (3.27), the coefficients G_{ij}^k are shown to be equal to the Christoffel symbols given in (2.7);

$$G_{ij}^k = \Gamma_{ij}^k. \tag{3.30}$$

In the dual manner to (3.26), we obtain, for the moving frame (s^a, s^i),

$$J_a s^b = -\sum_{c=1}^3 G_{ac}^b s^c - \sum_{k=1}^f G_{ak}^b s^k,$$

$$J_a s^i = -\sum_{c=1}^3 G_{ac}^i s^c - \sum_{k=1}^f G_{ak}^i s^k,$$

$$\xi_i s^a = -\sum_{c=1}^3 G_{ic}^a s^c - \sum_{k=1}^f G_{ik}^a s^k, \tag{3.31}$$

$$\xi_i s^j = -\sum_{c=1}^3 G_{ic}^j s^c - \sum_{k=1}^f G_{ik}^j s^k.$$

3.3   The IRC equation

Now that we have finished geometric setting up, we are to treat the potential. We denote by $\tilde{U}$ the potential function defined on the center-of-mass system X. Since the $\tilde{U}$ is invariant under the $SO(3)$ action, it determines a function U on the internal space M through

$$\tilde{U}(x) = U \circ \pi(x), \tag{3.32}$$

which is the potential we have treated in Sec.2. The distinction between $\tilde{U}$ and U makes sense in what follows. The IRC equation in the center-of-mass system X is described as

$$\frac{dx_\alpha}{\dfrac{1}{m_\alpha}\dfrac{\partial \tilde{U}}{\partial x_\alpha}} = d\tau, \quad \alpha = 1, \ldots, N. \tag{3.33}$$

Since the tangent map $d\pi$ has the property that $d\pi(J_a) = 0$, $d\pi(\xi_i) = \partial/\partial q^i$, one verifies that

$$J_a(U \circ \pi) = 0, \quad \xi_i(U \circ \pi) = \frac{\partial U}{\partial q^i}, \quad a = 1,2,3, \ i = 1, \ldots, f. \tag{3.34}$$

From (3.22) and (3.34), the gradient of $\tilde{U}$ is expressed as

$$\frac{\partial \tilde{U}}{\partial x} = \sum_{i=1}^{f} s^i \frac{\partial U}{\partial q^i}. \tag{3.35}$$

Eqs. (3.34) and (3.35) are put together along with (3.23,25) to give rise to the IRC equation (2.5) together with

$$\sum_{\alpha=1}^{N} m_\alpha x_\alpha \times \frac{dx_\alpha}{d\tau} = 0 \tag{3.36}$$

along the meta-IRC in X, $i.e.$, along a generic solution to (3.33). Equation (3.36) means that during the motion along the meta-IRC in X the molecule does not rotate. We note also that (3.36) is equivalent to $\omega^a = 0$ along the meta-IRC in X, $i.e.$,

$$\omega^a = \sum_{b=1}^{3} \theta_b^a d\phi^b + \sum_{i=1}^{f} \beta_i^a dq^i = 0. \tag{3.37}$$

If q^i are given as the meta-IRC in X, Eq.(3.37) provides ordinary differential equations for ϕ^a, so that a moving frame is determined, which is called an IRC frame.

3.4   Hessian matrix at the equilibrium position

We are now in a position to discuss the Cartesian Hessian matrix which is defined to be

$$H_{\beta\alpha} = \frac{\partial}{\partial x_\beta} \left(\frac{\partial \tilde{U}}{\partial x_\alpha} \right). \tag{3.38}$$

It is to be noted here that the matrix elements $H_{\beta\alpha}$'s are not scalars, but take the 3×3 matrix form. At an equilibrium position, from (3.35) and (3.38) the matrix elements turn out to be expressed as

$$H_{\beta\alpha} = \sum_{i,j=1}^{f} H_{ij} s_{\beta}^{j} s_{\alpha}^{i}, \tag{3.39}$$

where H_{ij} is given by (2.6) with $\partial U / \partial q^i = 0$, and $s_{\beta}^{j} s_{\alpha}^{i}$'s are dyadics. This matrix is referred to as the Cartesian force constant matrix.

Eigenvectors $V_n = (V_n^{\alpha})$, $n = 1, \ldots, f$, for $H_{\beta\alpha}$ are given in terms of the eigenvectors $v_n = (v_n^i)$ for H_{ij} by

$$V_n^{\alpha} = \sum_{i=1}^{f} B_i^{\alpha} v_n^i, \tag{3.40}$$

which satisfy

$$\sum_{\beta=1}^{N} (H_{\alpha\beta} | V_n^{\beta}) = \mu_n m_{\alpha} V_n^{\alpha}, \tag{3.41}$$

together with the normalization conditions

$$\sum_{\alpha=1}^{N} m_{\alpha} (V_n^{\alpha} | V_m^{\alpha}) = \delta_{nm}. \tag{3.42}$$

In generic situation, the Cartesian Hessian matirx $(H_{\alpha\beta})$ has six zero eigenvalues, three of which are associated with translation and the other three with rotation. In our case, the center-of-mass system is treated, the Hessian matrix has three zero eigenvalues which are linked with rotation. In fact, for vectors B_a^{α} linked with rotation (see (3.18)), one has

$$\sum_{\beta=1}^{N} (H_{\alpha\beta} | B_a^{\beta}) = 0, \quad a = 1, 2, 3. \tag{3.43}$$

The Taylor expansion of $\tilde{U}$ at the equilibrium position is now described as

$$\tilde{U}(\Delta x) - \tilde{U}_0 = \frac{1}{2} \sum_{\alpha,\beta=1}^{N} (\Delta x_{\alpha} | H_{\alpha\beta} | \Delta x_{\beta}). \tag{3.44}$$

In view of (3.16), we express the small increment Δx as

$$\Delta x = \sum_{a=1}^{3} B_a \hat{\omega}^a + \sum_{i=1}^{f} B_i \Delta q^i, \tag{3.45}$$

where $\hat{\omega}^a$ are given by (3.13) with $\Delta \phi^a$ and Δq^i substituted for $d\phi^a$ and dq^i, respectively. Inserting (3.45) into (3.44) and using (3.39), we obtain

$$\tilde{U}(\Delta x) - \tilde{U}_0 = \frac{1}{2} \sum_{i,j=1}^{f} H_{ij} \Delta q^i \Delta q^j, \tag{3.46}$$

which is equal to the Taylor expansion of U given in (2.11).

3.5   Hessian matrix along the meta-IRC

Making use of (3.22) along with (3.31) and (3.34), we obtain the matrix elements in the form

$$
H_{\beta\alpha} = -\sum_{j=1}^{f}(\sum_{a,k} s_{\beta}^{a}s_{\alpha}^{k}G_{ak}^{j} + \sum_{a,c} s_{\beta}^{a}s_{\alpha}^{c}G_{ac}^{j} + \sum_{i,c} s_{\beta}^{i}s_{\alpha}^{c}G_{ic}^{j})\frac{\partial U}{\partial q^{j}} + \sum_{i,j=1}^{f} s_{\beta}^{i}s_{\alpha}^{j}H_{ij}, \quad (3.47)
$$

where H_{ij} is defined by (2.6). From this it follows that Eq.(3.43) holds no longer, implying that the vectors B_{a}^{β} are no longer associated with zero eigenvalues. The Taylor expansion of $\tilde{U}$ is now expressed as

$$
\begin{aligned}
\tilde{U}(\Delta x) &- \tilde{U}_{0} \\
&= \sum_{\alpha=1}^{N}(\frac{\partial\tilde{U}}{\partial x_{\alpha}}|\Delta x_{\alpha}) + \frac{1}{2}\sum_{\alpha,\beta=1}^{N}(\Delta x_{\beta}|H_{\beta\alpha}|\Delta x_{\alpha}) \\
&= \sum_{i=1}^{f}\frac{\partial U}{\partial q^{i}}\Delta q^{i} + \frac{1}{2}\sum_{i,j=1}^{f} H_{ij}\Delta q^{i}\Delta q^{j} \\
&\quad -\frac{1}{2}\sum_{i,j=1}^{f}(\hat{\omega}^{a}G_{ak}^{j}\Delta q^{k} + \Delta q^{i}G_{ic}^{j}\hat{\omega}^{c} + \hat{\omega}^{a}G_{ac}^{j}\hat{\omega}^{c})\frac{\partial U}{\partial q^{j}}.
\end{aligned} \quad (3.48)
$$

This is different from the Taylor expansion, (2.14), of U. Put another way, at a generic position, the Taylor expansions of $\tilde{U}$ and U are not equal. We have to be reminded here that the potential $\tilde{U}$ is defined on the center-of-mass system X and U on the internal space M. However, if we restrict the small increment Δx_{α} to the vibrational one, that is, if we set $\hat{\omega}^{a} = 0$ in (3.45), the Taylor expansion (3.48) becomes equal to the Taylor expansion (2.14) for U.

As was pointed out in the last paragraph of Subsec.3.1, rotational and vibrational motions can not be separated. With the non-separability taken into account, geometric setting for classical molecular dynamics has been set up in [12].

4   Quantum mechanics

The reaction path theory is not a quantum theory. The quantum theory we have set up in [6] starts with the description of the kinetic energy density in terms of rotational and vibrational vectors;

$$
\frac{\hbar^{2}}{2}\sum_{\alpha=1}^{N}\frac{1}{m_{\alpha}}\left(\frac{\partial\Psi^{*}}{\partial x_{\alpha}}\Big|\frac{\partial\Psi}{\partial x_{\alpha}}\right) = \frac{\hbar}{2}\sum_{a,b=1}^{3}(I^{-1})_{ab}J_{a}\Psi^{*}J_{b}\Psi + \frac{\hbar}{2}\sum_{i,j=1}^{f}a^{ij}\xi_{i}\Psi^{*}\xi_{j}\Psi. \quad (4.1)
$$

The volume element dV on the center-of-mass system X is given by the product of the volume elements dG on $SO(3)$ and dM on M, which are defined, respectively, to be

$$dV = dG\,dM,$$

$$dG = \sin\phi^2 d\phi^1 d\phi^2 d\phi^3,$$

$$dM = J_{\mathrm{int}}dq^1 \cdots dq^f, \quad J_{\mathrm{int}} := \sqrt{\left(\frac{\sum_{\alpha=1}^{N} m_\alpha}{\prod_{\alpha=1}^{N} m_\alpha}\right)^3 \det(I_{ab})\det(a_{ij})}.$$

(4.2)

Since the kinetic energy is the integral of (4.1) with respect to dV, the kinetic energy operator is determined by the integration by parts applied to the kinetic energy integral.

To obtain the kinetic energy operator on the internal space, we have to use another notion from differential geometry. However, for motions of zero angular momentum, the operator can be expressed in the usual manner as

$$\frac{\hbar^2}{2} \frac{1}{J_{\mathrm{int}}} \sum_{i,j=1}^{f} \frac{\partial}{\partial q^i} \left(a^{ij} J_{\mathrm{int}} \frac{\partial}{\partial q^j}\right),$$

(4.3)

which was given in our previous article [6]. For motions of non-zero angular momentum, we have to choose a unitary representation of $SO(3)$ and thereby to associate a complex vector bundle over the internal space, a notion from differential geometry. Briefly, the differential operators $\partial/\partial q^i$ in (4.3) are replaced by the covariant differential operators to define the kinetic energy operator on the internal space, which acts on cross sections in the complex vector bundle. See [13,14] for instance.

5   Examples

We are to give examples which illustrate the IRC and meta-IRC discussed in Sec.2.

Fig. 1. A reaction proceeds from a reactant (R) through the transition state (TS) to a product (P), following the direction of the gradient vector of the potential energy U. The intrinsic reaction coordinate (IRC) designates the reaction pathway on M. All the other reaction pathways that lead to the same reaction product are called meta-IRC's.

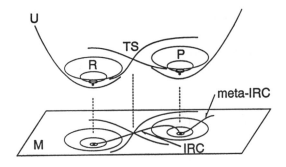

Figure 1 shows a potential energy surface U together with three critical points, R, TS, P, and level curves. The IRC designates the reaction pathway $R \to TS \to P$ on the internal space M.

For excited state chemical reactions, some alternative routes which lead to the same P may be selected according to the IRC-equation (2.5). Generic solutions to (2.5) are called meta-IRC's, the set of which defines the cell structure of the chemical reaction system [8].

Fig. 2. A chemical reaction of a water molecule in the excited state relaxes the bending vibration into the ground state, a final product. The meta-IRC shows the way how the reaction takes place.

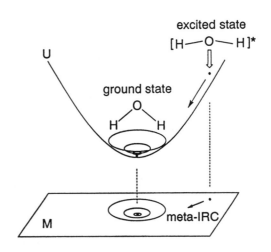

For an example of meta-IRC's, see Fig.2, which shows that an excited water molecule H_2O is relaxing into a ground-state water molecule along a chemical reaction path, i.e., along a meta-IRC. This process describes the relaxation of the bending vibration. The C_{2v} symmetry is assumed during the relaxation, so that only two geometrical parameters, the OH bond distance r and the HOH angle ϕ, suffice for the description of the internal states. The data for drawing the figure are as follows: The equilibrium geometry of H_2O is assumed to be $r_{eq} = 0.9572$Å, $\phi_{eq} = 104°32'$. The potential energy U is assumed, around the equilibrium position, to take the form of Eq. (2.11), i.e., to be expressed as a quadratic form in Δq^i with $q^1 = r_1 = r$, $q^2 = r_2 = r$, $q^3 = \phi$. The present meta-IRC is the solution to Eq. (2.5) with the initial excited state prescribed by $\Delta r = 0.0$Å and $\Delta \phi = 5.0°$ in the vicinity of the equilibrium position. The GF parameters used are as follows [1]: $H_{11} = H_{22} = 8.454$md/Å, $H_{33} = 0.6971$mdÅ/rad^2, $H_{13} = H_{23} = 0.224$md/rad, and $H_{12} = -0.100$md/Å. In integrating the IRC equation, the accumulation time of reaction, τ, is used [8], the unit of which is here set to be sec^{2}amu/kg $= 1.66057 \times 10^{-27}$sec^2. The results of the calculation are shown in Fig. 3(a,b,c). The change in the parameter Δr and in $\Delta \phi$ are drawn in Fig. 3(a), in view of which one finds that the $\Delta \phi$ decreases monotonously to the equilibrium value $\Delta \phi_{eq} = 0$, and the Δr remains virtually zero, although it slightly diminishes at least at the first stage, and finally approaches the equilibrium value $\Delta r_{eq} = 0$. The change in the potential energy U along the meta-IRC is drawn in Fig. 3(b). The value of U at the equilibrium point is, of course, set equal to zero: $U_0 = 0$. The unit of the U is kcal/mol.

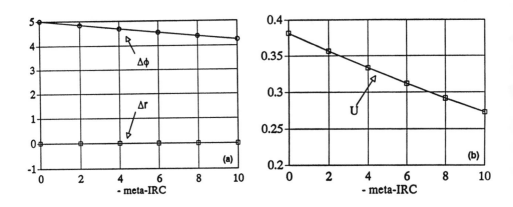

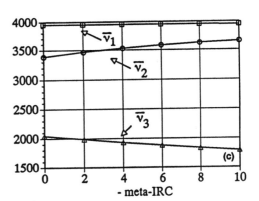

Fig. 3. Characteristics of the meta-IRC starting at $\Delta r = 0.0\text{Å}$ and $\Delta\phi = 5.0°$ in the vicinity of the equilibrium state of H_2O: the change (a) in the geometrical parameters Δr and $\Delta\phi$, (b) in the potential energy U, and (c) in the vibrational frequencies.

Figure 3(c) shows the change in the wave numbers of the vibrational frequencies across the meta-IRC, $\bar{\nu}_n = (1/2\pi c)\sqrt{\mu_n}$ with c the speed of light and $n = 1, 2, 3$, the unit of the wave numbers being cm^{-1}. The first, second, and third modes are the antisymmetric stretching, symmetric stretching, and bending modes, respectively, which tend to 3942.2, 3832.0, and 1648.3 at the equilibrium point, respectively. It should be noted that the third mode converges to the smallest vibrational frequency at the equilibrium point, in accordance with the mathematical statement called the Stable Limit Theorem [8].

If the molecule is rotating, the rotational energy [1] is added to U to form an effective potential energy, which is given by $U_{eff} = U + E_{rot}$. We are allowed to consider the meta-IRC associated with the effective potential. We assume in turn that a chemical reaction starts at the excited states with $\Delta r = 0.0 \text{Å}$ and $\Delta\phi = 70.0°$. In the case of the angular momentum $J = 1$, three rotational modes, B_1, B_2, B_3, are present; the B_1 state corresponds to the rotation around the C_2 axis, one of the two principal axes of inertia on the molecular plane, and the B_2 state to that around the other principal axis of inertia. The B_3 state corresponds to the rotation around remaining principal axis of inertia perpendicular to the molecular plane.

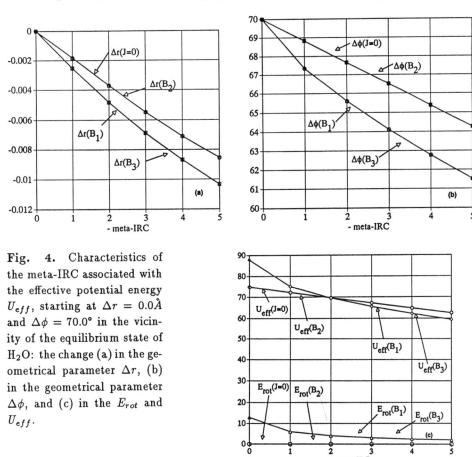

Fig. 4. Characteristics of the meta-IRC associated with the effective potential energy U_{eff}, starting at $\Delta r = 0.0 \text{Å}$ and $\Delta\phi = 70.0°$ in the vicinity of the equilibrium state of H_2O: the change (a) in the geometrical parameter Δr, (b) in the geometrical parameter $\Delta\phi$, and (c) in the E_{rot} and U_{eff}.

We observe: The Δr, again, slightly diminishes at the first stage, as shown in Fig. 4 (a), but it reaches the equilibrium value $\Delta r_{eq} = 0$ in the end. Further, the $B_{1,3}$ states affect the Δr more than the B_2 state does. The rotational influence on $\Delta\phi$ is shown in Fig. 4 (b). In Fig. 4 (c) are shown the change in the values of E_{rot} and in U_{eff} along the meta-IRC.

References

1. E.B. Wilson, J.C. Decius, and P.C. Cross, *Molecular Vibrations*, McGraw-Hill, New York, 1955.

2. K. Fukui, J. Phys. Chem. **74**, 4161 (1970).

3. A. Tachibana, in *Conceptual Trends in Quantum Chemistry*, eds. J.-L. Calais and E.S. Kryachko, Kluwer Academic, 1994.

4. F. Fukui, Acc. Chem. Res. **14**, 363 (1981).

5. A. Guichardet, Ann. Inst. Henri Poincaré, **40**, 329 (1984).

6. A. Tachibana and T. Iwai, Phys. Rev. A**33**, 2262 (1986).

7. A. Tachibana and K. Fukui, Theor. Chim. Acta (Berl.), **49**, 321 (1978).

8. A. Tachibana and K. Fukui, Theor. Chim. Acta (Berl.), **51**, 189 (1979).

9. A. Tachibana and K. Fukui, Theor. Chim. Acta (Berl.), **51**, 275 (1979).

10. A. Tachibana and K. Fukui, Theor. Chim. Acta (Berl.), **57**, 81 (1980).

11. A. Tachibana, Theor. Chim. Acta (Berl.), **58**, 301 (1981).

12. T. Iwai, Ann. Inst. Henri Poincaré, **47**, 199 (1987).

13. T. Iwai, J. Math. Phys., **28**, 964 (1987)

14. T. Iwai, J. Math. Phys., **28**, 1315 (1987)

THE INVARIANCE OF THE REACTION PATH DESCRIPTION IN ANY COORDINATE SYSTEM

WOLFGANG QUAPP

Mathematisches Institut, Universität Leipzig
D-04109 Leipzig, Germany

1. Introduction

Reaction paths are a widely used concept in theoretical chemistry. It is evident that the invariance problem, which was mathematically solved a long time ago (cf. the report given in Ref. [1]), penetrates again and again the discussions in this field (see Ref. [2]). We give both the non-invariant and the invariant definitions with respect to the choice of the particular coordinate system for two important kinds of chemical reaction pathways (RP), namely, steepest descent lines (SDP) and gradient extremal (GE) curves.
Note: The invariance problem is trivial for stationary points because the gradient is zero at those points. Invariance problems arise from a non-vanishing gradient.

The geometrical arrangement of the nuclei of a molecule or any chemical species in the 3-dimensional space R^3 can be computed in a definite mathematical way, along the lines of quantum chemistry to obtain the electronic energy of exactly this shape of the molecule. If we change some parts (coordinates) of the molecular structure, cf. **Figure 1**, we will get another energy. Thus, the potential energy surface (PES) emerges as the result of pure mathematical computations as a picture of a surface over the configuration space of the molecule. The geometry of every molecular structure clearly corresponds to a particular molecular electronic energy and these energies are independent from the kind of coordinates in the configuration space of the molecule. If we accept the independence from coordinate system, we can go to the next step: We define by a pure mathematical concept a "pathway" of changing the molecular structure from one special point of its configuration space to another point. Again, a definite energy of the molecule belongs to any point along this hypothetical pathway of the molecular rearrangement in the 3-dimensional real space, R^3. A picture of overlaid structures in R^3 of a molecular reactive change is given in **Figure 1**. From a mathematical point of view, it is clear that we can define this pathway as being independent from the choice of coordinates in the configuration space of the molecular system. And, indeed, this is the case! Each atom describes its own pathway in the 3D Cartesian space, and the total movement of N atoms of the molecular system defines the migration of a point in a configuration space R^{3N}, the migration of the so-called system point. We have yet to define this pathway.

D. Heidrich (ed.), The Reaction Path in Chemistry:
Current Approaches and Perspectives, 95–107.
© 1995 *Kluwer Academic Publishers. Printed in the Netherlan.*

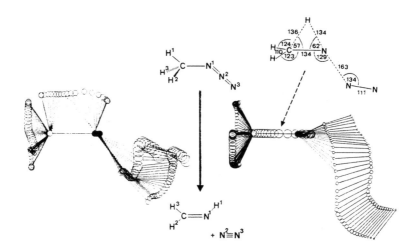

Figure 1. Overlaid molecular structure of a chemical reaction path, courtesy [3]. Left: view onto the H^1CN^1 plane; right: view in the H^1CN^1 plane. Broken bonds (- - -) depict the position of the H^1 at the saddle point.

Perhaps, the constant confusion concerning the invariance problem (cf. [2]) comes from the fact that the usual concepts for defining reaction paths use the properties of the PES in a concrete coordinate system. But, again, from a purely mathematical point of view, a change in the coordinate system by means of a definite transformation formula, can always be compensated for by changing the method for the computation of the reaction path through an inverse transformation formula. In Scheme 1 we

1. Find an adequate coordinate system for representation of PES	2. Give device for the calculation of paths on this surface	3. Result: The reaction path in the configuration space

1'. Coord.transform.: PES representation is changed	2'. Give device for calculation of paths using inverse formulas	3. Result: The same reaction path in the configuration space

1'. Coord.transform.: PES representation is changed	2. Use the original calculation of paths as in the first row	3'. Result: A different, non-invariant reaction path

Scheme 1. Different approaches for reaction path calculations

demonstrate an overview of different possibilities: First, we placed the intrinsic choice of coordinates in the first row of the scheme. For any change of the coordinate system, we have to choose row two or three. If we use the second row of scheme 1,

$\boxed{1'} \rightarrow \boxed{2'} \rightarrow \boxed{3}$, the question remains: what is the intrinsic, the genuine coordinate system for $\boxed{1} \rightarrow \boxed{2} \rightarrow \boxed{3}$?

In general, this is, and remains, a question of convention. It depends on the purpose of the investigation. For instance, a mass-weighted Cartesian system is well suited, if we are searching for chemical reaction pathways. It is an isoinertial system, and it is useful for dynamic calculations as a natural continuation of the spectroscopic treatments of vibrations and force constants of a molecule [4,5].

2. Mathematical Background

2.1. COVARIANT AND CONTRAVARIANT COORDINATES [6]

$\{e_i\}$ should be a basis in $\mathbf{R}^3$; for any vector $\mathbf{A}$, there are numbers A^i forming in this basis the sum

$$\mathbf{A} = \sum_{i=1}^{3} A^i e_i, \tag{1}$$

which is composed of the three summands, the *components* of $\mathbf{A}$. A vector can be an abstract object in $\mathbf{R}^3$ existing also without a basis $\{e_i\}$. But for computations in the sense of Cartesian geometry, we introduce a basis. The numbers A^i are called *coordinates* of $\mathbf{A}$. The scalar product of a vector $\mathbf{A}$ with a second vector $\mathbf{B} = \sum B^i e_i$ can formally be defined by

$$\mathbf{A} \cdot \mathbf{B} = (\sum_{i=1}^{3} A^i e_i)(\sum_{j=1}^{3} B^j e_j) = \sum_{i,j=1}^{3} A^i B^j (e_i \cdot e_j). \tag{2}$$

In addition to the products of the coordinates $A^i B^j$, we must acknowledge the corresponding products of the basis vectors. This product is the trivial $e_i \cdot e_j = \delta_{ij}$ in case of orthogonal unit vectors. The product is more complicated for curvilinear coordinates, and we write it with a symbol

$$e_i \cdot e_j = g_{ij} = g_{ji}, \quad i = 1, \ldots, 3. \tag{3}$$

The matrix of functions $\mathbf{g}$ of the right hand side is called the metric tensor of the basis vectors. Thus, we have the scalar product

$$\mathbf{A} \cdot \mathbf{B} = \sum_{i,j=1}^{3} A^i B^j g_{ij}. \tag{4}$$

It is a bilinear form in A^i and in B^j. If we use it as a linear form of $\mathbf{B}$, we get

$$\mathbf{A} \cdot \mathbf{B} = \mathbf{A} \cdot (\sum_{j=1}^{3} B^j e_j) = \sum_{j=1}^{3} (\mathbf{A} \cdot e_j) B^j. \tag{5}$$

We put $\mathbf{A} \cdot e_j =: A_j$, and get $\mathbf{A} \cdot \mathbf{B} = \sum_{j=1}^{3} A_j B^j$. The objects A_j with a subscript j define the vector $\mathbf{A}$ as well as the coordinates A^i. Hence, they are called "covariant" coordinates, and the old A^i, in contrast to this name, are termed the "contravariant" coordinates.

In a 2-dimensional plane we can illustrate these two types of coordinates by simple comparisons of the corresponding projections on the axes, see **Figure 2**.

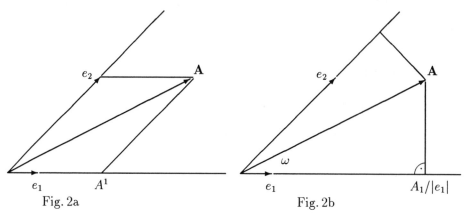

Fig. 2a   Fig. 2b

Figure 2. Contravariant (a) and covariant (b) coordinates in a plane

In contravariant coordinates, the vector **A** is, with $A^1=3$, $A^2=1$, calculated by the vector addition

$$A = 3e_1 + 1e_2. \tag{6}$$

The $A^i e_i$ is named the e_i-component of vector **A** (and not its coordinate). **A** is composed of its components.

In the covariant coordinates, the picture changes a little: Per definition we have $A_1 = \mathbf{A} \cdot e_1 = |A|\,|e_1|\,cos(\omega)$. If $|e_1| = 1$, then the coordinate A_1 is the orthogonal projection of **A** on the e_1 axis. But if $|e_1| \neq 1$, we have to divide A_1 by $|e_1|$, and get the projection

$$|\mathbf{A}|\,cos(\omega) = \frac{A_1}{|e_1|}. \tag{7}$$

The characterization of the two different coordinates is determined by the behaviour under a transformation of the basis vectors e_i into e_i'.

$$e_i' = \sum_{j=1}^{3} \alpha_i^j e_j, \tag{8}$$

where the matrix (α_i^j) is the (3×3) matrix of the transformation.

The covariant coordinates A_i are defined by $A_i = \mathbf{A} \cdot e_i$, so, it follows that

$$A_i' = \mathbf{A} \cdot e_i' = \mathbf{A} \cdot \sum_{j=1}^{3} \alpha_i^j e_j = \sum_{j=1}^{3} \alpha_i^j (\mathbf{A} \cdot e_j) = \sum_{j=1}^{3} \alpha_i^j A_j, \tag{9}$$

Eq.(9) says the A_i change in the same way as the basis vectors e_i. They are equally variable, i.e. *co-variant*. It is easy to calculate that the A^i change with the inverse transformation of $(\beta_i^j)\cdot(\alpha_j^k) = (\delta_i^k)$ by

$$e_i = \sum_{j=1}^{3} \beta_i^j e_j'. \tag{10}$$

The A^i are named *contravariant*. Covariant coordinates can be represented by contravariant coordinates (and vice versa)

$$A_j = \mathbf{A}\cdot e_j = \left(\sum_{i=1}^{3} A^i e_i\right)e_j = \sum_{i=1}^{3} A^i(e_i\cdot e_j) = \sum_{i=1}^{3} A^i g_{ij} = \sum_{i=1}^{3} g_{ij} A^i. \tag{11}$$

The work is done by a system of linear equations at any point where the matrix (g_{ij}) gives the transformation device. The inverse way uses the inverse matrix, designated by (g^{ij}):

$$A^i = \sum_{i=1}^{3} g^{ij} A_j. \tag{12}$$

The last two operations describe the shift down or the shift up of the index.

2.2. GRADIENT VECTOR AND HESSIAN MATRIX

The energy functional $E(x^1,\ldots,x^n)$ is a scalar depending on the $n = 3N$ coordinates x^i of the N nuclei of the molecule. Each nucleus has 3 coordinates (x,y,z) summed as the contravariant vector $\mathbf{x} = (x^1,\ldots,x^n)^T$. Each derivation of $E(\mathbf{x})$ to x^i yields a coordinate of a vector, termed the gradient vector of E. $\partial E/\partial x^i$ takes the ith place in

$$\mathbf{GradE} := \left(\frac{\partial E}{\partial x^1},\ldots,\frac{\partial E}{\partial x^n}\right)^T.$$

The coordinates should change, in keeping with eqs.(8) and (10), with the $(3N \times 3N)$ matrices $(\beta_i^j),(\alpha_j^i)$, as

$$x'^j = \sum_{j=1}^{3} \beta_i^j x^i, \qquad x^i = \sum_{i=1}^{3} \alpha_j^i x'^j, \tag{13}$$

If we transform the gradient, we get

$$\frac{\partial E}{\partial x^i} = \sum_{k} \frac{\partial E}{\partial x'^k}\frac{dx'^k}{dx^i} = \sum_{k} \beta_i^k \frac{\partial E}{\partial x'^k}, \qquad \frac{\partial E}{\partial x'^j} = \sum_{k} \frac{\partial E}{\partial x^k}\frac{dx^k}{dx'^j} = \sum_{k} \alpha_j^k \frac{\partial E}{\partial x^k}, \tag{14}$$

an inverse formula in comparison to eq.(13). Hence, the **Grad E**$=(G_1,\ldots,G_n)^T$ is a covariant vector [1,7]. Correspondingly, for eq.(12) we get its contravariant coordinates by

$$G^i = \sum_{j=1}^{3} g^{ij} G_j. \tag{15}$$

A further derivation of **Grad E** to any x^k gives a two-dimensional field of combinations of $\partial^2 E/\partial x^i \partial x^k$ with i and k. It is usually arranged in a matrix, named the Hessian matrix (after Otto Hesse, 1811-1874). In general, co- or contravariant characteristic cannot be assigned to the only partial derivatives of this matrix under a coordinate transformation, because there are mixed terms coming out of the chain rule. The new terms are connected with the coordinate system and can be compressed in special symbols. The matrix

$$H_{ij} = (\frac{\partial^2 E}{\partial x^i \partial x^j} - \sum_k^n \frac{\partial E}{\partial x^k}\Gamma_{ij}^k), \qquad i,j = 1,\ldots,n \tag{16}$$

with

$$\Gamma_{ij}^k = \frac{1}{2}g^{kl}(\frac{\partial g_{jl}}{\partial x^i} + \frac{\partial g_{il}}{\partial x^j} - \frac{\partial g_{ij}}{\partial x^k}) \quad \text{and} \quad g^{kl}g_{lm} = \delta_m^k, \tag{17}$$

shows the character of a two-fold covariant tensor. The functions Γ_{ij}^k are the Christoffel symbols (after Elwin Bruno Christoffel, 1829-1900). In the Cartesian coordinates, the metric elements, g_{ij}, are all constants, the Christoffel symbols are zero, and **H** reduces to the second order partial derivatives only. This simple matrix can be the initial matrix for a general definition of the Hessian tensor, in the general case of curvilinear coordinates, which we develop with eq.(16), cf. [8,9].

3. Invariant and Non-Invariant Definition of Reaction Paths of PES

3.1. PATH OF STEEPEST DESCENT

We assume a curve $\mathbf{c}(\mathbf{x}(t))=(c^1(t).\ldots,c^n(t))$ in the configuration space $\mathbf{R}^n$ where the tangent vector of $\mathbf{c}(t)$ should point at any value of the parameter t in the direction of the negative gradient vector of E. The curve $\mathbf{c}(t)$ is described by contravariant coordinates, $c^i(t)$, being itself simple functions of t. Thus, their derivatives to t are dc^i/dt (also contravariant coordinates). In any equation, we can only compare objects of the same covariant or contravariant character. We have to use the contravariant form of the gradient vector [1]

$$\frac{dc^i}{dt} = -\sum_{j=1}^n g^{ij}G_j = -G^i, \quad i = 1,\ldots,n \tag{18}$$

Both sides of eq.(18) are contravariant vectors and change according to the same rule. Mathematically, a solution of eq.(18) yields a curve invariant from the actual coordinate system.
When making the ansatz

$$\frac{dc^i}{dt} = -G_i, \quad i = 1,\ldots,n \tag{19}$$

one deals with two different kinds of vectors on both sides of the equation. Thus, we could not expect similar behaviour on both sides of the equation under coordinate transformations. We can compute a solution curve of eq.(19) with analytical formulas, if possible, or by numeric methods. However, a computation derived from these methods would make the curve of eq.(19), the steepest descent of E in the actual coordinates, a non-invariant line (see [10,11] for examples, and [1] for an extended analysis of the problem). Eq.(18) can be utilized to calculate a RP if the saddle point of the PES is found beforehand. This "intrinsic" steepest descent path (IRC) [12] leads to the next minimum. However, in general, it does not follow the valley floor line (see **Figure 4** of Ref.[13]). The reason is that the valley floor represents an asymptote of all gradient curves descending to the reaction valley from its side slopes [14]. We need a criterion that allows us to distinguish the floor line points of a valley from all other points. This is found in the gradient extremal equation [15–18].

3.2. GRADIENT EXTREMALS

For a valley floor, the ansatz of a defining curve results from the idea that the gradient norm of the PES, if proved along an equipotential level, should be minimal. In Cartesian coordinates, this leads to the equation

$$\mathbf{HG} = \lambda \mathbf{G}, \qquad (20)$$

where $\mathbf{H}$ is the Hessian matrix, $\mathbf{G}$ is the gradient vector, and λ is an eigenvalue of the Hessian matrix. Eq.(20) means that the gradient $\mathbf{G}$ itself is an eigenvector of the Hessian. Note: The character of the system of eqs.(20) is different from that of the system (18). The steepest descent system is a system of differential equations of the first order allowing an integration constant. Thus, its solution line can be fitted to arbitrary starting points. We can draw a net of gradient lines over a region of the configuration space. Eq.(20) has, in general, isolated solution curves. They do not form a field of neighbouring lines [19].

The n rows of the matrix eq.(20) become in long form

$$\sum_j (H_{ij} - \lambda \delta_{ij}) G_j = 0, \quad i = 1, \dots, n. \qquad (21)$$

If we assume the Hessian in full contravariant character corresponding to eq.(16), we can transform eq.(21) onto new coordinates x_i'. We get

$$0 = \sum_{k,l} \sum_j (H'_{kl} \frac{dx'^k}{dx^i} \frac{dx'^l}{dx^j} - \lambda \delta_{ij})(\sum_m G'_m \frac{dx'^m}{dx^j}) =$$

$$\sum_{l,m,k} (H'_{kl} G'_m \frac{dx'^k}{dx^i} (\sum_j \frac{dx'^l}{dx^j} \frac{dx'^m}{dx^j})) - \lambda \sum_k G'_k \frac{dx'^k}{dx^i} =$$

$$\sum_{l,m,k} (H'_{kl} G'_m g'^{lm} - \lambda G'_k) \frac{dx'^k}{dx^i} = 0, \qquad i = 1, \dots, n.$$

Because $\frac{dx'^k}{dx^i} \neq 0$, the transformation should not be singular. We get the general, invariant gradient extremal equation, cf. [16]

$$\sum_{l,m}^n g^{lm} H_{kl} G_m = -\lambda G_k, \quad k = 1, \dots, n. \qquad (22)$$

Again, it follows from the tensor character of this equation that any solution curve is invariant under coordinate transformations.

4. Discussion

4.1. STEEPEST DESCENT PATHS

We have shown (cf. Figure 4 in ref.[13]) that SDP and GE are different curves in the general curvilinear case. It seems that GE curves describe valley bottoms well. Hence, no steepest descent path can be assumed to be the valley line if it is a curved line! Only the GE along a straight floor line coincides with the SDP. However, there is another interesting intersection between SDP and GE [20,21], but only at special points. If the curvature of a SDP turns from positive to negative value (or vice versa) then it meets a GE at exactly that point where its curvature is zero. This gives a second device to construct GE curves: Connect the points of different SDP with zero curvature [20].

Remark: Only in the points where SDP and GE meet, i.e. where the gradient is per definition an eigenvector of $\mathbf{H}$, the tangent to a SDP is also an eigenvector of $\mathbf{H}$. At all other points this statement would be nonsense (or, in other words, would yield a very crude approximation).

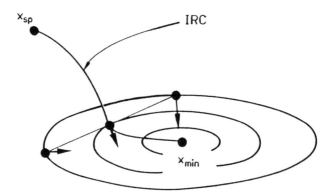

Figure 3. Model PES with a steepest descent path (IRC) which is not a valley floor line

It is an easy exercise to see that a SDP meets a minimizer along the eigenvector to the smallest eigenvalue [11]. Nearly all SDP within a minimum catchment basin have this property. We can choose the special SDP starting at a saddle point. This special SDP also converges to the next minimum in direction of the smallest eigenvector. It is the so-called intrinsic reaction coordinate (IRC) of Fukui [7,12,22]. Chemistry is interested in such a connection-line. A connection between a saddle point and the corresponding reactant or product structure yields an image of the possible change of the molecular system during a reaction even if, for instance, a classical trajectory of the system point follows another pathway on the PES to reach the saddle point after excitation from a minimum.

Because the SDP does not, in general, describes the valley floor, the SDP is a poor model to imagine vibrations orthogonal to this path. In **Figure 3** we symbolize this situation [1]. A SDP, coming from a saddle point, goes down to the next minimum somewhere on the slope of the PES. How should we imagine a vibration orthogonal to this descent? Of course, if the PES is convex, the tangent to an equipotential line shows an ascent of energy in both directions. In an harmonic approximation we could be misslead into thinking that the ansatz

$$E(s,t) = IRC(s) + \omega(s)t^2 \tag{23}$$

would be valid. Here, s is the arc length of the SDP and t is a coordinate directed orthogonal to s, measuring along the tangent of an equipotential line. **Figure 3** shows the uselessness of this ansatz (cf. [23]) for the definition of vibrations orthogonal to the "IRC valley" in dynamic treatments.

4.2. STRUCTURE OF VALLEYS

The most severe topological objects of a PES are the stationary points: minima and saddle points of any index between 1 and (n-1). Here, the gradient of E is zero, and this property overrides all regular coordinate transformations.

In the usual case a reaction path is going from a min_1 to the saddle and to min_2 along a valley of the PES. What happens with the valley under coordinate transformation? In most cases, a valley is a dominant topological feature of the PES. But note, this is not the problem in defining an invariant pathway. If we choose a mass-weighted Cartesian coordinate system to be the intrinsic one and define a steepest descent or a gradient extremal in this system, we can then define these curves in any other system: As in Scheme 1 along row 3, or along row 2 to get the old one. In the former case, under a non-invariant calculation, the new curve is a proper SDP or GE of the PES in the new coordinates. But we do not know how the valley is changed. In the latter case, the new curve can cross the isopotential lines of the new PES in a skew manner, along an invariant calculation [10].

A further problem is the knowledge of the neighbourhood of a special curve defined by a GE. We are interested in the curvature properties ω orthogonal to the reaction path line analogous to eq.(23). Many things can happen. For example, a valley coming down from the saddle point can bifurcate, in which case, its continuation is at least threefold: Two valleys with a ridge inbetween should result [11,24-27]. We cannot determine a bifurcation point at the slope of the PES by means of SDP because the differential equation system (18) is an autonomous one. Its solution cannot bifurcate outside of a critical point [27]. Taking bifurcations into consideration, we need the new tool of GE equations [27,28]. Nevertheless, a definition of a GE (and any RP) is a *one-dimensional* concept. It does not answer the question of how to explore its neighbourhood if we go the non-invariant way of a coordinate transformation. We can use any kind of parametrization, for example by the arc lenght s, for a SDP [11], or the GE [29]. But this says nothing to the approximation of the PES in the neighborhood. We can also approximate the remainder of the (n-1) coordinates of the PES in a "non-physical" coordinate choice (see a warning from Truhlar [30]). In **Figure 4**, we give a rough impression of how such a non-physical approach can result in an artificial valley-ridge inflection. In Figure 4 top row, an IRC is the

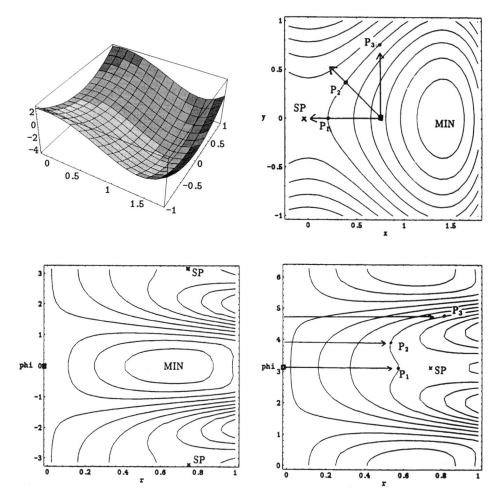

Figure 4. Top row: model PES E(x,y)=x²(-4+2x²)+3y², with a trivial IRC=MinimumGE along the x axis. At bottom: the same PES in non-physical coordinates (see text).

x axis and arrows parallel to the y axis always meet the equipotential lines in a convex manner. Thus, the IRC is a valley line throughout. In Figures 4 bottom row a curvilinear coordinate transformation is done with a simple polar coordinate transformation at point [0.75, 0.0] depicted by a small quadrat □:

$$x = r \cos \phi + 0.75, \qquad y = r \sin \phi.$$

The transformation changes the character of the PES totally. The bottom row left shows the ϕ=0 line in the center. So, the view along the IRC to the MIN is quite normal. The former valley continues to be a valley. However, in the bottom row right, we obtain a ridge. In the center, the ϕ=180° line is shown in order to get a view onto the SP. Only shortly before the saddle point does the character change to a valley.

The three angles $90°$, $135°$, and $180°$ are pointed out by arrows top right and bottom right. The point P_2 is a normal point on the level line in the x,y coordinates, but it becomes a ridge point in the r,ϕ coordinates after the coordinate transformation.

4.3. COVARIANCE

The definitions of eq.(18) for an invariant SDP and particularly of eq.(22) for an invariant GE follow a main concept of differential geometry and modern physics. If we have any geometrical object in a space and we can show its tensor field behaviour, then we can transform partial derivatives (not forming a tensor field itself) into a tensor field by adding new terms [31].

Definition: E is a scalar field, G_j is a covariant tensor field, and G^j a contravariant tensor field. We define *covariant differentiation* by

$$\nabla_i E = \frac{\partial E}{\partial x^i}, \quad i,j = 1,\ldots,n, \tag{24}$$

$$\nabla_i G_j = \frac{\partial G_j}{\partial x^i} - \sum_s^n \Gamma_{ij}^s G_s, \quad i,j = 1,\ldots,n, \tag{25}$$

$$\nabla_i G^j = \frac{\partial G^j}{\partial x^i} + \sum_s^n \Gamma_{is}^j G^s, \quad i,j = 1,\ldots,n. \tag{26}$$

We have the mathematical proposition that the covariant differentiation (24)-(26) defines tensor fields of the corresponding index picture. Its proof follows the simple mathematical rule of thumb: "Tensor calculus is an application of the chain rule." With eq.(8), we have

$$g'_{kl} = \sum_{i,j} \alpha_k^i \alpha_l^j g_{ij}, \tag{27}$$

$$g'^{kl} = \sum_{i,j} \beta_i^k \beta_j^l g^{ij}, \tag{28}$$

and we obtain by differentiation

$$\Gamma'^r_{st} = \sum_{i,j,k} \alpha_s^i \alpha_t^j \beta_k^r \Gamma_{ij}^k + \sum_l \frac{\partial \alpha_t^l}{\partial x^s} \beta_l^r. \tag{29}$$

Thus, the Γ_{ij}^k are not transformed tensorially. An additional term appears. This term compensates in definitions (24)-(26) of the covariant differentiation for the troublesome terms and yields, for example, for (25) the rule

$$\nabla'_k G'_l = \sum_{i,j}^n \alpha_k^i \alpha_l^j \nabla_i G_j. \tag{30}$$

Any further covariant differentiation for more general tensor fields uses the following method:

(i) Besides the given tensor field $T_{i_1,\ldots,i_n}^{j_1,\ldots,j_n}$ consider the product $T_{i_1}\ldots T_{i_n}T^{j_1}\ldots T^{j_n}$.

(ii) Compute the covariant derivation of the product by formally applying the rules of eq.(25) or eq.(26).

(iii) The covariant differentiation of the given tensor field is defined analogously to (ii).

We obtain, for example, a covariant differentiation of the Hessian tensor eq.(16) by

$$\nabla_k H_{ij} = \frac{\partial H_{ij}}{\partial x^k} - \sum_s^n \Gamma_{ki}^s H_{sj} - \sum_s^n \Gamma_{kj}^s H_{is}, \qquad i,j,k = 1,\ldots,n, \qquad (31)$$

which we could possibly need in the invariant calculation of bifurcation points of GE [20,28].

Acknowledgments

WQ thanks Dietmar Heidrich and Olaf Imig for long and fruitful discussions. The work was made possible by the financial support of the Deutsche Forschungsgemeinschaft.

References

1. W. Quapp and D. Heidrich, *Theor. Chim. Acta* **66**, 245 (1984).

2. See special issue of *J. Chem. Soc. Faraday Trans.* **90**, No.12 (1994).

3. H. Bock and R. Dammel, *Angew. Chem. (Int. Ed.)* **26**, 504 (1987).

4. G. Herzberg, Molecular Spectra and Molecular Structure, Vol.II, Krieger Publ., Malabar, 1991

5. E. B. Wilson, J. C. Decius , and P. C. Cross, Molecular Vibrations, McGraw-Hill Comp., New York, 1955.

6. G. Eisenreich, Lineare Algebra und Analytische Geometrie, Akademie-Verl., Berlin, 1989.

7. A. Tachibana and K. Fukui, *Theor. Chim. Acta* **49**, 321 (1978).

8. D. Gromoll, W. Klingenberg, and W. Meyer, Riemannsche Geometrie im Großen, *Lecture Notes in Mathematics* **55**, pp.89, 1968.

9. A. Banerjee and N. P. Adams, *Int. J. Quantum Chem.* **43**, 855 (1992).

10. P. G. Mezey, Potential Energy Hypersurfaces, Elsevier, Amsterdam, 1987.

11. D. Heidrich, W. Kliesch, and W. Quapp, Properties of Chemically Interesting Potential Energy Surfaces, *Lecture Notes in Chemistry* **56**, Springer, Berlin, 1991.

12. K. Fukui, *J. Phys. Chem.* **74**, 4161 (1970).

13. W. Quapp, O. Imig, and D. Heidrich, this book.

14. M. V. Basilevski, *Chem. Phys.* **24**, 81 (1977).

15. S. Pancíř, *Coll. Czech. Chem. Comm.* **40**, 1112 (1975).

16. M. V. Basilevski and A. G. Shamov, *Chem. Phys.* **60**, 347 (1981).

17. M. V. Basilevski, *Chem. Phys.* **67**, 337 (1982).

18. D. K. Hoffman, R. S. Nord, and K. Ruedenberg, *Theor. Chim. Acta*, **69**, 265 (1986).

19. W. Quapp, *Theor. Chim. Acta* **75**, 447 (1989).

20. J.-Q. Sun and K. Ruedenberg, *J. Chem. Phys.* **98**, 9707 (1993).

21. D. J. Rowe and A. Ryman, *J. Math. Phys.*, **23**, 732 (1982).

22. L. L. Stacho and M. I. Ban, *Theor. Chim. Acta* **84**, 535 (1993).

23. B. A. Ruf and W. H. Miller, *J. Chem. Soc., Faraday II* **84**, 1523 (1988).

24. A. Tachibana, H. Fueno, I. Okazaki, and T. Yamabe, *Int. J. Quantum Chem.* **42**, 929 (1992).

25. T. Taketsuga and T. Hirano, *J. Chem. Phys* **99**, 9806 (1993).

26. H. B. Schlegel, *J. Chem. Soc., Faraday Trans.* **90**, 1569 (1994).

27. W. Quapp, *J. Chem. Soc., Faraday Trans.* **90**, 1607 (1994).

28. O. Imig, D. Heidrich, and W. Quapp, in preparation, (1995).

29. H. B. Schlegel, *Theor. Chim. Acta* **83**, 21 (1992).

30. D. Truhlar, *J. Chem. Soc., Faraday Trans.* **90**, 1608 (1994).

31. E. Zeidler, Nonlinear Functional Analysis and its Application IV, Application to Mathematical Physics, Springer, New York, 1988.

SECOND-ORDER METHODS FOR THE OPTIMIZATION OF MOLECULAR POTENTIAL ENERGY SURFACES

TRYGVE HELGAKER and KENNETH RUUD
Department of Chemistry
University of Oslo
P. O. Box 1033, Blindern
N-0315 Oslo, Norway

PETER R. TAYLOR
San Diego Supercomputer Center
P. O. Box 85608, San Diego
CA 92186-9784 USA

1 Introduction

The optimization of *ab initio* Born-Oppenheimer potential energy surfaces is an important subject in quantum chemistry. Methods for the efficient localization and characterization of stationary points of the molecular potential energy surface are an essential tool in studies of molecular structure and reactivity. Of particular importance are methods for the determination of minima and first-order saddle points, corresponding to molecular equilibrium structures and transition states.

Over the years, many techniques have been devised that allow the stationary points to be located in a reasonable amount of computer time (compared to the time required for the calculation of the energy of a single point on the potential energy surface). In particular, techniques for minimizations have developed to a stage where the equilibrium structure of large molecules can be calculated routinely for accurate wave functions. In one way or another, these techniques are modifications or applications of standard methods for minimizations, as described in the monographs by Fletcher [1], by Gill, Murray and Wright [2], and by Dennis and Schnabel [3]. In particular, the so-called quasi-Newton methods have become popular for minimizations of potential energy surfaces. With a suitable choice of coordinates and initial Hessian, these methods converge in a reasonable number of iterations. In addition to the energy, each iteration requires the evaluation of the molecular gradient only. Molecular gradients are today available for all important classes of electronic *ab initio* wave functions and may be calculated in about the same amount of time that is required for the energy.

The localization of first-order saddle points corresponding to molecular transition states is a much more difficult problem. Indeed, transition state searches cannot yet be called routine. There are several reasons for this situation. First, there is apparently little interest in the subject of saddle points outside the area of quantum chemistry. Thus, the above

D. Heidrich (ed.), The Reaction Path in Chemistry:
Current Approaches and Perspectives, 109–136.
© *1995 Kluwer Academic Publishers. Printed in the Netherlands.*

monographs contain no information on methods for the localization of saddle points. Those methods that have been developed within the field of quantum chemistry therefore have not been able to draw on extensive experience from other fields. Second, searches of first-order saddle points are inherently more difficult than minimizations since there is no simple measure of progress. Clearly, the surface should be climbed uphill in one direction and downhill in all others, but it is not easy to formulate a simple criterion that allows us to distinguish the uphill and downhill directions. Third, the user often has little information or intuition on which to base his initial guess of the transition state—considerably less than for minimizations. These circumstances combine to make the determination of transition states a much more difficult problem than the localization of equilibrium structures.

Nevertheless, over the last ten years or so quite a few useful transition state methods have been developed and implemented in standard *ab initio* electronic structure programs. In our experience, the more useful of these require the evaluation of the exact molecular Hessian in each iteration—in addition to the energy and the gradient. With such techniques, it is possible to determine "conformational" saddle points representing conformational rearrangements rather straightforwardly, but the determination of "reactive" saddle points—representing bond breaking and bond formation—remains a difficult task, at least for systems containing more than a few atoms. Indeed, the successful exploration of reactive transition states is still to a large extent a matter of trial and error, being critically dependent on the initial guess. It is perhaps appropriate here to cite Havlas and Zahradník, who in a survey of optimization techniques in 1984 noted [4]: "Localization of a saddle point is easy to make only in laboratories other than our own. For us it has always been a troublesome job." Although much progress has been made since then, it is still true that the determination of saddle points sometimes resembles art more than technique, relying more on inspiration than on routine.

We review in this paper some methods for the optimization of minima and first-order saddle points of *ab initio* Born-Oppenheimer molecular potential energy surfaces. Although standard quasi-Newton methods are covered, our emphasis throughout is on second-order methods, requiring the knowledge of the exact molecular gradient and Hessian at each step. Our reasons for this choice are twofold. First, although for minimizations the ubiquitous first-order quasi-Newton methods are often the most economical, difficult situations do arise where quasi-Newton methods do not converge in a reasonable number of steps. Second, first-order methods are not particularly well suited for saddle points since the gradients and updated Hessians do not in general provide the information needed to identify the appropriate uphill and downhill directions. Thus, the development of robust, globally convergent techniques for the routine optimization of saddle points requires the use of second-order methods. Indeed, there are some indications that even anharmonicity information may be required for the development of black-box methods. We shall here discuss some schemes for saddle points and illustrate their usefulness by a few selected examples. For other reviews of methods for optimization of potential energy surfaces, see Schlegel [5, 6], Head, Weiner, and Zerner [7] and Head and Zerner [8].

2 General Considerations

2.1 PARAMETRIZATION AND NOTATION

Mathematically our task is to identify and characterize stationary points $\mathbf{x}^*$ of a smooth non-linear function $f(\mathbf{x})$ of a set of parameters $\mathbf{x}$. Physically, the function $f(\mathbf{x})$ corresponds to the potential energy surfaces (PES) of some molecular system and $\mathbf{x}$ is a set of coordinates describing the molecular configuration. Different choices of coordinates can be made. In many cases one may simply choose to work with a redundant set of Cartesian coordinates, possibly mass-weighted. In other cases, it may be more advantageous to parametrize the PES in terms of a set of redundant or non-redundant internal coordinates (bond lengths and bond angles). Although we do not discuss the choice of coordinates in much depth, we shall briefly return to this subject in Section 4.5.

It is appropriate here to introduce some notation. The gradient is a column vector whose elements are the first derivatives of $f(\mathbf{x})$

$$\mathbf{g}(\mathbf{x}) = \nabla f(\mathbf{x}) \tag{1}$$

and the Hessian is a matrix whose elements are the second derivatives with respect to $\mathbf{x}$

$$\mathbf{G}(\mathbf{x}) = \nabla\nabla^T f(\mathbf{x}) \tag{2}$$

where $\nabla\nabla^T$ is a dyadic product. The Hessian is symmetric and may therefore be diagonalized

$$\mathbf{G}(\mathbf{x})\mathbf{v}_i(\mathbf{x}) = \lambda_i(\mathbf{x})\mathbf{v}_i(\mathbf{x}) \tag{3}$$

to yield a set of real eigenvalues $\lambda_i(\mathbf{x})$ and an associated set of orthonormal eigenvectors $\mathbf{v}_i(\mathbf{x})$

$$\mathbf{v}_i^T(\mathbf{x})\mathbf{v}_i(\mathbf{x}) = \delta_{ij} \tag{4}$$

We may thus write the Hessian in the spectral representation as

$$\mathbf{G}(\mathbf{x}) = \sum_i \lambda_i(\mathbf{x})\mathbf{v}_i(\mathbf{x})\mathbf{v}_i^T(\mathbf{x}) \tag{5}$$

and the gradient may be written in the following way

$$\mathbf{g}(\mathbf{x}) = \sum_i \phi_i(\mathbf{x})\mathbf{v}_i(\mathbf{x}) \tag{6}$$

where the projections of the gradient on the eigenvectors are given by

$$\phi_i(\mathbf{x}) = \mathbf{v}_i^T(\mathbf{x})\mathbf{g}(\mathbf{x}) \tag{7}$$

In the following we shall often expand the function in a Taylor series around some "current" point $\mathbf{x}_c$:

$$f(\mathbf{x}) = f(\mathbf{x}_c + \mathbf{s}) = f_c + \mathbf{g}_c^T\mathbf{s} + \frac{1}{2}\mathbf{s}^T\mathbf{G}_c\mathbf{s} + O(s^3) \tag{8}$$

Here the subscript c indicates evaluation at $\mathbf{x}_c$, and we have also introduced $\mathbf{s} = \mathbf{x} - \mathbf{x}_c$.

2.2 STATIONARY POINTS

The stationary points (optimizers $\mathbf{x}^*$) of interest are either minima (representing equilibrium configurations) or first-order saddle points (representing transition states). Sufficient conditions for minima are vanishing a gradient $\mathbf{g}(\mathbf{x})$ and a positive definite Hessian $\mathbf{G}(\mathbf{x})$, i.e., zero slopes and positive curvatures in all directions. The corresponding conditions for transition states are a vanishing gradient and a Hessian with one and only one negative eigenvalue, i.e., zero slopes and positive curvatures in all directions except one (the reaction mode). The number of negative eigenvalues of the Hessian is referred to as the Hessian index. Thus minima have zero Hessian index and transition states have Hessian index one.

Although a positive definite Hessian is a sufficient condition for a minimum, it is not a necessary one since a so-called weak minimum may exist with one or several zero eigenvalues. Obviously, in a redundant coordinate system all minima are weak since for any molecular geometry we may continuously change the redundant (Cartesian or internal) coordinates without affecting the potential energy. In the following, we assume that zero eigenvalues always arise as a result of a redundant parametrization of the PES. Zero eigenvalues arising from redundant parametrizations are easily dealt with by a suitable projection technique.

2.3 STRATEGIES FOR OPTIMIZATION

Roughly speaking, optimizations proceed in two stages. In the first stage the purpose is to take us from some initial guess of the optimizer to a point in its neighborhood. In this global part of the search we may for example start with a Hessian with incorrect index and our immediate goal is to locate a region with correct curvature. In the final, local stage of the optimization our purpose is to determine the exact location of the optimizer starting from a point in its immediate vicinity. A useful algorithm should handle both stages successfully. It should converge globally (i.e., from any starting point) and be fast in the local region.

All practical methods start by constructing a *local model* of the function in the vicinity of the current estimate x_c of x^*. These models should accurately represent the function in some region around x_c, they should be easy to construct and yet flexible enough to guide us towards the stationary point. In the *local region* we may expect the model to represent the function accurately in the neighborhood of x^*. Therefore, in this region the search is rather simple: We take a step to the optimizer of the local model, construct a new model and repeat until convergence. Techniques for the local region are discussed in Section 3.

In the *global region* the model should guide us in the right general direction towards x^*. This is relatively easy in minimizations since any step that reduces the function may be considered a step in the right direction. Global strategies for minimizations are treated in Section 4. In saddle point optimizations it is harder to judge the quality of a global step. Since a saddle point is a minimum in some directions and a maximum in others, the strategy is to identify these directions and take a step that increases and decreases the function accordingly. Such methods are discussed in Section 5.

It should be noted that no practical methods exist that are guaranteed to locate the global minimum of a PES. In fact, none of the methods discussed in this review make any attempt at identifying the global minimum as opposed to the local minima of a PES. However, some strategies have been developed for selectively locating the physically more important regions of potential energy surfaces (simulated annealing and so on), but these have found little use in *ab initio* quantum chemistry and are not treated here. Therefore, whenever a minimum or a saddle point has been located, its chemical significance should be further scrutinized based on additional criteria. For example, a minimum may be too high in energy or too shallow to be of interest. Similarly, the significance of a first-order saddle point should always be carefully considered. In particular, if some reaction or rearrangement is studied, one should always ascertain that the saddle point in question in fact connects the two minima of interest. Also, one should always consider the possibility that another saddle point of lower energy may connect the same two minima.

Our discussion is restricted to methods for unconstrained optimization, although there are many situations where constrained searches may be useful—for example, we may wish to minimize the energy for a fixed bond length or a fixed dihedral angle. Such problems can be dealt with using techniques discussed elsewhere, mostly simple extensions of the unconstrained techniques discussed here [1, 2, 9]. One class of constrained optimizations is particularly easy to deal with: symmetry restricted optimizations, where the molecule is assumed to belong to a certain point group. For example, much computational effort is saved by carrying out the minimization of the benzene molecule in the C_s point group, disregarding all modes that break this symmetry. Nevertheless, it should always be kept in mind that a stationary point located and characterized within a given symmetry may no longer have the desired Hessian index when the symmetry constraint is lifted. In particular, it may turn out that a minimum located within a given symmetry becomes a saddle point when symmetry breaking modes are considered. Thus—while it may be useful and sensible to carry out a search of stationary points with symmetry constraints imposed, these constraints should be lifted in the final analysis of the stationary point.

3   The Local Region

In the local region the situation is particularly simple, and in most situations Newton's method with approximate or exact Hessians will work well. Even so, there are situations where even Newton's method fails locally due to small eigenvalues, large anharmonicities or strong couplings between the modes. In the following we assume that this is not the case.

3.1   NEWTON'S METHOD

In the local region we proceed by constructing a model that contains within its region of validity (in some loose sense) the optimizer. This is the idea behind Newton's method, where we expand the surface to second order in displacements from the current point:

$$m_{SO}(\mathbf{s}) = f(\mathbf{x}_c) + \mathbf{g}_c^T \mathbf{s} + \frac{1}{2}\mathbf{s}^T \mathbf{G}_c \mathbf{s} \tag{9}$$

To locate the stationary point of the second-order (SO) model Eq. (9), we differentiate this model and set the result equal to zero. We obtain a linear set of equations

$$\mathbf{G}_c \mathbf{s} = -\mathbf{g}_c \tag{10}$$

which has a unique solution

$$\mathbf{s} = -\mathbf{G}_c^{-1}\mathbf{g}_c \tag{11}$$

provided $\mathbf{G}_c$ is non-singular. The second-order model with a non-singular Hessian thus has one and only one stationary point. This is a minimum if $\mathbf{G}_c$ is positive definite. It should be noted that Newton's method works equally well for minima and saddle points.

Sufficiently close to the optimizer, Newton's method converges quadratically. To see what this means, let $\mathbf{x}_k$ be a sequence of points converging to $\mathbf{x}^*$ (obtained, for example, by a sequence of Newton steps)

$$\lim_{k \to \infty} \mathbf{x}_k = \mathbf{x}^* \tag{12}$$

and let $\mathbf{e}_k$ be the error in $\mathbf{x}_k$

$$\mathbf{e}_k = \mathbf{x}_k - \mathbf{x}^* \tag{13}$$

Convergence towards $\mathbf{x}^*$ is said to be quadratic if for some (preferably) small number a

$$\lim_{k \to \infty} \frac{|\mathbf{e}_{k+1}|}{|\mathbf{e}_k|^2} = a \tag{14}$$

Thus, quadratic convergence implies that the number of correct figures in $\mathbf{x}_c$ doubles in each iteration, clearly a desirable property. A proof of quadratic convergence for Newton's method is given in textbooks on optimization [1, 3].

3.2   HESSIAN UPDATE METHODS

Newton's method requires the calculation of the gradient and the Hessian in each iteration. Often the Hessian is computationally much more demanding than the gradient. This is for example true in optimizations of potential energy surfaces, where the molecular Hessian requires an effort several times that of the gradient and the energy. We may therefore wonder whether it is possible to develop a less expensive method by replacing the exact Hessian $\mathbf{G}_c$ by some approximation $\mathbf{B}_c$. This will obviously detract from the quadratic convergence of Newton's method, but provided the approximate Hessian is carefully constructed, the increase in number of iterations should be modest and lead to an overall faster convergence (measured in CPU time rather than in iterations) towards the stationary point.

The most successful of such approaches are the Hessian update methods. The idea behind these methods is to use an approximate Hessian which in each iteration is improved upon (updated) based on our knowledge of gradients at nearby points. The update techniques are designed to determine an approximate Hessian $\mathbf{B}_+$ at

$$\mathbf{x}_+ = \mathbf{x}_c + \mathbf{s}_c \tag{15}$$

in terms of the Hessian $\mathbf{B}_c$ at $\mathbf{x}_c$, the gradient difference

$$\mathbf{y}_c = \mathbf{g}_+ - \mathbf{g}_c \tag{16}$$

and the step vector $\mathbf{s}_c$. Expanding around $\mathbf{x}_+$ gives

$$\mathbf{y}_c = \mathbf{B}_+\mathbf{s}_c + O\left(s_c^2\right) \tag{17}$$

which shows that the gradient difference $\mathbf{y}_c$ contains a component of the finite-difference approximation to the exact Hessian along the direction $\mathbf{s}_c$. This finite-difference information together with some structural characteristics of the exact Hessian are used to form the Hessian updates described below.

Based on the finite-difference formula Eq. (17), all Hessian updates are required to fulfil the quasi-Newton condition

$$\mathbf{y}_c = \mathbf{B}_+\mathbf{s}_c \tag{18}$$

and to possess the property of hereditary symmetry, i.e., $\mathbf{B}_+$ is symmetric if $\mathbf{B}_c$ is symmetric. These requirements are fulfilled by the Powell-symmetric-Broyden (PSB) update given by

$$\mathbf{B}_+ = \mathbf{B}_c + \frac{\left(s_c^T s_c\right)\mathbf{T}_c s_c^T + \left(s_c^T s_c\right) s_c \mathbf{T}_c^T - \left(\mathbf{T}_c^T s_c\right) s_c s_c^T}{\left(s_c^T s_c\right)^2} \tag{19}$$

where

$$\mathbf{T}_c = \mathbf{y}_c - \mathbf{B}_c\mathbf{s}_c \tag{20}$$

Notice that the construction of the updated Hessian involves simple matrix and vector multiplications of gradient and step vectors.

It is often desirable that the approximate Hessian is positive definite so that the quadratic model has a minimum. To ensure this we may use the Broyden-Fletcher-Goldfarb-Shanno (BFGS) update given by

$$\mathbf{B}_+ = \mathbf{B}_c + \frac{\mathbf{y}_c\mathbf{y}_c^T}{\mathbf{y}_c^T\mathbf{s}_c} - \frac{\mathbf{B}_c\mathbf{s}_c\mathbf{s}_c^T\mathbf{B}_c}{\mathbf{s}_c^T\mathbf{B}_c\mathbf{s}_c} \tag{21}$$

which under certain weak conditions on the step vector has the property of hereditary positive definiteness, i.e., if $\mathbf{B}_c$ is positive definite, $\mathbf{B}_+$ is positive definite. As we shall see, this is useful for minimizations even when the exact Hessian has directions of negative curvature. It is then more appropriate to speak of $\mathbf{B}_c$ as an effective rather than approximate Hessian.

Use of updated Hessians destroys the attractive quadratic convergence of Newton's method Eq. (14). However, with the update schemes presented above, the resulting convergence is superlinear in the local region

$$\lim_{k\to\infty} \frac{|e_{k+1}|}{|e_k|} = 0 \tag{22}$$

which is still quite satisfactory. There are other Hessian updates but for minimizations the BFGS update is the most successful. In the global region, Hessian update techniques are usually combined with line search (see Section 4.1) and the resulting minimization algorithms are called quasi-Newton methods. In saddle point optimizations we must allow the approximate Hessian to become indefinite and the PSB update is therefore more appropriate.

4   Global techniques for minimization

Newton's method and the Hessian update schemes are in general very robust in the local region of the optimization. In the global part of the search, however, the second-order model does not accurately represent the true surface in the region of the optimizer. Indeed, sufficiently far away from the true minimizer, the Newton step may lead us *away* from the minimizer. This behavior is easily understood by considering a simple function such as the Gaussian distribution

$$f\left(x\right) = \exp\left(-x^2\right) \tag{23}$$

which has a maximum at $x^* = 0$. The Newton step at x is seen to be given by

$$s_c = \frac{x_c}{2x_c^2 - 1} \tag{24}$$

and the next estimate of the optimizer $x^* = 0$ is therefore

$$x_+ = \frac{2x_c^3}{2x_c^2 - 1} \tag{25}$$

From a consideration of Eq. (25) we see that the Newton step converges only for $|x_c| < 1/2$ and that it oscillates between $+1/2$ and $-1/2$ for $|x_c| = 1/2$. For larger values, Newton's method diverges and for $|x_c| = 1/\sqrt{2}$ it gives infinite steps due to the presence of an inflection point. This behavior is illustrated in the Fig. 1.

From the figure we see that the divergence away from $x^* = 0$ is very slow for large values of $|x_c|$. It should be noted, however, that the Gaussian function $\exp(-x^2)$ has global minima at $|x| = \infty$. Thus for large x, Newton's method converges towards these two minima, but very slowly so. The slow rate of convergence arises since for large x the third and higher derivatives are much larger than the first and second derivatives.

Before leaving this simple example, we notice that for the function Eq. (23) Newton's method converges *cubically* towards the optimizer $x^* = 0$, see Eq. (25). This occurs since the Gaussian is an even function in x around x^* which implies that its third derivative vanishes at x^*.

Several techniques have been developed to deal with the non-convergence of Newton's method in the global region. Broadly speaking, these techniques all work by restricting the step in some sense. Within the chosen restriction, a step is sought which fulfils some simple criteria, designed to ensure that each step leads to an acceptable reduction in the function. We shall here consider a few such methods, in particular the line-search and trust-region methods.

4.1   LINE SEARCHES AND QUASI-NEWTON METHODS

By far the most popular technique for optimization in the global region—at least in connection with updated Hessians—are the line search methods. The idea behind these schemes is that although the Newton or quasi-Newton step may not be satisfactory and therefore must be discarded, it still contains useful information. In particular, we may use the step to provide a direction for a one-dimensional minimization of the function. We then carry out

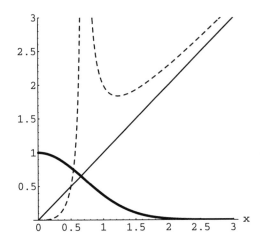

Figure 1: Newton's method for the function $f(x) = exp(-x^2)$. The bold line represents the function $f(x)$, the straight line is $|x_c|$, and the dashed line is $|x_+|$.

a search along the Newton step until an acceptable reduction in the function is obtained and the result of this line search becomes our next step.

All line searches start by defining a descent direction. Consider all vectors $\mathbf{z}$ that fulfil the condition

$$\mathbf{z}^T \mathbf{g}_c < 0 \tag{26}$$

Since

$$\left. \frac{df\,(\mathbf{x}_c + t\mathbf{z})}{dt} \right|_{t=0} = \mathbf{z}^T \mathbf{g}_c < 0 \tag{27}$$

there must be a positive number t such that

$$f\,(\mathbf{x}_c + t\mathbf{z}) < f\,(\mathbf{x}_c) \tag{28}$$

and $\mathbf{z}$ is therefore said to be a descent direction. The negative gradient obviously is a descent direction (often referred to as the steepest-descent direction) as is the Newton or quasi-Newton step

$$s_N = -\mathbf{B}_c^{-1} \mathbf{g}_c \tag{29}$$

provided the Hessian is positive definite

$$s_N^T \mathbf{g}_c = -\mathbf{g}_c^T \mathbf{B}_c^{-1} \mathbf{g}_c < 0 \tag{30}$$

It is for this reason the positive definite BFGS update Eq. (21) is preferred over the PSB update Eq. (19) for minimizations. The positive definite Newton step is usually a better direction than steepest descent since it takes into account features of the function further away from $\mathbf{x}_c$ than does steepest descent.

Given the direction of search $\mathbf{z}_c$ at $\mathbf{x}_c$, we must find a satisfactory step along this direction. It would seem that the best is to minimize $f\,(\mathbf{x}_c + t\mathbf{z}_c)$ with respect to t and take the step

$$s_c = x_c + t^* z_c \qquad (31)$$

where t^* is the minimizer. However, such exact line searches are expensive and not used in practice. Instead inexact or partial line searches are carried out to generate an acceptable point s_c along z_c. By acceptable we mean for example a point that fulfils the condition

$$f\left(x_c + t z_c\right) < f\left(x_c\right) + \frac{1}{2} t g_c^T z_c \qquad (32)$$

for some $0 < t \leq 1$. The parameter t may be determined by interpolation. We first try the full Newton step. If this is not acceptable, a smaller step is obtained by interpolation and tested. This backtracking process is repeated until an acceptable step is found. For more information on line searches, see [1, 2, 3].

Line searches are often used in connection with Hessian update formulas and provide a relatively stable and efficient method for minimizations. However, line searches are not always successful. For example, if the Hessian is indefinite there is no natural way to choose the descent direction. We may then have to revert to steepest descent although this step makes no use of the information provided by the Hessian. It may also more generally be argued that backtracking from the Newton step never makes full use of the available information since the Hessian is used only to generate the direction and not the length of the step.

4.2   TRUST-REGION MINIMIZATION

A different approach to optimization in the global region is based on the trust-region concept. In this approach, we retain the strategy from the local region of minimizing in each step a model surface, but we impose the additional requirement that the search for the minimizer of the local model should only be carried out in some restricted region around the expansion point. In other words, we accept that our model can only be trusted to represent the true surface accurately in some local region—*the trust region*—and we proceed carefully by forbidding ourselves to take a step out of this region. Therefore, in each iteration we determine the global minimizer of the model surface within the trust region and then take the step to this point. In the global part of the search, the minimizer of the local model will be on the boundary of the trust region. Close to the true minimizer, the minimizer of the local model will be inside this boundary and the step will correspond to the Newton step.

To apply this idea in practice, we cannot specify our trust region in great detail and for convenience we assume that it has the shape of a hypersphere $|s| \leq h$ where h is the trust radius. Expanding to second order in the displacements, we obtain the so-called restricted second-order (RSO) model:

$$m_{RSO}\left(s\right) = f\left(x_c\right) + g_c^T s + \tfrac{1}{2} s^T G_c s \quad \text{for } s^T s \leq h^2 \qquad (33)$$

Before discussing the trust-region minimization itself, we consider the stationary points of the RSO model. Of course, for minimizations we are only interested in the global minimizer of the local model, but for later reference we here give a discussion of all stationary points of the RSO model.

We first notice that the RSO model must have several stationary points. If the Newton step Eq. (11) is shorter than the trust radius $|s_N| < h$, then the RSO model has a stationary point in the interior. It also has at least two stationary points on the boundary $|s| = h$. To see this we introduce the Lagrangian

$$L\left(s, \mu\right) = m_{SO}\left(s\right) - \frac{1}{2} \mu \left(s^T s - h^2\right) \qquad (34)$$

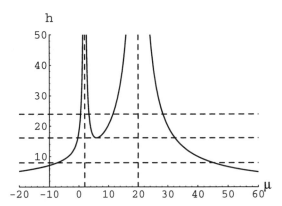

Figure 2: The step length (solid line) as a function of the level-shift parameter μ for the function $x^2 + 10y^2$ of at $(x, y) = (-15, 10)$. The positions of the two peaks are determined by the eigenvalues $(2, 20)$ and their widths by the gradient at the point of expansion $(-30, 200)$. The three dashed horizontal lines correspond to different choices of the trust radius h: 8, 16.13 and 24.

where μ is an undetermined multiplier. Differentiating this expression and setting the result equal to zero, we obtain

$$\mathbf{s}(\mu) = -(\mathbf{G}_c - \mu\mathbf{1})^{-1}\mathbf{g}_c \qquad (35)$$

where the level shift parameter μ is chosen such that the step is to the boundary

$$\sqrt{\mathbf{g}_c^T(\mathbf{G}_c - \mu\mathbf{1})^{-2}\mathbf{g}_c} = h \qquad (36)$$

To see the solutions to Eq. (36) more clearly, we have plotted in Fig. 2 its left- and right-hand sides as functions of μ for the simple function

$$f(x) = x^2 + 10y^2 \qquad (37)$$

at $(x, y) = (-15, 10)$, where the gradient is given by $(-30, 200)$ and the (constant) Hessian eigenvales are $(2, 10)$. The step length function has poles at the eigenvalues as can be seen from Eq. (36) in the diagonal representation

$$\sum_i \frac{\phi_i^2}{(\lambda_i - \mu)^2} = h^2 \qquad (38)$$

where λ_i are the eigenvalues and ϕ_i the components of the gradient along the eigenvectors. In Fig. 2, the eigenvalues determine the positions of the peaks and the gradient their widths.

The solutions to Eq. (36) are found at the intersections of the left- and right-hand sides. Since the step goes to infinity at the eigenvalues and to zero at infinity there are at least two solutions: one with $\mu < \lambda_1$ (the smallest eigenvalue) and another with $\mu > \lambda_n$ (the largest eigenvalue). We may also have stationary points in each of the $n-1$ regions $\lambda_k < \mu < \lambda_{k+1}$. For large h we have two solutions in each region, for small h there may be no solution. If the model is constructed around a stationary point, the peaks are infinitely narrow and we have 2n stationary points on the boundary (two along each eigenvector direction) in addition to the point in the interior. In Fig. 2 we see that the function $f(x)$ in Eq. (37) has two stationary points on the boundary for $h < 16.13$ and four stationary points on the

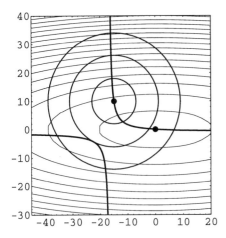

Figure 3: Levenberg-Marquardt trajectories on the surface $x^2 + 10y^2$ obtained by expansion around $(-15, 10)$. The trajectories intersect the trust-region boundaries at their stationary points. The concentric circles represent boundaries of trust regions of radius 8, 16.13, and 24.

boundary for $h > 16.13$. For $h \approx 16.13$ we have one maximum, one minimum and one inflection point on the boundary.

We may regard Eq. (35) as defining a set of trajectories $\mathbf{s}(\mu)$ as a function of μ on the energy surface $f(\mathbf{x})$. These trajectories are referred to as the Levenberg-Marquardt trajectories and are depicted in Fig. 3 for the function $f(x, y)$ given by Eq. (37). The Levenberg-Marquardt trajectories depend on the point of expansion—the trajectories in the figure are obtained by expansion around $(-15, 10)$. The stationary points on the trust-region boundaries correspond to the intersections of the concentric circles and the Levenberg-Marquardt trajectories in Fig. 3, in the same way as the stationary points correspond to the intersections of the step length function and the horizontal lines in Fig. 2. Note that on a (non-degenerate) second-order surface it is always possible to generate n disjoint trajectories by choosing μ in the different intervals between the eigenvalues of the Hessian. Close to a minimum, one and only one of these passes through the expansion point and the stationary point, as illustrated in Fig. 3.

It may be shown that if $\mu < \lambda_1$, then the solution to Eq. (36) is the global minimum on the boundary. In the diagonal representation the step may be written

$$\sigma_i = -\frac{\phi_i}{\lambda_i - \mu} \tag{39}$$

where σ_i is the component of the step along the i'th eigenvector. We see that by selecting the solution $\mu < \lambda_1$ we take a step opposite the gradient in each mode.

The other solutions to Eq. (36) correspond to stationary points where the function is increased in some directions and reduced in others. For example, if we select a solution in the region $\lambda_2 < \mu < \lambda_3$ then the step is toward the gradient of the first two modes and opposite the gradient of all higher modes. The second-order change in the function may be written

$$\Delta m_{RSO} = \sum_i \frac{\phi_i^2 \left(\mu - \frac{1}{2}\lambda_i\right)}{(\lambda_i - \mu)^2} \tag{40}$$

and we see that the model decreases along all modes for which $\lambda_k > 2\mu$ and increases along all others. This completes our discussion of stationary points of the RSO model.

The strategy for *trust-region* or *restricted step* minimization should now be obvious. In each iteration the gradient and Hessian are calculated and a step is taken to the global minimizer of the RSO model. Thus, in the global region a step is taken to the boundary of the trust region

$$s(\mu) = -(G_c - \mu 1)^{-1} g_c \tag{41}$$

since this contains the minimizer of the model. In the local region the model has a minimizer inside the trust region and we take the Newton step

$$s(0) = -G_c^{-1} g_c \tag{42}$$

The method therefore reduces to Newton's method in the local region with its rapid rate of convergence. Notice that in each iteration we always first try the Newton step Eq. (42), resorting to the level-shifted step Eq. (41) only if the Newton step is larger than the trust radius.

When solving Eq. (41) the level shift parameter μ must be chosen such that $s(\mu)$ corresponds to the global minimizer on the boundary. From the above discussion we know that μ must be smaller than the lowest Hessian eigenvalue. Also, μ must be negative since otherwise the step becomes longer than the Newton step. The exact numerical value of μ may be found by bisection or interpolation.

The trust radius h reflects our confidence in the SO model. For highly anharmonic functions the trust region must be set small, for quadratic functions it is infinite. Clearly, during an optimization we must be prepared to modify h based on our experience with the function. This is accomplished by means of a feedback mechanism. In the first iteration some arbitrary but reasonable value of h is assumed. In the next iteration, h is updated based on a comparison between the predicted reduction in $f(x)$ and the actual reduction. If the ratio between actual and predicted reductions

$$R_c = \frac{f_+ - f_c}{g_c^T s_c + \frac{1}{2} s_c^T G_c s_c} \tag{43}$$

is close to one, h is increased. If the ratio is small the radius is reduced. If the ratio is negative the step is rejected, the trust radius reduced and a new step is calculated from Eq. (41). For details on the updating of the trust radius, see for example [10, 11].

The trust-region method is usually implemented with the exact Hessian. An approximate Hessian may also be used but, unless the initial Hessian has been calculated accurately and the topology of the surface has not changed much, an updated Hessian does not contain enough information about the function to make the trust region reliable in all directions. The trust-region method provides us with the possibility to carry out an unbiased search in all directions at each step. In general, an updated Hessian does not contain the information necessary for such a search. Updated Hessians are therefore more useful in connection with line search methods, as discussed above.

It is interesting to note the difference between the RSO trust-region method and Newton's method with line search. In the trust-region method we first choose the size of the step (the trust radius), and then determine the direction of the step (constrained minimization on the boundary of the trust region). In Newton's method with line search we first choose the direction of the step (the Newton direction) and next determine the size of the step (constrained minimization along the Newton direction). The trust-region method is more robust (guaranteed convergence for smooth and bounded functions) and has no problems with indefinite Hessians. It is more conservative than line search since the size of the step is predetermined. However, line search requires additional energy calculations and is not equally well suited for handling indefinite Hessians.

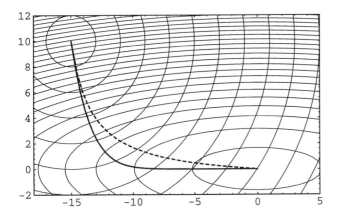

Figure 4: Steepest-descent (full line) and Levenberg-Marquardt (dashed line) trajectories starting at $(-15, 10)$ and leading to the minimum $(0, 0)$ on the surface $x^2 + 10y^2$.

4.3 QUADRATIC STEEPEST DESCENT

We now turn to a method closely related to the Levenberg-Marquardt trust-region scheme. Rather than searching for the global optimizer of the restricted local model, we seek to trace out on the second-order model surface the steepest-descent path as defined by the equation

$$\frac{d\mathbf{x}(t)}{dt} = -\mathbf{g}(\mathbf{x}) \tag{44}$$

Expanding the gradient to first order in the displacements, we obtain the following differential equation

$$\frac{d\mathbf{x}(t)}{dt} = -\mathbf{g}_c - \mathbf{G}_c \mathbf{x}(t) \tag{45}$$

which may be solved to yield

$$\mathbf{s}_{SD}(t) = -\mathbf{G}_c^{-1}\left[1 - \exp\left(-\mathbf{G}_c t\right)\right]\mathbf{g}_c \tag{46}$$

for the steepest-descent step on a quadratic surface. Clearly, for a downhill step, t should be a positive number.

We now have a method closely related to the Levenberg-Marquardt trust-region method. In each iteration, we first investigate the Newton step by trying $t = \infty$. If this step is too large (i.e., larger than the trust radius), we search for a finite value of t that gives us a step to the boundary of the trust region. The trust region is updated in each iteration in the same way as in the trust-region minimization scheme. As for the trust-region minimization method, quadratic steepest descent does not suffer from problems related to the presence of negative eigenvalues and close to $\mathbf{x}^*$ it turns naturally into Newton's method. Quadratic steepest descent has been discussed by several authors, mostly in connection with reaction path following [12, 13, 14] but also for minimizations of electronic wave functions [15]. Different formulations have been given by Gonzalez and Schlegel [16] and by Sun and Ruedenberg [17, 18, 19, 20].

It is of some interest to compare the steepest-descent path Eq. (46) with the Levenberg-Marquardt trajectory Eq. (35) which may be rewritten as

$$\mathbf{s}_{LM}\left(t\right) = -\mathbf{G}_c^{-1}\left[1 + (t\mathbf{G}_c)^{-1}\right]^{-1}\mathbf{g}_c \tag{47}$$

This expression highlights the similarities between the second-order steepest-descent trajectory Eq. (46) and the Levenberg-Marquardt trajectory. In particular, we notice that both trajectories end at the minimum of the second-order surface for $t = \infty$. For small steps, we expand around $t = 0$ and obtain

$$\mathbf{s}_{LM}\left(t\right) = -\left(1 - \mathbf{G}_c t + \mathbf{G}_c^2 t^2 \cdots\right)\mathbf{g}_c t \tag{48}$$

$$\mathbf{s}_{SD}\left(t\right) = -\left(1 - \frac{1}{2}\mathbf{G}_c t + \frac{1}{6}\mathbf{G}_c^2 t^2 \cdots\right)\mathbf{g}_c t \tag{49}$$

As expected, both steps are opposite the gradient for small t. Thus, the level-shifted Newton method and quadratic steepest descent differ only in the interpolation between small steps (which are opposite the gradient) and large steps (which are equivalent to the Newton step). The steepest-descent and Levenberg-Marquardt trajectories starting at $(-15, 10)$ for the function Eq. (37) are compared in Fig. 4. Clearly, the steepest-descent line intersects the contour lines at right angles, while the Levenberg-Marquardt line intersects the circles centered at $(-15, 10)$ (representing trust-region boundaries) at their global minima. We notice that steepest-descent trajectories have a tendency to minimize in one direction at a time and therefore take a more indirect route towards the minimizer. This effect should be larger for a function with a greater difference between the two eigenvalues. Note also that for a fixed distance away from the starting point (for example at the intersections of the trajectories with one of the concentric circles in the figure), the Levenberg-Marquardt trajectory is lower in energy than the steepest-descent trajectory.

4.4   RATIONAL FUNCTION MINIMIZATION

The trust region was introduced since the second-order model is a good approximation to the function only in some region around the expansion point. The resulting restricted second-order model has stationary points on the boundary of the trust region and possibly a stationary point in the interior. These points are used to construct globally convergent optimization algorithms. There is another way to introduce restrictions on the step lengths in the global part of an optimization. The *rational function (RF) model* is given by

$$m_{RF}\left(\mathbf{s}\right) = f\left(\mathbf{x}_c\right) + \frac{\mathbf{g}_c^T\mathbf{s} + \frac{1}{2}\mathbf{s}^T\mathbf{G}_c\mathbf{s}}{1 + \mathbf{s}^T\mathbf{S}\mathbf{s}} \tag{50}$$

which may be written in the form

$$m_{RF}\left(\mathbf{s}\right) = f\left(\mathbf{x}_c\right) + \frac{1}{2}\frac{\left[\,\mathbf{s}^T \quad 1\,\right]\begin{bmatrix}\mathbf{G}_c & \mathbf{g}_c \\ \mathbf{g}_c^T & 0\end{bmatrix}\begin{bmatrix}\mathbf{s} \\ 1\end{bmatrix}}{\left[\,\mathbf{s}^T \quad 1\,\right]\begin{bmatrix}\mathbf{S} & 0 \\ 0 & 1\end{bmatrix}\begin{bmatrix}\mathbf{s} \\ 1\end{bmatrix}} \tag{51}$$

where the metric $\mathbf{S}$ is a symmetric matrix. This model is bounded since large elements in the numerator are balanced by large elements in the denominator. Also, to second order the RF and SO models are identical since

$$\left(1 + \mathbf{s}^T\mathbf{S}\mathbf{s}\right)^{-1} = 1 - \mathbf{s}^T\mathbf{S}\mathbf{s} + O\left(s^4\right) \tag{52}$$

Therefore, in the RF model we have added higher order terms to the SO model to make it bounded. The explicit form of the RF model depends on the matrix $\mathbf{S}$ which should reflect the anharmonicity of the function. Usually we do not have sufficient information to specify $\mathbf{S}$ in detail and we simply take it to be the unit matrix multiplied by a scalar S. We note that the rational function model Eq. (50) is closely related to the idea of Padé approximations.

To find the stationary points of the RF model we differentiate Eq. (51) and set the result equal to zero. We arrive at the eigenvalue equations

$$\begin{bmatrix} \mathbf{G}_c & \mathbf{g}_c \\ \mathbf{g}_c^T & 0 \end{bmatrix} \begin{bmatrix} \mathbf{s} \\ 1 \end{bmatrix} = \nu \begin{bmatrix} \mathbf{S} & 0 \\ 0 & 1 \end{bmatrix} \begin{bmatrix} \mathbf{s} \\ 1 \end{bmatrix} \tag{53}$$

Solving these equations we obtain $n + 1$ eigenvectors and eigenvalues

$$\nu = \frac{\begin{bmatrix} \mathbf{s}^T & 1 \end{bmatrix} \begin{bmatrix} \mathbf{G}_c & \mathbf{g}_c \\ \mathbf{g}_c^T & 0 \end{bmatrix} \begin{bmatrix} \mathbf{s} \\ 1 \end{bmatrix}}{\begin{bmatrix} \mathbf{s}^T & 1 \end{bmatrix} \begin{bmatrix} \mathbf{S} & 0 \\ 0 & 1 \end{bmatrix} \begin{bmatrix} \mathbf{s} \\ 1 \end{bmatrix}} = 2\Delta m_{RF}(\mathbf{s}) \tag{54}$$

corresponding to the $n + 1$ stationary points of the RF model. From Eq. (54) we see that the minimum belongs to the lowest eigenvalue. In general, the k'th eigenvalue belongs to a saddle point of index $k - 1$. Notice that the eigenvalues Eq. (54) give the change in the model Eq. (51) rather than in the function $f(\mathbf{x})$ when the step is taken. The stationary points of the RF model do not necessarily correspond to stationary points of $f(\mathbf{x})$ but they are useful for constructing globally convergent optimization algorithms.

Since the coefficient matrix of the eigenvalue equations Eq. (53)

$$\mathbf{G}_c^+ = \begin{bmatrix} \mathbf{G}_c & \mathbf{g}_c \\ \mathbf{g}_c^T & 0 \end{bmatrix} \tag{55}$$

has dimension $n + 1$ and contains the Hessian in the upper left corner it is referred to as the augmented Hessian. The $n + 1$ eigenvalues of the augmented Hessian $\mathbf{G}_k^+$ bracket the Hessian eigenvalues

$$\lambda_1^+ \leq \lambda_1 \leq \lambda_2^+ \leq \lambda_2 \leq \cdots \leq \lambda_n \leq \lambda_{n+1}^+ \tag{56}$$

The eigenvalues of Eq. (53) coincide with the eigenvalues of the augmented Hessian only when $\mathbf{S}$ equals unity.

To compare the RSO and RF models, we expand Eq. (53) and obtain

$$\mathbf{s} = -(\mathbf{G}_c - \mu \mathbf{1})^{-1} \mathbf{g}_c \tag{57}$$

$$\mathbf{g}_c^T \mathbf{s} = \frac{\mu}{S} \tag{58}$$

where

$$\mu = \nu S \tag{59}$$

Inserting Eq. (57) in Eq. (58), we find

$$-\mathbf{g}_c^T (\mathbf{G}_c - \mu \mathbf{1})^{-1} \mathbf{g}_c = \frac{\mu}{S} \tag{60}$$

or in the diagonal representation

$$\sum_i \frac{\phi_i^2}{\mu - \lambda_i} = \frac{\mu}{S} \tag{61}$$

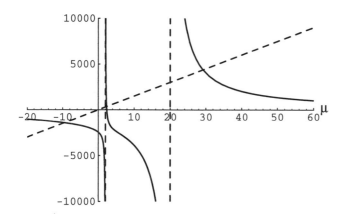

Figure 5: Solutions for the step-length parameter μ for the rational-function model for the function $x^2 + 10y^2$ at $(-15, 10)$ and with $S = 1/150$.

Plotting the left- and right-hand sides of this equation as functions of μ for the function Eq. (37) at $(-15, 10)$ and with $S = 1/150$, we obtain Fig. 5. The left-hand side goes to infinity at the Hessian eigenvalues. We have $n + 1$ intersections, one for each stationary point of the rational model. Changing the metric S changes the slope of the straight line and therefore the intersections.

Once μ has been determined we calculate the step from the modified Newton equations Eq. (57). Therefore, the RF and RSO steps are calculated in the same way. The only difference is the prescription for obtaining the level shift. In the RSO approach μ reflects the trust radius h, in the RF model μ reflects the metric S. By varying h and S freely the same steps are obtained in the two models.

Close to a stationary point $\mathbf{g}_c$ vanishes. One of the eigenvalues of the augmented Hessian Eq. (55) then goes to zero and the remaining approach those of $\mathbf{G}_c$. The zero-eigenvalue step becomes the Newton step and the remaining n steps become infinite and parallel to the Hessian eigenvectors.

A global strategy for minimization may now be based on the RF model. In each iteration the equations

$$\begin{bmatrix} \mathbf{G}_c & \mathbf{g}_c \\ \mathbf{g}_c^T & 0 \end{bmatrix} \begin{bmatrix} \mathbf{s} \\ 1 \end{bmatrix} = \nu \begin{bmatrix} \mathbf{S} & 0 \\ 0 & 1 \end{bmatrix} \begin{bmatrix} \mathbf{s} \\ 1 \end{bmatrix} \tag{62}$$

are solved for the lowest eigenvalue ν and the corresponding eigenvector. This gives a step

$$\mathbf{s}(\nu\mathbf{S}) = -(\mathbf{G}_c - \nu\mathbf{S})^{-1}\mathbf{g}_c \tag{63}$$

to the minimizer of the rational function. This step is opposite the gradient in each mode if $\nu S < \lambda_1$. In the local region, ν goes to zero and the step approaches the Newton step. The parameter S may be used for step size control in the same way as h in trust-region minimizations. However, S is usually set to one and the step is simply scaled down if it is unsatisfactory. The RF or augmented-Hessian method was used for optimization of electronic wave functions by Banerjee and Grein [21] and by Yarkony [22]. It was applied to surface studies by Banerjee et al. [23].

4.5 FIRST- AND SECOND-ORDER METHODS COMPARED

We have seen that the methods for optimization of molecular geometries can be divided into two broad classes: second-order methods which require the exact gradient and Hessian in each iteration, and first-order (quasi-Newton) methods which require the gradient only. Both methods are in widespread use, but the first-order methods are more popular since analytical energy gradients are available for almost all electronic structure methods, whereas analytical Hessians are not. Also, the simpler first-order methods usually perform quite well, converging in a reasonable number of iterations in most cases.

We illustrate in this section the usefulness of the different methods of optimization by carrying out sample calculations on a few selected systems. In particular, we compare the performance of the first- and second-order methods. We shall see that although the simpler first-order methods in many cases represent the most cost-effective approach, difficult situations do arise where the performance of the second-order methods is superior. However, it is pertinent to discuss first two areas which are particularly important for first-order methods, namely the choice initial Hessian and the choice of coordinate system.

The choice of initial Hessian can critically influence the convergence of first-order methods. One (somewhat expensive) possibility is to calculate the exact Hessian at the starting geometry. This initial Hessian is then updated at each geometry step using one of the quasi-Newton schemes; if the optimization proceeds satisfactorily the Hessian is not reevaluated until the minimum is reached and vibrational frequencies are to be calculated. Although there are no formal grounds for using the updated Hessians for, e.g., trust-region updates, we may expect that for well-behaved systems this approach could simply be substituted for the true second-order method. Such a scheme is implemented in the program GRADSCF [24] and functions very satisfactorily. However, this approach still requires that the full Hessian can be calculated when required.

True quasi-Newton schemes attempt to sidestep any calculation of the exact Hessian. In such cases, an approximate Hessian must be estimated by the program or by the user. How should such an estimate be obtained? One attractive possibility is to utilize experimental data on vibrational frequencies, or to transfer estimates from one system to a related system. Here is where the choice of coordinate system can be crucial. In general, such estimates are likely to provide information only about, in effect, the normal modes of vibration of a system: they specify only the diagonal elements of the Hessian *in a particular coordinate system*. There are few sources of empirical information about coupling force constants, for example, and few sources about force constants expressed in Cartesian coordinates. Also Hessian information in Cartesian coordinates is rarely transferable from one molecule to another. Hence, for first-order methods the choice of coordinate system in which the optimization is performed is strongly influenced by the need to obtain Hessian information.

In the second-order methods we have described, the choice of coordinate system was not made explicit. From a quantum-chemical perspective, analytical derivatives are most conveniently computed in Cartesian (or symmetry-adapted Cartesian) coordinates. Indeed, second-order methods are not particularly sensitive to the choice of coordinate system and second-order implementations based on Cartesian coordinates usually perform quite well. As we discussed above, however, if the Hessian is to be estimated empirically, a representation in which the Hessian is diagonal, or close to diagonal, is highly desirable. This is certainly not true for Cartesian coordinates: some set of "internal" coordinates that better resemble normal coordinates would be required. Two related choices are popular. The first choice is the internal coordinates suggested by Wilson, Decius and Cross [25], which comprise bond stretches, bond angle bends, motion of a bond relative to a plane defined by several atoms, and torsional (dihedral) motion of two planes, each defined by a triplet of atoms. Commonly, the molecular geometry is specified in Cartesian coordinates, and a linear transformation between Cartesian displacement coordinates and internal displacement coordinates is either supplied by the user or generated automatically. Less often, the (curvilinear) transformation from Cartesian coordinates to internals may be computed. The second choice is "Z-matrix" coordinates, popularized by a number of semiempirical and nonempirical quantum chemistry programs, in which the geometry and internal coordinate system is specified in terms of bond lengths, bond angles, and torsional angles. The Cartesian coordinates can be generated straightforwardly from the Z-matrix and so can a linear transformation connecting displacements in the two representations. The Z-matrix procedure favors nonsymmetrical or low symmetry systems — specification of the geometry for high symmetry species usually requires some contortion and the use of dummy atomic centers.

Armed with a set of internal coordinates (that is, most commonly a linear transformation between Cartesian displacements and internal displacements), and some initial guess at the internal coordinate Hessian (commonly generated from mean experimental frequencies for different types of vibration, or from prior experience), various first-order or approximate second-order schemes can be used to optimize the geometry. However, some molecules, particularly those that contain closed rings of atoms, present additional problems. The first requirement with ring systems is usually to include all of the ring bonds as internal coordinates, augmented with as many ring angles as are nonredundant. (Pulay and coworkers [26, 27] have had some success with the use of redundant coordinates in optimizations.) This "closing the ring" is usually straightforward in a program in which internal coordinates are specified by the user. However, Z-matrix coordinates are not as convenient for this purpose, and again it may be necessary to define dummy centers or to resort to other devices in order to close a ring. Indeed, it has been suggested that Cartesian coordinates (even with the difficulty of choosing an initial Hessian) may be preferable [28, 29]. We may note in passing that (even when using second-order methods) internal coordinates offer one significant advantage: they are more convenient when one or more coordinates are to be fixed during an optimization. Such a constraint is almost always expressed in terms of internal coordinates (freezing certain lengths or angles).

Concluding our discussion of coordinates, we would like to note that *any* method that relies on the construction of model surfaces to the true surface will to some extent be sensitive to the choice of coordinates. A good choice of coordinates will extend the region in which the model is valid, minimizing the unaccounted effects of higher-order derivatives and thus reducing the number of iterations needed. This is true for second-order methods where a good choice of internal coordinates can improve convergence, but it is much more critical for first-order methods where an unfortunate choice of coordinates can make the convergence intolerably slow.

To summarize experience with these different approaches to optimization, we consider some published results for several first-order optimizations and compare them to results obtained with our second-order trust-region scheme. The molecules chosen are from the list used by Baker and Hehre [28] in a comparison of Cartesian and Z-matrix coordinates, and later by Schlegel [30] in comparisons with other choices of internal coordinates. All calculations are minimal basis self-consistent field, and starting geometries were obtained from a molecular mechanics force field. The second-order trust-region minimizations have been carried out using the HERMIT-SIRIUS-ABACUS program system [31, 32, 33].

A ring system which is reasonably well behaved is the molecule 2-fluorofuran in Fig. 6. Optimization of the geometry of this system requires about seven steps for first-order

Figure 6: 2-Fluorofuran

schemes, irrespective of whether internal, Cartesian, or a mixed internal/Cartesian set of coordinates is used. Our second-order scheme requires four steps, so in this case there is apparently nothing to be gained by using the more elaborate scheme. We may point out, though, that the use of a second-order scheme allows extremely stringent convergence criteria to be used. The four steps taken ensure convergence of the molecular gradient norm to 10^{-5}: since the optimization is now in the quadratic basin another step would double the number of figures converged in the gradient. It is also worth noting that one of the internal coordinates specified in the Z-matrix methods is the distance between O and and the carbon atom to which the fluorine is attached, as recommended by Schlegel in a discussion of ring internal coordinates. This is not a naively obvious choice and demonstrates the care that must be taken to choose appropriate internal coordinates for first-order optimization schemes.

A less tractable case is bicyclo[2.2.2]octane, see Fig. 7. Here the use of Z-matrix co-

Figure 7: Bicyclo[2.2.2]octane

ordinates as recommended by Schlegel results in 11 steps being taken in the geometry optimization. A Z-matrix in which no device is used to close the rings requires 25 steps. An optimization in Cartesian coordinates requires 19 steps, and a mixture of internal and Cartesian coordinates is somewhat more effective, requiring 14 steps. The second-order scheme requires six steps.

Finally, we may consider the protonated histamine molecule in Fig. 8. This is asserted

$$HN \diagdown N \qquad NH_3{}^+$$

Figure 8: Protonated histamine

to be a difficult case because of the slow optimization of the side chain. Certainly this is true for the first-order methods. Even with the use of Schlegel's recommended ring coordinates 42 steps are required for a first-order method. Mixed Cartesian and internal coordinates perform somewhat worse, at 47 steps, while in Cartesians alone the optimization does not converge in 100 steps! However, this molecule is not significantly more challenging for the second-order method than the others considered here (or elsewhere); convergence of the geometry is obtained in seven steps. If we assume that calculating an analytical Hessian is three times more expensive than computing the analytical gradient, it is clearly substantially more cost-effective in this case to employ the second-order method. Even if the Hessian were ten times as expensive as the gradient, it would still be cost-effective to use the second-order method here. This illustrates the power of second-order methods: convergence is obtained in relatively few steps even for difficult systems.

It is our experience that from (say) a molecular mechanics-derived starting geometry, a trust-region-based second-order optimization will converge in five to seven steps, and only in rare cases will it require as many as nine. From a more accurate starting geometry, such as one obtained from an *ab initio* calculation in a smaller basis, convergence is typically obtained in two or three steps. It is quite likely that for well-behaved systems, and/or some creativity on the part of the user, first-order methods can rival or surpass these results in terms of overall cost-effectiveness. However, the second-order method displays this convergence behavior in essentially *all* cases, and requires no special efforts on the part of the user. Hence if one deals often with difficult cases, the second-order methods are likely to be the most cost-effective.

5   Strategies for Saddle Point Optimizations

We now turn to methods for first-order saddle points. As already noted, saddle points present no problems in the local region provided the exact Hessian is calculated at each step. The problem with saddle point optimizations is that in the global region of the search, there are no simple criteria that allow us to select the step unambiguously. Thus, whereas for minimization methods it is often possible to give a proof of convergence with no significant restrictions on the function to be minimized, no such proofs are known for saddle-point methods, except, of course, for quadratic surfaces. Nevertheless, over the years several useful techniques have been developed for the determination of saddle points. We here discuss some of these techniques with no pretence at completeness.

5.1   LEVENBERG-MARQUARDT TRAJECTORIES

From our discussion of the restricted second-order model it became apparent that by letting the level-shift parameter of the restricted Newton step run over all real numbers, we trace out a set of trajectories—the Levenberg-Marquardt trajectories—on the second-order model surface. The trust-region minimization scheme may be seen as a one-dimensional search along one particular Levenberg-Marquardt trajectory, namely the trajectory generated by selecting the level shift parameter smaller than the smallest eigenvalue of the Hessian.

There have been several attempts at developing algorithms that allow us to trace out the other Levenberg-Marquardt trajectories on the second-order surface. Thus, by selecting a level-shift parameter between the first and second Hessian eigenvalues, we may take a step that initially at least increases the function along the lowest eigenvalue and minimizes it along all others. This procedure may indeed lead to a transition state, but in this way we always increase the function along the lowest mode. If we wish to increase it along a higher mode this can only be accomplished in a somewhat unsatisfactory manner by coordinate scaling. Nevertheless, such methods, pioneered by Cerjan and Miller [34], have been used by several authors with considerable success, see for example [10, 11, 35, 36].

A problem with this approach is that the Levenberg-Marquardt trajectories are not unique or global in the sense that there exists for each point on the energy surface a complete set of Levenberg-Marquardt trajectories, different from the trajectories belonging to other points. In a sense, the Levenberg-Marquardt trajectories belong more to the second-order model than to the true surface. An interesting question then arises as to whether or not there exists a set of lines on the potential energy surface that are global (independent of local models) and also connect stationary points. Given the existence of such lines, we can imagine the development of an algorithm that allows us to trace in an automatic and systematic manner these lines between the different stationary points of the energy surface.

In fact, one such set are the steepest-descent lines that connect saddle points and minima. By tracing steepest-descent lines down from a first-order saddle point we end up in a minimum (provided the surface is bounded). Unfortunately, steepest-descent lines are of little use to us since they cannot be used for uphill walks. There exists, however, a different set of lines that are global, connect stationary points and also can be climbed uphill. These are the gradient extremals, to which we now turn our attention.

5.2 GRADIENT EXTREMALS

To develop a method to locate saddle points by selectively following one eigenvector we note that at a stationary point all n components of the gradient in the diagonal representation are zero:

$$\phi\left(\mathbf{x}_{SP1}\right) = \begin{bmatrix} 0 \\ 0 \\ \vdots \\ 0 \end{bmatrix} \tag{64}$$

These n conditions define a point in n-dimensional space. We now move away from the stationary point in a controlled manner by relaxing only one of these conditions. For example, we may no longer require the second component of the gradient to be zero. We are then left with $n - 1$ conditions, which define a line in n-dimensional space passing through the stationary point Eq. (64):

$$\phi\left[\mathbf{x}_{GE}\left(t\right)\right] = \begin{bmatrix} 0 \\ \phi\left(t\right) \\ \vdots \\ 0 \end{bmatrix} \tag{65}$$

We call this line a gradient extremal (GE) [37, 38, 39, 40, 41, 42]. Unless the function increases indefinitely, the gradient element $\phi\left(t\right)$ must eventually become zero or approach zero. It is therefore reasonable to expect that by following a GE we sooner or later hit a new stationary point

$$\phi\left(\mathbf{x}_{SP2}\right) = \begin{bmatrix} 0 \\ 0 \\ \vdots \\ 0 \end{bmatrix} \tag{66}$$

If we start at a minimum we may expect this new stationary point to be a minimum, although this is by no means guaranteed. This observation is the basis for the GE algorithm: Transition states are determined by carrying out one-dimensional searches along GEs starting at a minimum. Since n GEs pass through each stationary point, there are $2n$ directions along which we may carry out such line searches.

The condition that only one component of the gradient differs from zero means that the gradient is an eigenvector of the Hessian at GEs:

$$\mathbf{G}\left(\mathbf{x}\right)\mathbf{g}\left(\mathbf{x}\right) = \mu\left(\mathbf{x}\right)\mathbf{g}\left(\mathbf{x}\right) \tag{67}$$

We obtain the same equations by optimizing the squared norm of the gradient in the contour subspace where $f\left(\mathbf{x}\right)$ is equal to a constant k. Differentiating the Lagrangian

$$L\left(\mathbf{x}, \mu\right) = \mathbf{g}^{T}\left(\mathbf{x}\right)\mathbf{g}\left(\mathbf{x}\right) - 2\mu\left[f\left(\mathbf{x}\right) - k\right] \tag{68}$$

and setting the result equal to zero we arrive at Eq. (67). This means, for example, that the gradient extremal belonging to the lowest eigenvalue may be interpreted as a valley floor.

To develop a method for tracing gradient extremals we return to the restricted second-order model Eq. (33). In this model the Hessian is constant

$$\mathbf{G}_{SO}\left(\mathbf{x}\right) = \mathbf{G}_{c} \tag{69}$$

and the gradient is given by

$$\mathbf{g}_{SO}(\mathbf{x}) = \mathbf{g}_c + \mathbf{G}_c \mathbf{s} \tag{70}$$

Inserting Eqs. (69) and (70) in the gradient extremal equation Eq. (67), we obtain

$$(\mu \mathbf{1} - \mathbf{G}_c)(\mathbf{g}_c + \mathbf{G}_c \mathbf{s}) = \mathbf{0} \tag{71}$$

If μ is different from all Hessian eigenvalues we obtain the Newton step

$$\mathbf{g}_c + \mathbf{G}_c \mathbf{s} = \mathbf{0} \tag{72}$$

which takes us to the stationary point on the SO model. This is trivially a gradient extremal point. We therefore set μ equal to the k'th Hessian eigenvalue and obtain

$$(\mu_k \mathbf{1} - \mathbf{G}_c)(\mathbf{g}_c + \mathbf{G}_c \mathbf{s}) = \mathbf{0} \tag{73}$$

which has the solutions

$$\mathbf{x}_k(t) = -\mathbf{P}_k \mathbf{G}_c^{-1} \mathbf{g}_c + t \mathbf{v}_k \tag{74}$$

Here $\mathbf{v}_k$ is the eigenvector

$$\mathbf{G}_c \mathbf{v}_k = \mu_k \mathbf{v}_k \tag{75}$$

and $\mathbf{P}_k$ the projector

$$\mathbf{P}_k = \mathbf{1} - \mathbf{v}_k \mathbf{v}_k^T \tag{76}$$

Therefore the k'th GE is a straight line passing through the stationary point of the model in the direction of the k'th eigenvector.

Equation (74) forms the basis for a second-order saddle point algorithm [43]. In each iteration we first identify the mode to be followed (*the reaction mode*), then take a projected Newton step to minimize the function along all other modes (*the transverse modes*), and finally take a step along the reaction mode until we reach the boundary of the trust region. If the projected Newton step takes us out of the trust region, we minimize the transverse mode on the boundary instead and do not take a step along the reaction mode. In the local region $\mathbf{x}^*$ lies within the trust region and we take the unprojected Newton step.

Although the above algorithm may be useful in some situations, it has not proved to be stable enough for routine uphill walks. Indeed, Sun and Ruedenberg have recently pointed out that a stable algorithm for tracing gradient extremals requires a knowledge of the third derivatives of the potential energy surfaces [42]. Therefore, an algorithm based on a second-order model is not satisfactory. We do not go into any details of their algorithm, but merely note that only the third derivatives in the direction of $\mathbf{g}_c$ are needed—and these may readily be obtained from a simple finite difference scheme. Their algorithm has not been tried on *ab initio* potential energy surfaces.

5.3   TRUST-REGION IMAGE MINIMIZATION

In the GE method the minimization of the transverse modes comes first and the maximization of the reaction mode second. We now describe a method that weights minimization and maximization equally. We assume the existence of an image function [44] with the following properties. If the function to be optimized $f(\mathbf{x})$ has the following gradient and eigenvalues at $\mathbf{x}$

$$\phi(\mathbf{x}) = \begin{bmatrix} \phi_1(\mathbf{x}) \\ \phi_2(\mathbf{x}) \\ \vdots \\ \phi_n(\mathbf{x}) \end{bmatrix} \qquad \lambda(\mathbf{x}) = \begin{bmatrix} \lambda_1(\mathbf{x}) \\ \lambda_2(\mathbf{x}) \\ \vdots \\ \lambda_n(\mathbf{x}) \end{bmatrix} \tag{77}$$

the gradient and eigenvalues of the image function $\overline{f}(\mathbf{x})$ are

$$
\overline{\phi}(\mathbf{x}) = \begin{bmatrix} -\phi_1(\mathbf{x}) \\ \phi_2(\mathbf{x}) \\ \vdots \\ \phi_n(\mathbf{x}) \end{bmatrix} \quad \overline{\lambda}(\mathbf{x}) = \begin{bmatrix} -\lambda_1(\mathbf{x}) \\ \lambda_2(\mathbf{x}) \\ \vdots \\ \lambda_n(\mathbf{x}) \end{bmatrix} \tag{78}
$$

Hence, in the diagonal representation the gradient and Hessian are identical except for opposite signs in the lowest mode. Therefore, a first-order saddle point of the function coincides with a minimum of the image and we may determine the transition state by minimizing the image function.

To minimize the image function we use a second-order method since in each iteration the Hessian is needed anyway to identify the mode to be inverted (*the image mode*). Line search methods cannot be used since it is impossible to calculate the image function itself when carrying out the line search. However, the trust-region RSO minimization requires only gradient and Hessian information and may therefore be used. In the diagonal representation the step Eq. (41) becomes

$$
\mathbf{s}(\mu) = -\frac{\phi_1}{\lambda_1 + \mu}\mathbf{v}_1 - \sum_{i \neq 1}\frac{\phi_i}{\lambda_i - \mu}\mathbf{v}_i \tag{79}
$$

The only difference between the steps along the image mode and the transverse modes is the sign of the level shift. The level shift $\mu < -\lambda_1$ is determined such that the step is to the boundary of the trust region. Equation (79) forms the basis for the *trust-region image minimization* (TRIM) method for calculating saddle points [45].

To compare the image and GE methods, notice that in the diagonal representation the quadratic model may be written

$$
m_{SO}(\mathbf{s}) = f(\mathbf{x}_c) + \sum_i m_i(\sigma_i) \tag{80}
$$

where

$$
m_i(\sigma_i) = \phi_i\sigma_i + \frac{1}{2}\lambda_i\sigma_i^2 \tag{81}
$$

In RSO minimizations we minimize Eq. (80) within the trust region. If instead we wish to maximize the lowest mode and minimize the others we may use the model

$$
\overline{m}_{SO}(\mathbf{s}) = f(\mathbf{x}_c) - m_1(\sigma_1) + \sum_{i \neq 1} m_i(\sigma_i) \tag{82}
$$

since by reducing $-m_1(\sigma_1)$ we increase $m_1(\sigma_1)$. Equation (82) is the SO model of the image function. We can now see the difference between TRIM and the GE method as presented here. In the GE algorithm we first minimize the transverse modes and then maximize the reaction mode. In the TRIM method we minimize and maximize simultaneously by introducing an auxiliary function Eq. (82). The underlying idea of the GE method are lines connecting stationary points. The idea behind the TRIM method is an auxiliary function (the image function) whose minima coincide with the saddle points of the original function.

An image function does not exist for all functions. It exists by definition for all quadratic functions. It also exists trivially for functions in one variable since an image is obtained simply by changing the sign of the function. To analyze this is greater detail, we follow Sun and Ruedenberg [46] and note that the gradient and Hessian for the image function may be written in the form

$$
\overline{\mathbf{g}}(\mathbf{x}) = \left[1 - 2\mathbf{v}_1(\mathbf{x})\mathbf{v}_1^T(\mathbf{x}) \right]\mathbf{g}(\mathbf{x}) \tag{83}
$$

$$\overline{\mathbf{G}}\left(\mathbf{x}\right) = \left[1 - 2\mathbf{v}_1\left(\mathbf{x}\right)\mathbf{v}_1^T\left(\mathbf{x}\right)\right]\mathbf{G}\left(\mathbf{x}\right) \qquad (84)$$

The relationship between the gradient and Hessian elements is for the true potential energy surface given by

$$\frac{\partial g_j}{\partial x_i} = G_{ij} \qquad (85)$$

which follows directly from the definitions of gradients and Hessians. For the image gradient and Hessians, however, the following relationship holds

$$\frac{\partial \overline{g}_j\left(\mathbf{x}\right)}{\partial x_i} = \overline{G}_{ij} - 2\frac{\partial\left[v_{1j}\left(\mathbf{x}\right)\mathbf{v}_1^T\left(\mathbf{x}\right)\right]}{\partial x_i}\mathbf{g}\left(\mathbf{x}\right) \qquad (86)$$

In general, the second term vanishes only for functions $f\left(\mathbf{x}\right)$ with constant eigenvectors (quadratic functions). For other than quadratic functions, the second term will contain non-zero elements and thus no image function exists for such functions. Nevertheless, the image concept has proven useful for formulating an algorithm.

5.4   RATIONAL FUNCTION MODE FOLLOWING

In the GE algorithm we select one eigenvector as the reaction mode and follow this towards the transition state. A similar mode following technique has also been developed within the RF framework. However, the RF model Eq. (50) is not sufficiently flexible for mode following. Instead we use the model [23, 47]

$$m_{RF}\left(\mathbf{x}\right) = f\left(\mathbf{x}_c\right) + \frac{m_1\left(\sigma_1\right)}{1 + \sigma_1^2} + \frac{\sum_{i \neq 1} m_i\left(\sigma_i\right)}{1 + \sum_{i \neq 1} \sigma_i^2} \qquad (87)$$

where we have separated the model in two parts representing the reaction mode and the transverse modes. Each term is divided by a quadratic to make it bounded. The two metrics in Eq. (87) are set to unity. Note that two heuristic parameters appear in Eq. (87) rather than one as in the methods previously discussed.

The model Eq. (87) has $2n$ stationary points. To see this we differentiate the model and obtain two independent sets of equations

$$\begin{bmatrix} \lambda_1 & \phi_1 \\ \phi_1 & 0 \end{bmatrix}\begin{bmatrix} \sigma_1 \\ 1 \end{bmatrix} = \nu_R\begin{bmatrix} \sigma_1 \\ 1 \end{bmatrix} \qquad (88)$$

$$\begin{bmatrix} \lambda_2 & 0 & \cdots & 0 & \phi_2 \\ 0 & \lambda_3 & \cdots & 0 & \phi_3 \\ \vdots & \vdots & \ddots & \vdots & \vdots \\ 0 & 0 & \cdots & \lambda_n & \phi_n \\ \phi_2 & \phi_3 & \cdots & \phi_n & 0 \end{bmatrix}\begin{bmatrix} \sigma_2 \\ \sigma_3 \\ \vdots \\ \sigma_n \\ 1 \end{bmatrix} = \nu_T\begin{bmatrix} \sigma_2 \\ \sigma_3 \\ \vdots \\ \sigma_n \\ 1 \end{bmatrix} \qquad (89)$$

Equation (88) has two solutions for the reaction mode and Eq. (89) has n solutions for the transverse modes. To determine a transition state we choose the maximum of Eq. (88) and the minimum of Eq. (89). If the step becomes too big it is scaled down.

The methods discussed in this section all work by the same principle. The Hessian is diagonalized and the reaction mode is identified. A step is then taken such the function is increased along the reaction mode and decreased along the transverse modes. The methods differ in the way this maximization and minimization is carried out.

5.5 EXAMPLES OF TRANSITION STATE OPTIMIZATIONS

We have focussed our attention on methods for automated searches of transition states: methods that follow specific vibrational modes or gradient extremal paths, and methods that transform the surface topology so that the desired optimization becomes a minimization on the image surface. The ideal method should be globally convergent and allow us to initiate "uphill" walks in various directions from an optimized minimum energy structure. In this way, it might be hoped that all or most chemically interesting transition states would be identified. Unfortunately, this attractively simple approach has some deficiencies in practice. Thus, for mode-following methods, we must identify a particular vibrational mode at each step along the walk: unless very small steps are taken this may not be possible when the modes change substantially along the walk. For gradient extremals there is no requirement that an extremal walk that begins at a minimum will necessarily pass over a saddle point, and a stable algorithm must take into account the anharmonicity of the surface along the path. Image-surface minimization proves in practice to have a very large radius of convergence to the *lowest* saddle point on a given surface—this may or may not be chemically interesting. Thus although second-order optimization methods allow us to implement these techniques for locating transition states, their use in some automatic, "black-box" implementation is not always successful.

A major factor that complicates such black-box approaches is the dimensionality of the potential energy surface. For triatomic systems, for example, there may be only one or two transition states, both of chemical interest, and easily found by for example following different vibrational modes of the reactant or product. Even for four-atom systems there are usually not many saddle points. Clearly, as the size of the molecule increases the number of stationary points of any Hessian index will increase, but this is not really the issue. The problem is that for larger systems, many of the saddle points (especially low-lying ones) will be transition states on a pathway corresponding to some conformational change in the molecule. Thus a number of uphill walks from a minimum will arrive at conformational transition states that are uninteresting from the point of view of chemical reactions. In such cases a method such as TRIM, for example, which shows excellent convergence to the lowest saddle point on a potential surface, will almost always locate only a conformational transition state, even when the starting geometry is far away. The automated location of reactive transition states, beginning from optimized minima, is a goal not yet realized for molecules with more than about four atoms.

Let us give a few examples of how conformational transition states may be determined in an automated manner. The calculations have been carried out with HERMIT-SIRIUS-ABACUS [31, 32, 33] at the self-consistent field level using a minimal basis set. A simple example is provided by PH_5, whose equilibrium structure is a trigonal bipyramid of D_{3h} symmetry. Starting at equilibrium and applying the TRIM method without symmetry constraints, we arrive after nine iterations at a square pyramidal transition state of C_{4v} symmetry. This structure represents the barrier to a Berry pseudo-rotation connecting two equivalent D_{3h} equilibrium structures, as confirmed by walking down the other side of the barrier.

For a more complicated case, consider the structure of cyclohexane, which exists in two conformational mimima, the chair structure of D_{3d} symmetry and the twist-boat structure of D_2 symmetry. The chair conformer is stabilized with respect to the twisted boat by about 25 kJ/mole in a minimal basis. Starting at the D_{3d} equilibrium structure, the TRIM method reaches a first-order saddle point of C_2 symmetry in 16 iterations and one backstep, again with no symmetry constraints applied. The activation energy is 49 kJ/mole. Minimizing from this structure, we reach the D_2 twist-boat conformation in 10 second-order iterations. Thus we have been able to walk between the two stable conformers of cyclohexane in an automated manner. The uphill and downhill walks both require only a modest number of iterations, considering that they started far away in the global region.

We may explore the surface further by initiating another TRIM walk, this time from the twist-boat structure. After 10 iteration we reach a new barrier top, only 3.7 kJ/mole above the twisted boat. This saddle point corresponds to the C_{2v} boat conformer of cyclohexane, previously thought to represent a minimum configuration. Minimizing down from this barrier, we reach another twist-boat structure, different but equivalent to the one we started from. The detected C_{2v} structure thus represents the barrier to an interconversion between two twist-boat equilibrium conformers.

We would like to stress that all stationary points found for cyclohexane were determined without interference from the user, just by letting the algorithm start from a minimum. In general, however, there is no need to start TRIM at a minimum. Often, it may be more useful or even necessary to start the search away from equilibrium in order to "catch" some suspected transition state which would otherwise be hidden from the algorithm due to the presence of another saddle point with a larger radius of convergence.

These examples demonstrate that it is indeed possible to determine transition states in an automated manner, even for quite large molecules. However, all transition states found above are conformational and do not involve the breaking or the formation of chemical bonds. Reactive transition states are harder to find by automated uphill walks from minima, although some examples have appeared in the literature. For example, in a study of N_2H_2 using Levenberg-Marquardt trajectories, three equilibrium structures (cis -, trans -, and iso -diazene) were determined along with three conformational and three reactive transition states, dissociating in $N_2 + H_2$ and $N_2H + H$ [48].

If we relax the condition that the algorithms should be globally convergent, then quasi-Newton methods may be quite useful and cost-effective, as demonstrated by Culot et al. [49]. These authors use a trust-region based quasi-Newton method with an exact initial Hessian and converge in a modest number of iterations (ten to twenty) provided the initial guess is a reasonable one.

In practice, whatever transition-state optimization method is used, the crucial need is almost always a good starting guess. Obviously, if the guess is lucky enough to lie within the quadratic basin of the saddle point, the problem is straightforward to solve. This is seldom the case in most applications, and perhaps the most important factor is experience: a "nose" for a good starting structure. A good example is provided by Walch's work on the reaction of CH with N_2 [50], which can be contrasted with unsuccessful attempts by Martin and Taylor [51] for the same reaction. Simple methods such as "linear synchronous transit" walks—linear interpolation between reactant and product geometries—can sometimes be used to generate a useful starting guess. Nevertheless, we must reiterate that, even with the convergence advantages formally conferred by second-order methods, successful transition-state optimizations are still largerly a matter of skill and experience (especially the choice of initial guess). Much work remains to be done before such optimizations can be as automated as second-order methods for locating minima.

References

[1] R.Fletcher. *Practical Methods of Optimization, 2nd ed.* Wiley, Chichester, 1987.

[2] P.E.Gill, W.Murray, and M. H. Wright. *Practical Optimization.* Academic, London, 1981.

[3] J. E. Dennis, Jr. and R. B. Schnabel. *Numerical Methods for Unconstrained Optimization and Nonlinear Equations.* Prentice-Hall, Englewood Cliffs, 1983.

[4] Z. Havlas and R. Zahradník. *Int.J.Quantum Chem.*, 26:607, 1984.

[5] H.B.Schlegel. *Adv.Chem.Phys.*, 69:249, 1987.

[6] H. B. Schlegel. In J. Bertrán and I.G.Csizmadia, editors, *New Theoretical Concepts for Understanding Organic Reactions*, page 33. Kluwer, Dordrecht, 1989.

[7] J.D.Head, B.Weiner, and M.C.Zerner. *Int.J.Quantum Chem.*, 33:177, 1988.

[8] J.D.Head and M.C.Zerner. *Adv.Quantum Chem.*, 20:239, 1989.

[9] J.Baker. *J.Comp.Chem.*, 13:240, 1992.

[10] J.Simons, P.Jørgensen, H.Taylor, and J.Ozment. *J. Phys. Chem.*, 87:2745, 1983.

[11] J.Nichols, H.Taylor, P.Schmidt, and J.Simons. *J.Chem.Phys.*, 92:340, 1990.

[12] P.Pechukas. *J.Chem.Phys.*, 64:1516, 1976.

[13] M.Page and J.W.McIver, Jr. *J.Chem.Phys.*, 88:922, 1988.

[14] J.Ischtwan and M.A.Collins. *J.Chem.Phys.*, 89:2881, 1988.

[15] R.N.Camp and H.F.King. *J.Chem.Phys.*, 77:3056, 1982.

[16] C.Gonzalez and H.B.Schlegel. *J.Chem.Phys.*, 90:2154, 1989.

[17] J.-Q.Sun and K.Ruedenberg. *J.Chem.Phys.*, 99:5257, 1993.

[18] J.-Q.Sun and K.Ruedenberg. *J.Chem.Phys.*, 99:5269, 1993.

[19] J.-Q.Sun and K.Ruedenberg. *J.Chem.Phys.*, 99:5276, 1993.

[20] K.Ruedenberg and J.-Q.Sun. *J.Chem.Phys.*, 100:6101, 1994.

[21] A.Banerjee and F.Grein. *Int.J.Quantum Chem.*, 10:123, 1976.

[22] D.R.Yarkony. *Chem.Phys.Lett.*, 77:634, 1981.

[23] A.Banerjee, N.Adams, J.Simons, and R.Shepard. *J. Phys. Chem.*, 89:52, 1985.

[24] A.Komornicki. "GRADSCF, an *ab initio* program system, developed at Polyatomics Research Institute".

[25] E.Bright Wilson, Jr., J.C.Decius, and P.C.Cross. *Molecular Vibrations. The Theory of Infrared and Raman Vibrational Spectroscopy.* McGraw-Hill, New York, 1955.

[26] G.Fogarasi, X.F.Zhou, P.W.Taylor, and P.Pulay. *J.Am.Chem.Soc.*, 114:8191, 1992.

[27] P.Pulay and G. Fogarasi. *J.Chem.Phys.*, 96:2856, 1992.

[28] J.Baker and W.J.Hehre. *J.Comp.Chem.*, 12:606, 1991.

[29] J.Baker. *J.Comp.Chem.*, 14:1085, 1993.

[30] H.B.Schlegel. *Int. J. Quantum Chem. Symp.*, 26:243, 1992.

[31] T.Helgaker, P.R.Taylor, K.Ruud, O.Vahtras, and H.Koch. "HERMIT, a molecular integral program".

[32] H.J.Aa.Jensen and H.Ågren. "SIRIUS, a program for calculation of MCSCF wave functions".

[33] T.Helgaker, K.L.Bak, P.Dahle, H.J.Aa.Jensen, P.Jørgensen, R.Kobayashi, H.Koch, K.Mikkelsen, J.Olsen, K.Ruud, P.R.Taylor, and O.Vahtras. "ABACUS, a second-order MCSCF molecular property program".

[34] C.J.Cerjan and W.H.Miller. *J.Chem.Phys.*, 75:2800, 1981.

[35] D.T.Nguyen and D.A.Case. *J. Phys. Chem.*, 89:4020, 1985.

[36] H.J.Aa.Jensen, P.Jørgensen, and T.Helgaker. *J.Chem.Phys.*, 85:3917, 1986.

[37] J.Pancíř. *Coll. Czech. Chem. Commun.*, 40:1112, 1975.

[38] J.Pancíř. *Coll. Czech. Chem. Commun.*, 42:16, 1977.

[39] M.V.Basilevsky and A.G.Shamov. *Chem.Phys.*, 60:347, 1981.

[40] D.K.Hoffman, R.S.Nord, and K.Ruedenberg. *Theor.Chim.Acta*, 69:265, 1986.

[41] W. Quapp. *Theor.Chim.Acta*, 75:447, 1989.

[42] J.-Q.Sun and K.Ruedenberg. *J.Chem.Phys.*, 98:9707, 1993.

[43] P.Jørgensen, H.J.Aa.Jensen, and T.Helgaker. *Theor.Chim.Acta*, 73:55, 1988.

[44] C.M.Smith. *Theor.Chim.Acta*, 74:85, 1988.

[45] T.Helgaker. *Chem.Phys.Lett.*, 182:503, 1991.

[46] J.-Q.Sun and K.Ruedenberg. *J.Chem.Phys.*, 101:2157, 1994.

[47] J.Baker. *J.Comp.Chem.*, 7:385, 1986.

[48] H.J.Aa.Jensen, P.Jørgensen, and T.Helgaker. *J.Am.Chem.Soc.*, 109:2895, 1987.

[49] P.Culot, G.Dive, V.H.Nguyen, and J.M.Ghuysen. *Theor.Chim.Acta*, 82:189, 1992.

[50] S.P.Walch. *Chem.Phys.Lett.*, 208:214, 1993.

[51] J.M.L.Martin and P.R.Taylor. *Chem.Phys.Lett.*, 209:143, 1993.

GRADIENT EXTREMALS AND THEIR RELATION TO THE MINIMUM ENERGY PATH

WOLFGANG QUAPP*, OLAF IMIG^c, DIETMAR HEIDRICH^c

*Mathematisches Institut,
^{c}Institut für Physikalische und Theoretische Chemie,
Universität Leipzig,
D-04109 Leipzig, Germany

1. Introduction: Defining and Tracing of Reaction Paths

The concept of the reaction path (RP) of potential energy surfaces (PES) has gained increasing importance in theoretical chemistry [1,2]. Qualitatively, the RP is a curve in the configuration space of the atoms forming the chemical system which connects two minimizers of the PES along points of minimal energy in comparison to neighbouring points. The energy profile over the reaction path should be a "valley floor" leading via a point of highest energy, the saddle point of index 1 of the PES. This point corresponds to the transition structure of the "transition state theory". The fundamental problem in handling PES is the problem of dimensionality. Molecules with a number of atoms more than N=4 force an overwhelming number of net points in the dimension n=3N-6. The RP concept is a promising way out. It requires finding an algorithm for chemically reasonable one-dimensional curves of the PES determinable by differential properties of the PES – gradient and Hessian matrix – without knowledge of the whole, or of large parts of the PES. Slope and curvature of $E(\mathbf{x})=E(x^1,\ldots,x^n)$ can be calculated from the gradient vector, $\mathbf{g}(\mathbf{x})=\nabla E(\mathbf{x})$, and from the Hessian matrix (second derivatives, $\mathcal{H}(\mathbf{x})=\nabla\nabla^T E(\mathbf{x})$, of the PES), respectively. [We shall denote geometrical vectors in the configuration space and column matrices of their Cartesian coordinates by boldface lower-case letters, second order tensors and square matrices of their components by caligraphic upper-case letters. Scalars are often denoted by greek letters.]

All parametrizations s in $\mathbf{x}(s)=(x^1(s),\ldots,x^n(s))^T$ of the reaction path are called *reaction coordinate s* when forming the abscissa in the diagrams used in chemistry, see **Figure 1**. Different mathematical definitions of the reaction path may produce different curves. From the point of view of a simple chemical application, this is not necessarily disturbing. However, it is of particular importance when working beyond transition state theory.

D. Heidrich (ed.), The Reaction Path in Chemistry:
Current Approaches and Perspectives, 137–160.
© 1995 *Kluwer Academic Publishers. Printed in the Netherlands.*

In this chapter, the problem of defining (and tracing) a chemically meaningful RP of PES is reinvestigated in the light of the gradient extremal (GE). GEs of n-dimensional hypersurfaces, $E=E(x^1,\ldots,x^n)$, are curves defined by the condition that the gradient, ∇E, is an eigenvector of the Hessian matrix, $\nabla\nabla^T E$. The relationship between the conventional steepest descent path (SDP) and the GE is analyzed using the curvature of SDPs. A classification of the solutions of the gradient extremal equation is given for the two-dimensional case. GE pathway tracing can be used as a tool to study the valley paths. The tracing is depicted below by means of model surfaces. Singular points of GE – where valleys or ridges appear, dissipate, or bifurcate – are included in the analysis. In addition, the description of the valley floor line by curvature extrema of the contour lines is discussed.

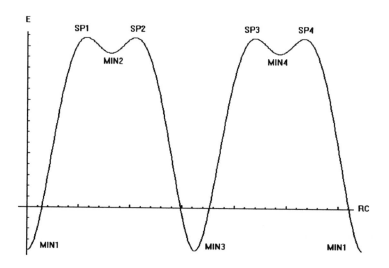

Figure 1. Energy profile over a reaction coordinate.

1.1. The Two Main Methods

(i) Steepest Descent Path (SDP)

– starting at a saddle point in the direction of negative main curvature,

– stepping in the direction of negative gradient vector and

– reaching a possible minimizer.

(ii) The least ascent path

– starting at a minimizer (saddle point), and following the direction of the smallest (negative) main curvature of the PES,

- following a valley floor gradient extremal (VF-GE) by proving the gradient along the equipotential surface E(x)=const,

- reaching a saddle point in suitable cases (or a minimum in the opposite direction) or crossing a second GE,

- obtaining suggestions for more complex valley structure, as blind valleys, branching valleys etc.

In general, both methods do not yield the same results. The second method (ii) produces additional curves which are other (artificial) solutions of the GE equation which do not characterize valley floors. It should be noted that there are some simplifications of (ii) when following a given eigenvector of the Hessian by the so-called stream bed method, see [3–5]. SDP and least ascent path (VF-GE) have each certain advantages as well as limitations.

2. The Method of Steepest Descent

If the SDP [6] is treated in a system of mass-weighted Cartesian coordinates [7], and if it starts at a saddle point then K. Fukui termed it "intrinsic reaction coordinate (IRC)" [8]. It is now accepted in many program packages as the favorable, mathematically defined RP in chemistry. First of all, we will look critically at this path's ability to characterize a valley floor line. We will assume to be working with mass-weighted coordinates.

Any steepest descent along an energy profile of the PES functional E(x) corresponds to a one-dimensional curve called SDP, x=x(t), in the configuration space of the chemical system, i.e. in the hyperplane below the surface. x(0) is the starting point. x(t) has to satisfy the system of differential equations [1,9]

$$\frac{d\mathbf{x}}{dt} = -\mathbf{g}(\mathbf{x}). \tag{1}$$

The solution curve converges, for $t \to \infty$, towards the minimizer $\mathbf{x}_{min}$ of E(x), if there is a minimum of the PES, at all. Because

$$-\left\|\frac{d\mathbf{x}}{dt}\right\|^2 = \mathbf{g}(\mathbf{x})^T \frac{d\mathbf{x}}{dt} = \frac{dE}{dt} \le 0,$$

we get a decrease of energy if t increases. When using the arclength s as parameter t in eq.(1), we get [1]

$$\frac{d\mathbf{x}}{ds} = -\frac{\mathbf{g}(\mathbf{x})}{\|\mathbf{g}(\mathbf{x})\|} = -\overset{\wedge}{\mathbf{g}}(\mathbf{x}). \tag{2}$$

$\overset{\wedge}{\mathbf{g}}(\mathbf{x})$ is the normalized gradient, and $d\mathbf{x}/ds$ the normalized tangent vector of the steepest descent curve x(s), if x(s) satisfies the system of eq.(2).

The curvature $\kappa(s)$ of $\mathbf{x}(s)$ is the absolute value of the curvature vector

$$\mathbf{k}(s) = \frac{d^2\mathbf{x}}{ds^2}.$$

The derivation of eq.(2) gives us the relation [10],

$$\mathbf{k}(s) = \frac{d^2\mathbf{x}}{ds^2} = [\mathcal{H} \cdot \hat{\mathbf{g}} - \hat{\mathbf{g}}\,(\hat{\mathbf{g}}^T \cdot \mathcal{H} \cdot \hat{\mathbf{g}})]/\|\mathbf{g}\|.$$

If the SDP is a straight line, i.e. $\mathbf{k}(s)=0$, then we obtain the important equation (see below)

$$\mathcal{H}(\mathbf{x}) \cdot \mathbf{g}(\mathbf{x}) = \lambda \mathbf{g}(\mathbf{x}) \tag{3}$$

with a real scalar function $\lambda(\mathbf{x})$ in all n equations of this system

$$\lambda(\mathbf{x}) = \hat{\mathbf{g}}\,(\mathbf{x})^T \cdot \mathcal{H}(\mathbf{x}) \cdot \hat{\mathbf{g}}\,(\mathbf{x}).$$

Eq.(3) means: Only if the curvature of $\mathbf{x}(s)$ is zero, then the gradient will be an eigenvector of the Hessian. This is an interesting relationship between the first derivatives of the PES, and the second derivatives forming the Hessian matrix. It holds only in very special points of a curvilinear SDP, and then it marks a prominent point of the PES [10,11]!

In practical calculations, the steepest descent method produces, with a finite step size, a minimizer in any exactness, where usually some numerically pathological cases should be avoided [12]. A simple algorithm for steps with step size $\mu > 0$ along the tangent of $\mathbf{x}(s)$ can be used

$$\mathbf{x}_{k+1} = \mathbf{x}_k - \mu\,\hat{\mathbf{g}}\,(\mathbf{x}_k). \tag{4}$$

Ref.[13] gives an actual review of a number of refined methods.
For a calculation of the IRC, we have to start at the saddle point $\mathbf{x}_0=\mathbf{x}(0)$, but the first step has to be forced in direction of the negative main curvature of the PES, because it is $\mathbf{g}(\mathbf{x}_0)=0$ at the saddle point, and eq.(4) would not start to work. After starting algorithm (4), it is sure to reach a minimizer of the PES. Sometimes numerical problems emerge through the zigzagging of the numerical SDP near the PES minimum, or in a flat part of the valley of the PES [12,14].

In much of the literature written on this subject, the IRC is interpreted as **the** minimum energy valley path and as such it is taken to be the valley floor line. If we define a valley path with the property that $E(\mathbf{x})$ should ascend in all directions orthogonal to the valley direction (i.e., orthogonal to the gradient vector) then the interpretation is valid in most cases (for a counterexample cf. Schlegel [15]). However, the interpretation of this property holds at **any** point of an equipotential line if this line is convex [9]. Therefore, the energy profile orthogonal to the gradient can not be a local criterion of the valley floor. It is beyond dispute that a local criterion of an IRC does not exist.
Another drawback of the SDPs is the impossibility to detect bifurcations of a valley path [1,15,16], whereas the reaction path definition by VF-GE gives us the possibility to determine bifurcations.

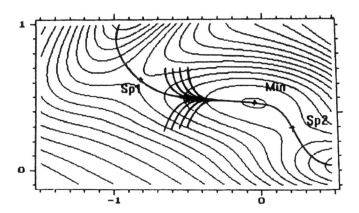

Figure 2. Diagram of contour lines (thin) of a 2D model potential surface, and of an IRC (bold-faced curve). A number of "non SP – SDPs" are illustrated which reach the MEP asymptotically.

The fact that the IRC starts at the saddle point is a further shortcoming. An inversion of the IRC concept is impossible (starting at the minimizer/ stepping in the direction of the least main curvature and along the solution of system (4) with $\mu < 0$.) The valley floor is an asymptote of all SDPs coming from the slopes of the PES. This is evident by inversion of the depicted path lines which would suddenly escape out of the bottom curve in **Figure 2**. Hence, we must find a local criterion for points lying on a valley floor line.

3. A Local Criterion for a Valley Floor Line as Preparation for Defining of GEs

Pancíř [17] and Basilevsky/Shamov [18] were the first who formulated local criteria for describing a valley floor line. Pancíř determined two conditions which he assumed to be obviously given:

(i) The energy must increase along all directions perpendicular to the direction of the valley floor line.

(ii) The curvature of the energy surface along the direction of the valley must be less than the curvature along any other direction.

Pancíř came to the conclusion that a path satisfying (i) and (ii) should be a sequence of points where the gradient **g** is an eigenvector of the Hessian, $\mathcal{H}$. Ruedenberg et al. [19] called curves which are formed by such points "*gradient extremals*". However, we will recognize below that the GE, defined in this manner, does only approximately satisfy requirements (i) and (ii) in the general case.

We will now characterize points forming a GE in the configuration space of a chemical system. The measure for the ascent of the n-dimensional PES functional $E(\mathbf{x})$ is the absolute value of $\mathbf{g}$, or, more suitable for calculations, the functional

$$\sigma(\mathbf{x}) = \frac{1}{2}\|\mathbf{g}(\mathbf{x})\|^2.$$

We search for a point $\mathbf{x}$ giving $\sigma(\mathbf{x})$ an extreme value under a variation along the equipotential hypersurface $E(\mathbf{x})=$const. A necessary condition for an extreme value is the zero of the change of $\sigma(\mathbf{x})$ along all directions orthogonal to $\mathbf{g}(\mathbf{x})$. In other words, the projection of $\nabla\sigma(\mathbf{x})$ onto all directions orthogonal to the gradient vanishes [19]

$$\mathcal{P} \cdot \nabla\sigma(\mathbf{x}) = \mathbf{0} \tag{5}$$

with

$$\mathcal{P} = \mathbf{1} - {}^g\mathcal{P}, \quad \text{and} \quad {}^g\mathcal{P}_{ij} = \frac{g_i g_j}{\|\mathbf{g}(\mathbf{x})\|^2}. \tag{6}$$

$\mathcal{P}$ is the projection matrix projecting any vector onto a direction orthogonal to the gradient. In contrast, ${}^g\mathcal{P}$ projects onto the gradient direction itself. ${}^g\mathcal{P}$ is the dyadic product of the normed gradient vector with itself, and, therefore, it is of rank 1. Then, $\mathcal{P}$ is of rank (n-1). It is (proof by calculation)

$$\nabla\sigma(\mathbf{x}) = \mathcal{H}(\mathbf{x}) \cdot \mathbf{g}(\mathbf{x}). \tag{7}$$

With the help of eqs.(6) and (7), we can write eq.(5) in the form

$$\mathcal{P} \cdot \nabla\sigma(\mathbf{x}) = \mathcal{H}(\mathbf{x}) \cdot \mathbf{g}(\mathbf{x}) - {}^g\mathcal{P} \cdot \mathcal{H}(\mathbf{x}) \cdot \mathbf{g}(\mathbf{x}) = \mathbf{0}.$$

The projection of any vector onto $\mathbf{g}(\mathbf{x})$ by ${}^g\mathcal{P}$ is, of course, parallel to $\mathbf{g}(\mathbf{x})$. It results the important relation

$$\mathcal{H}(\mathbf{x}) \cdot \mathbf{g}(\mathbf{x}) = \lambda(\mathbf{x})\mathbf{g}(\mathbf{x}) \tag{8}$$

with a factor $\lambda(\mathbf{x})$ which turns out to be an eigenvalue of the Hessian $\mathcal{H}(\mathbf{x})$. $\lambda(\mathbf{x})$ is a scalar. A point $\mathbf{x}$ which satisfies eq.(8) belongs to an extreme absolute value of $\mathbf{g}$ (slope of $E(\mathbf{x})$) along the equipotential hypersurface.

Note: So far, we do not know if this extremal of $|\mathbf{g}|$ is a minimum (corresponding to a VF-GE) or a maximum, or an inflection point. With these preliminary statements, we can now give the definition of a gradient extremal [19].

Definition: "A gradient extremal is a locus of points where the gradient of the PES is an eigenvector of the Hessian of the PES."

Eq.(8) defines a curve in configuration space of the molecule. It is a nonlinear system of n equations with n unknown variables $x^1, \ldots, x^n$, the rank of which is (n-1). To practically apply curve tracing, we need (n-1) equations. Knowledging the property of the extreme value of $\sigma(\mathbf{x})$ and eq.(8), we immediately have

$$GE_{i-1}(\mathbf{x}) = \nabla\sigma(\mathbf{x})^T \cdot {}^i\mathbf{v}(\mathbf{x}) = \frac{\partial\sigma(\mathbf{x})}{\partial^i\mathbf{v}} = 0, \quad i = 2, \ldots, n \tag{9}$$

with the ith eigenvector of $\mathcal{H}(\mathbf{x})$, ${}^i\mathbf{v}(\mathbf{x}) \neq \hat{\mathbf{g}}(\mathbf{x})$ for $i = 2, \ldots, n$ (we number ${}^1\mathbf{v}(\mathbf{x})$ $= \hat{\mathbf{g}}(\mathbf{x})$). This is possible, because the gradient is the eigenvector with the lowest eigenvalue, if eqs.(9) are fulfilled.) With eq.(6), it holds $\mathcal{P}\,{}^i\mathbf{v} = {}^i\mathbf{v}$, i=2,...,n. In eq.(9) a directional derivative of the scalar $\sigma(\mathbf{x})$ is used. It is the scalar product of the gradient vector of $\sigma(\mathbf{x})$ with the ith eigenvector. Hence, this derivation gives the one-dimensional ascent of the functional $\sigma(\mathbf{x})$ in the direction of the eigenvector ${}^i\mathbf{v}$. Each of the eqs.(9) requires that this ascent has to be zero on a GE. In short, eq.(9) is

$$\mathbf{GE}(\mathbf{x}) = \mathbf{0}, \qquad \mathbf{GE} : \mathbf{R}^n \to \mathbf{R}^{n-1}.$$

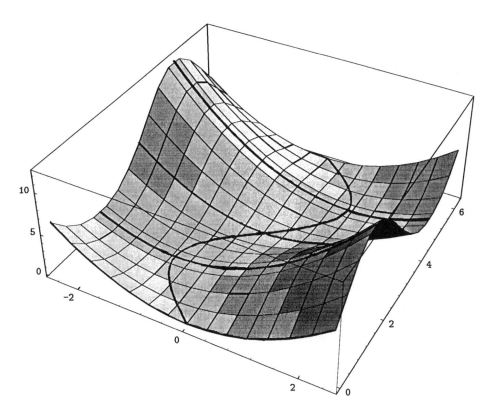

Figure 3. Model PES E(x,y) = [x + sin(y)]^2 for a coupled vibration of a large amplitude torsion and an internal rotation [20]. The bold-face curves are the energy profiles over the GEs in the configuration plane.

The (n-1) lines in the system of eqs.(9) produce for any point $\mathbf{x}$ the (n-1) components of a point $\mathbf{y}$ of an $\mathbf{R}^{n-1}$, on the right hand side. If $\mathbf{x}\in\mathbf{R}^n$ is a GE point, then the point $\mathbf{y}$ has to be the zero point of the $\mathbf{R}^{n-1}$. The GE system eq.(8) or (9) filters out all points of the configuration space having an extreme $\sigma(\mathbf{x})$ value under variation along an equipotential surface. We may find a valley floor path, or a crest of a ridge, if this $\sigma(\mathbf{x})$ is always a minimum.

In **Figure** 3, a model PES with the GE equation

$$\cos(y)\,\sin(y)\,[x + \sin(y)] = 0$$

is illustrated. It has a sinuslike VF-GE curve, x=−sin(y), at the zero floor which is intersected by straight, periodically changing GEs, y(x)=kπ/2, k=0,1,.... . For odd k, the GE goes from ridge to valley and vice versa (even values of k characterize flank lines as shown below). The model shows the torsional motion with a large amplitude and a coupling to an internal rotation [20].

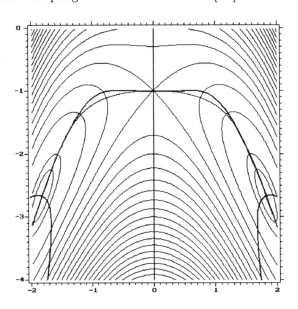

Figure 4. Contour lines of the 2D model PES E(x,y)= 2y + y² + yx² + 0.4x²; SDP from the SP: thin line, GE: bold-faced lines. SDP and GE curves are different and meet at the saddle.

Basilevsky's "mountaineer's algorithm" [18] beginning at a minimizer and going uphill to the saddle point, seems to be a useful interpretation of a VF-GE path and can be identified with a valley floor path. Mountaineer's algorithm:

(i) Start at a minimizer by a step in the direction of the least main curvature of the PES.

(ii) Search along a contour line (no variation in energy) to find the point with minimal $\sigma(\mathbf{x})$.

(iii) Step uphill in the direction of the gradient g; the "energy" expense of a step of constant length is minimal with respect of the variation along a contour line: It is a least ascent.

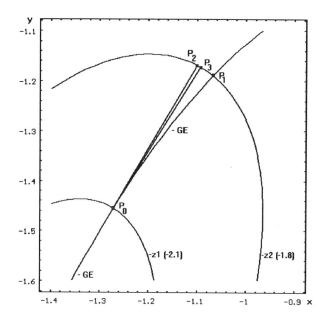

Figure 5. Enlargement of a region of Figure 4 showing the GE and the difference between the tangent vector $\overrightarrow{P_0P_3}$ of the GE at $P_0(x_o, y_o)$, and the gradient vector $\overrightarrow{P_0P_2}$.

(iv) Repeat (ii) – (iii) until a saddle point is reached.

If we execute (ii) and (iii) using infinitesimal steps, then we get the VF-GE curve. **Figure 4** shows a VF-GE path of a test potential leading from a minimizer to a saddle point. There is also a SDP with the same property. However, the two paths do not coincide. Along the VF-GE, $\sigma(\mathbf{x})$ has a minimum if we vary the test point along a contour line. In exactly these points the gradient vector of the PES is the eigenvector with the smallest eigenvalue of the Hessian of the PES. Here, the potential energy increases in the direction of the other (orthogonal) eigenvectors. However, this is **n o t** identical with the direction orthogonal to the tangent of the VF-GE curve, as it is shown in **Figure 5**. When following the VF-GE, the "Pancíř criteria" are **n o t** exactly fulfilled, because the direction (i.e. the tangent) of the GE is not identical with the gradient. SDP and VF-GE represent different curves if the SDP is curvilinear. They only coincide if the SDP is a straight line.

In **Figure 7**, we introduce another view of the relationship between SDPs and GE curves: On a model potential, an indicatrix of Dupin is presented in 121 raster points (cf. text books of differential geometry): It is constructed, in any point, by cutting E with a plane parallel to the tangential plane, rescaling the intersection curve, and taking its limit when the plane approach the tangential plane. In the case of the surface in **Figure 6**, all points are hyperbolic points and we obtain hyperbolas. The eigenvectors of the Hessian are the main directions of these hyperbolas. In every raster point, the two eigenvectors of the Hessian are drawn. Additionally,

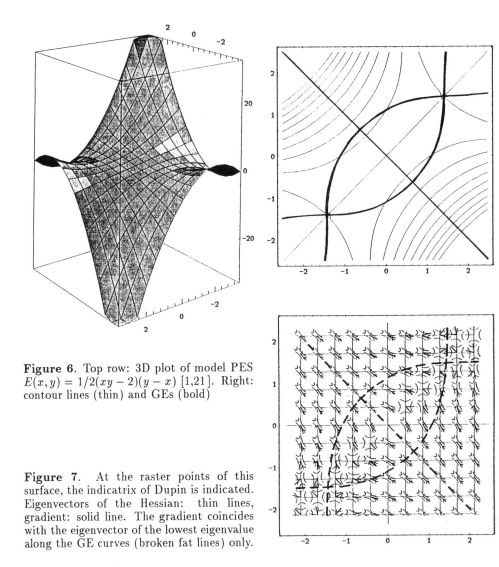

Figure 6. Top row: 3D plot of model PES $E(x,y) = 1/2(xy - 2)(y - x)$ [1,21]. Right: contour lines (thin) and GEs (bold)

Figure 7. At the raster points of this surface, the indicatrix of Dupin is indicated. Eigenvectors of the Hessian: thin lines, gradient: solid line. The gradient coincides with the eigenvector of the lowest eigenvalue along the GE curves (broken fat lines) only.

the negative gradient is given. On the GE curves the gradient coincides with the eigenvector direction of the lowest eigenvalue. This does not hold true for the IRC when going downhill from the saddle point.

Figure 8 shows an example for a "very oblique passing" of the contour lines by a VF-GE. In principle, the "skewness" of any GE versus the contour lines comes from the "individual" definition of all GE points on the corresponding contour line, which we may illustrate with **Figure 5** as follows: If a point P_0 of a GE is found on the line (z_1), then we will be able to start a new search to get a corresponding point P_1 on a next line (z_2) which also satisfies the GE equation. The new point P_1 may be found far away from the first point P_0. Thus, the connecting curve of P_0 and P_1,

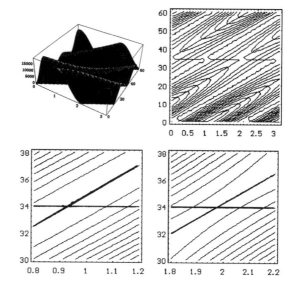

Figure 8. At the top row: Model PES for the torsion and inversion coordinate of acetaldehyde, CH_3CHO [19]. Bottom: Enlargement of the minimum and saddle point regions. The GE of interest is the horizontal one at 34°.

being the GE, obliquely crosses the contour lines. In contrast, the determination of SDPs does not use marked features of the contour lines. SDPs may start from any point of the PES.

Remark: The geometry optimization method to localize saddle points, based on finding the minima of the functional, may yield additional minima of $\sigma(\mathbf{x})$ where $\mathbf{g}\neq 0$ [22]. Along GE curves, this functional has extreme values in the (n-1) dimensional search region of the equipotential sub-hypersurface. If an extremum of σ additionally exists along the GE curve, this point should be indicated by the geometry optimization using σ. In this manner, a method for locating the inflection points on a valley floor, indicating the transition from an elliptic minimum bowl to a hyperbolic saddle point region, may be established.

The extrema of $\sigma(\mathbf{x})$, seen along a contour part, may also be maxima or shoulders. Thus, not all eigenvalues which belong to the eigenvectors orthogonal to the gradient have to be positive. And we must classify possible solutions of system (8).

4. Classification of GE

We search for extreme values of the functional

$$\sigma(\mathbf{x}) = \frac{1}{2}\|\mathbf{g}(\mathbf{x})\|^2$$

under a variation along an equipotential surface, i.e. orthogonal to the gradient vector $\mathbf{g}(\mathbf{x})$ of the PES. The gradient is then an eigenvector of $\mathcal{H}$. On the valley floor GE, all other eigenvectors belong to positive eigenvalues. The equations (8,9) are only neseccary criteria for a minimal value of $\sigma(\mathbf{x})$. In order to distinguish possible minima, maxima, or shoulders, we develop the $(n\text{-}1)n/2$ second derivatives of $\sigma(\mathbf{x})$ along all of the $(n\text{-}1)$ directions spreading the subspace perpendicular to the gradient vector. This produces a projected Hessian matrix of $\sigma(\mathbf{x})$:

$$\mathcal{X}(\mathbf{x}) = \left(\,({}^i\mathbf{v}^T \cdot \nabla)({}^j\mathbf{v}^T \cdot \nabla)\sigma(\mathbf{x}) \,\right) \tag{10}$$

where ${}^i\mathbf{v} \neq \hat{\mathbf{g}}$ are the eigenvectors of $\mathcal{H}$, $i=2,\ldots,n$, and, to reach the full n-dimensionality in the matrix, we put ${}^1\mathbf{v}=(0,\ldots,0)$ to be an n-dimensional zero vector. In matrix (10), a product of different eigenvectors ${}^i\mathbf{v}$ occurs. In eq.(6), the projector ${}^g\mathcal{P}$ is the dyadic product of the gradient. Any eigenvector, ${}^i\mathbf{v}$, gives an occasion to construct an analogous projection matrix onto its direction which is called ${}^i\mathcal{P}$. The matrix elements of $\mathcal{P}$ are then given by a formula which is different to eq.(5)

$$\mathcal{P} = \sum_{k=2}^{n} {}^k\mathcal{P} = \left(\sum_{k=2}^{n} {}^k v_i \, {}^k v_j \right).$$

This projected Hessian matrix of $\sigma(\mathbf{x})$ is symbolized by $\mathcal{X}$. Forming the derivatives in matrix (10), and sorting out the corresponding terms, the eigenvalues of the following matrix [19] have to be calculated

$$\mathcal{X}(\mathbf{x}) = \mathcal{P} \cdot (\mathcal{T} \cdot \mathbf{g}) \cdot \mathcal{P} + \mathcal{H}(\mathcal{H} - \lambda_1 \mathbf{I}) \tag{11}$$

where $\mathcal{P}, \mathcal{H}$, and λ_1 are defined above (eqs. 5,7,8), and $\mathcal{T}$ is the matrix of third derivatives of $E(\mathbf{x})$

$$\mathcal{T}_{rst} = \frac{\partial^3 E(\mathbf{x})}{\partial x^r \partial x^s \partial x^t} . \tag{12}$$

From the second term on the right hand side of eq.(11) we see that $\mathcal{X}(\mathbf{x})$ has eigenvalues which contain the values of the $\mathcal{H}(\mathbf{x})$, in the combination $\lambda_i(\lambda_i - \lambda_1)$, but, the eigenvalues of $\mathcal{X}(\mathbf{x})$ are different from $\mathcal{H}(\mathbf{x})$. They change in accordance with the term on the right hand side, i.e. by the third derivatives of the PES. The $(n\text{-}1)$ non-zero eigenvalues of $\mathcal{X}$ should be positive for a valley floor GE.

4.1. THE 2D CASE: A CONTRIBUTION TO THE CLASSIFICATION OF GEs

If $\mathbf{e}(\mathbf{x})$ is the unit vector of the tangent to the contour lines (perpendicular to the gradient $\mathbf{g}(\mathbf{x})$), we have

$$\mathbf{e}(\mathbf{x}) = \left(\frac{\partial E}{\partial x^2}, -\frac{\partial E}{\partial x^1} \right)^T \Bigg/ \left[\left(\frac{\partial E}{\partial x^1} \right)^2 + \left(\frac{\partial E}{\partial x^2} \right)^2 \right]^{1/2}. \tag{13}$$

Then the extremal equation for σ is

$$\frac{1}{2}(\mathbf{e}^T \cdot \nabla)(\nabla E)^2 = \mathbf{e}^T \cdot (\nabla\nabla E) \cdot (\nabla E) = 0, \tag{14}$$

thus

$$E_{xy}(E_x^2 - E_y^2) + (E_{yy} - E_{xx})E_x E_y = 0. \tag{15}$$

The subscripts indicate the corresponding partial derivatives. The two eigenvalues in the two main curvature directions, $\mathbf{e}$ and $\mathbf{g}$, are defined by the eigenvalue equation $\mathcal{H}\mathbf{v} = \lambda\mathbf{v}$. Hence, on the one hand, we have

$$\mathcal{H}\hat{\mathbf{g}} = \lambda\hat{\mathbf{g}}, \quad \text{thus} \quad \lambda = \hat{\mathbf{g}}^T\mathcal{H}\hat{\mathbf{g}} = (E_{xx}E_x^2 + 2E_{xy}E_xE_y + E_{yy}E_y^2)/(E_x^2 + E_y^2). \quad (16)$$

On the other hand, if $\lambda^\perp$ is the eigenvalue which belongs to $\mathbf{e}$, we find

$$\mathcal{H}\mathbf{e} = \lambda^\perp\mathbf{e}, \quad \text{thus} \quad \lambda^\perp = \mathbf{e}^T\mathcal{H}\mathbf{e} = (E_{xx}E_y^2 - 2E_{xy}E_xE_y + E_{yy}E_x^2)/(E_x^2 + E_y^2). \quad (17)$$

Normally, tracing the GE curve utilizes the direction of $\mathbf{g}$ for orientation, and $\mathbf{e}$

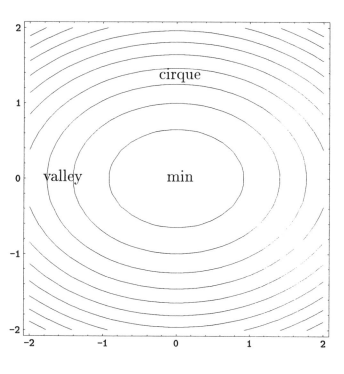

Figure 9. Model PES of an ellipsoide showing formations near a bowl-like valley and a cirque.

represents the other direction. $\lambda^\perp$ indicates wether the path is a valley or a cirque (positive $\lambda^\perp$), or wether it is a ridge or cliff (negative $\lambda^\perp$). In **Figure 9**, the formations of cirque and valley are visualized. A multiplication of the model PES by -1 changes a bowl into a summit, the valley into a ridge, the cirque into a cliff. The terms are used by Ruedenberg et al. [19]. These terms give us a good impression of whether the directions of gradient and the GE tangent are close together. If this is not the case, the terms are more or less good paraphrases. The direction of the GE tangent and the gradient may strongly differ and in an extreme case, they may be at a right angle to each other. In such a case, the gradient is the worst predictor. Also, a description using analogous mountain or valley formations can be misleading, see point 6 at the end of this chapter.

The distinction between a valley and a cirque on the one hand, and between a ridge and a cliff, on the other hand [19], depends on the sign of the second derivative of σ under a variation along the contour line. We have to prove

$$\lambda_\sigma^\perp = (\mathbf{e}^T \cdot \nabla)(\mathbf{e}^T \cdot \nabla) \cdot \sigma. \tag{18}$$

With the product rule of the differentiation we get

$$\lambda_\sigma^\perp = [\nabla \cdot \sigma] \cdot [(\mathbf{e}^T \cdot \nabla) \cdot \mathbf{e}] + \mathbf{e}^T \cdot [\nabla\nabla\sigma] \cdot \mathbf{e} \tag{19}$$

where differentiations do not act beyond the square brackets; and we calculate by a little algebra

$$\begin{aligned}
\lambda_\sigma^\perp &= -(\mathbf{e}^T \cdot \mathcal{H} \cdot \mathbf{e})(\mathbf{g}^T \cdot \mathcal{H} \cdot \mathbf{g})/(\mathbf{g}^T \cdot \mathbf{g}) + \mathbf{e}^T \cdot (\nabla\mathcal{H} \cdot \mathbf{g}) \cdot \mathbf{e} + (\mathbf{e}^T \cdot \mathcal{H}) \cdot (\mathcal{H} \cdot \mathbf{e}) \\
&= \lambda\lambda^\perp + \mathbf{e}^T \cdot (\nabla\nabla\nabla E \cdot \mathbf{g}) \cdot \mathbf{e} + (\lambda^\perp)^2 \\
&= \lambda^\perp(\lambda^\perp - \lambda) + \mathbf{e}^T \cdot (\nabla\nabla\nabla E \cdot \mathbf{g}) \cdot \mathbf{e}.
\end{aligned} \tag{20}$$

The second summand in eq.(20) is

$$\begin{aligned}
\mathbf{e}^T \cdot (\nabla\nabla\nabla E \cdot \mathbf{g}) \cdot \mathbf{e} \\
= (E_{xxx}E_x + E_{yyy}E_x)E_xE_y + E_{xxy}E_y(E_y^2 - 2E_x^2) + E_{xyy}E_x(E_x^2 - 2E_y^2).
\end{aligned}$$

We have two entities to characterize the GE: the eigenvalues $\lambda^\perp$ of $\mathcal{H}$ and $\lambda_\sigma^\perp$ of the Hessian $\mathcal{X}$ of σ, whose eigenvectors essentially point in the perpendicular direction to the GE curve. In addition to the previous classification [19], we consider the possibility of GEs to follow so-called flank lines. The flank line leads along the side where a valley and a ridge touch each other. The full GE line classification of the 2D case of a PES is outlined in Table 1.

Table 1. Classification scheme of GE curves of a two-dimensional PES

Type of GE	σ	$\lambda^\perp$	$\lambda_\sigma^\perp$
Valley line	Minimum	> 0	> 0
Cirque line	Maximum	> 0	< 0
Ridge line	Minimum	< 0	> 0
Cliff line	Maximum	< 0	< 0
Flank line	Minimum	≈ 0	< 0
	Maximum	≈ 0	> 0

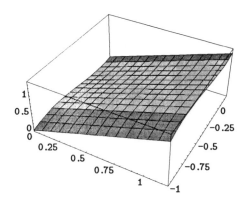

Figure 10. 2D model potential surface $E(x, y) = x + y^2(y + 1)$.

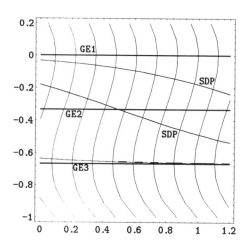

Figure 11. Contour lines with three SDPs and the gradient extremal GE$_2$, being the flank line, meet under a skew angle. (GE$_1$ and GE$_3$ are also SDPs).

4.2 Σαμπλε

We will treat the simple model function $E(x, y) = x + y^2(y + 1)$ given in Figures 10 and 11. If we calculate the gradient $\mathbf{g}(x, y) = (1, 3y^2 + 2y)^T$ and its norm functional $\sigma(x, y) = \frac{1}{2}(1 + (3y^2 + 2y)^2)$, the Hessian will be

$$\mathcal{H}(x, y) = \begin{pmatrix} 0 & 0 \\ 0 & 6y + 2 \end{pmatrix}$$

with eigenvalues $\lambda = 0$ and $\lambda^\perp = 6y + 2$. Furthermore

$$\nabla\sigma(x, y) = (0, y(2 + 3y)(2 + 6y))^T$$

and

$$\nabla\nabla^T\sigma(x, y) = \begin{pmatrix} 0 & 0 \\ 0 & 4 + 36y + 54y^2 \end{pmatrix}.$$

The eigenvector **e** becomes

$$\mathbf{e} = (3y^2 + 2y, -1)^T / [(3y^2 + 2y)^2 + 1]^{1/2}.$$

The GE eq.(9) (there is n-1=one equation) reads, after simplification

$$y(2 + 3y)(2 + 6y) = 0.$$

It has three solutions

$$VF - GE_1 : y = 0, \quad \text{it is a valley line with} \quad \lambda^\perp = 2, \quad \lambda_\sigma^\perp = 4,$$

$$GE_2 : \quad y = -\frac{1}{3}, \quad \text{it is a flank line with} \quad \lambda^\perp = 0, \quad \lambda_\sigma^\perp = -2,$$

$$GE_3 : \quad y = -\frac{2}{3}, \quad \text{it is a ridge line with} \quad \lambda^\perp = -2, \quad \lambda_\sigma^\perp = 4,$$

where $\lambda_\sigma^\perp = (54y^2 + 36y + 4)$, and any GE is a map $GE_i : \mathbf{R}^2 \to \mathbf{R}^1$. The assignment of the path types is done by help of Table 1. Because of the simplicity of the model PES, GE_1 (VF-GE) and GE_3 form two straight lines coinciding with the corresponding SDPs along the valley path, and along the ridge of the PES. On the flank line, GE_2, the valley touches the ridge; the contour lines change their curvature.

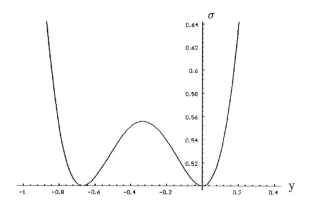

Figure 12. Value of σ along the contour line E=0.5 of Figures 10 and 11. The two minima correspond to MEP and ridge, where the maximum depicts the locus of the flank line.

However, we have a very different behaviour of the flank line GE_2 and any SDP intersecting this GE: The SDP meet at an angle of $18°$ shown in **Figure 11**. The SDPs inside the range $-2/3 < y < 0$ are given by analytic solutions of eq.(1)

$$y(x) = -\frac{2/3}{1 + exp[2(con - x)]},$$

and the constant, *con*, is an initial guess for the starting point of the SDPs. The three
SDPs in **Figure 11** are evaluated with con=1.5, 0.5, and -1.5. It is quite interesting
that the GE_2 is a straight line, too, in contrast to any SDP which forms a curvilinear
path when starting near the ridge and going to the valley in an asymptotic way. In
Figure 12, the value of $\sigma(x,y)$ is given along the contour line $E(x,y)=0.5$ being
$x=0.5-y^2(y+1)$. The two minima of $\sigma(x,y)$ correspond to valley and ridge of the
PES, and the maximum represents a point of the flank line, cf. Table 1.

5. Bifurcations and Turning Points of GE

Table 1 gives us the suitable tool for the PES analysis by GEs, especially to detect
inflection points and turning points. These are the points where $\lambda^{\perp}$ and $\lambda_{\sigma}^{\perp}$, respectively, change the sign, see [24].

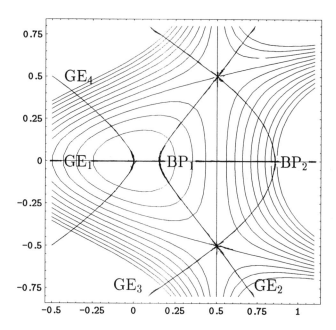

Figure 13. 2D model PES with two bifurcation points (BP) on GE_1, see text.

We will now analyze two more complex model PES. The first one is given in **Figure
13**: $E(x,y) = x^2 + (2 - 4x)y^2$. It shows a bifurcating path and the corresponding
bifurcation points (BP) of the GE_1 along the x axis. The zero of the coordinate
system is the minimizer of the PES, and on the line x=0.5 there are two saddle

points. The straight line at x=0.5 is, in the central part, an "inflection line" of the contour lines with a change from convex to concave behaviour. On this line, at the crossing point $(0.5, 0)$, there is the inflection point of the GE_1. In this point, $\lambda^\perp$ changes the sign from positive (for x < 0.5) to negative values (for x > 0.5). The GE_1 itself does not show any peculiarity at this point. However, in the region of the two bifurcation points, downhill and uphill, we find a particular behaviour of the GEs given by all of the four $\pm$ combinations of

$$y(x) = \pm \frac{\sqrt{2}}{4}\sqrt{3 - 6x + 8x^2 \pm \sqrt{9 - 28x + 20x^2 - 32x^3 + 64x^4}}.$$

The valley along the abscissa changes into a cirque formation at BP_1. At the inflection line, the cirque changes into a cliff. At the second BP_2, the cliff changes into a ridge. These are the usual phenomena when RP bifurcation occurs. So, in this case, there is no "valley ridge" inflection point. Valley and ridge are divided by cirque and cliff which touch each other along a contour line with zero curvature. Note an example of a direct contact between valley and ridge in Figure 16 below, cf.[16].

In **Figure 13**, the GE_2 is branching symmetrically to the x axis. Along a contour line near the zero point, the functional σ has two minima on the x axis (GE_1 follows a valley), and two maxima on a curve leaving off the y axis in a parabolic kind (GE_4). In BP_1, the σ along the x axis changes to a maximal value under variation along a contour line. GE_4 is originated by a maximal value of σ, too. Hence, between the two cirques characterized by maxima, there has to be a point (x,y) with a minimal value of $\sigma(x,y)$, and this forms GE_2. It characterizes the bifurcated valley floor. The two saddles, of course, are crossed by two valleys. These two valleys meet in BP_1 and go further downhill to the minimizer (0,0).

That is the usual way the simplest branching takes place: Somewhere at the slope in a BP a valley continues as a cirque, and two pitchfork like curved valleys deviate from this valley, more or less perpendicular to the left and to the right.

In BP_2, we have an analogous, but inverse picture. There, a ridge bifurcates into two ridges for the crossing GE_3 going through the two saddle points, and the GE_1 continues straight downhill as a cliff. It is interesting to think about the behaviour of the normal mode of a molecular vibration along the abscissa. What happens in the BP_1? Because a normal mode should stably exist in the direction of BP_1, we expect no particularity up to the cirque cliff inflection point (0.5, 0).

Let us discuss a further model PES shown in **Figure 14**. It is related to the real chemical system of silacyclopentane whith the ring-bending and the ring-twisting in the (x,y) plane [25]. This PES shows bifurcation and turning points of GEs. The contour lines are $150\,cm^{-1}$ apart. There are 4 minima (MIN_i) on the axes [(0.0, ±0.62361), ($\pm0.5,0.0$)], and there are 4 saddle points [SP_i: ($\pm0.402941, \pm0.214768$)] between, lying on a path connecting distorted forms of the molecule. The two deep minima represent $30°$ twist angles of the molecular ring. The VF-GE_1 surrounds the summit MAX_1. It is a saddle point of index 2. and the "barrier" to form a planar ring. The two minima on the y axis represent the most stable structures, while the two minima on the x axis are rather flat intermediates. The near saddle points are the bended configurations of the ring, having an energy about 2000 cm^{-1} above the twisted configurations. Perpendicular to the valley line VF-GE_1, we have another type of GEs in the minimum point regions, the cirque line GEs: This is the standard situation around stationary points. The GEs cross each other under a right angle. If one direction is the valley floor path, then the other main direction points at the

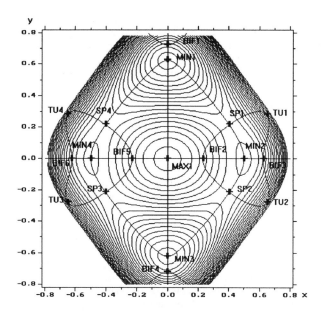

Figure 14. 2D model PES $E(x,y) = (2.5\ x^4 - 1.25\ x^2 + 2.25\ y^4 - 1.75\ y^2 + 9.5\ x^2 y^2)$ $\times 10^4$ for ring bending and ring twisting of silacyclopentane [25], with bifurcation points BIF_i, and turning points TU_i, see text. The profile over the cyclic VF-GE is given in Figure 1.

SPs to the cirque (or cliff, or ridge). Additionally, we find two straight line GEs: From MIN_1, one leads over the summit to MIN_3 beginning as a cirque line, changing into a cliff line, crossing the summit as cliff line, and changing back to a cirque line. Going through MIN_3, it continues as a cirque line up to a bifurcation point [BIF_4 at $(0.0, -0.7211)$]. There, it branches into two cirque lines and a straight valley line uphill. On the x axis from MIN_2, the GE leads uphill along a cirque line to the summit. Inside the cyclic MEP, it passes a branching point BIF_2 where it changes into the ridge of the summit. In this BIF, we have a ridge bifurcation leading to two saddle points.

Outside the cyclic MEP, a cirque line GE on the x axis also reaches a branching point [BIF_3 at $(0.6236, 0.0)$] where it splits into two flank lines and one ridge line. The two ascending flanks, however, show a strong peculiarity: They disappear in a "turning point" [TU $(0.6510, \pm 0.2804)$]. There, they meet the ascending ridge lines from the two neighbouring SPs. This is the usual pattern of an "end formation". However, the GE line does not end at an impasse. It changes its character to lead uphill, at a turning point TU, and continues downhill as new GE type [1,26,21].

Note: At a saddle point, the GEs usually meet at a right angle. This is not necessarily the case for general bifurcations, cf. the sinus surface in **Figure 3** where the VF-GE line and the flank line GE meet under a skew angle, and see also **Figure 8** where GEs meet in the stationary points under a skew angle.

If the dimension of the configuration space exceeds n=2, a possible classification scheme of GEs will become inpractical, because there are (n-1) main curvature directions perpendicular to the gradient which influence the classification. Thus, we will get mixed characters of the corresponding GE curves. Nevertheless, we can assume that pure valleys and pure cirques are the right objects for a MEP of a chemical reaction.

6. Discussion

"Some years ago" , in 1870, James Clerk Maxwell [27] gave a discussion of the so called lines of slope on the (two-dimensional) earth surface, now named SDPs:

" On Lines of Slope

Lines drawn so as to be everywhere at right angles to the contour-lines are called lines of slope. At every point of such a line there is an upward and a downward direction. If we follow the upward direction we shall *in general* reach a summit, and if we follow the downward direction we shall *in general* reach a bottom. In particular cases, however, we may reach a pass or a bar.

On Hills and Dales

Hence each point of the /../ surface has a line of slope, which begins at a certain summit and ends in a certain bottom. Districts whose lines of slope run into the same bottom are called Basins or Dales. Those whose lines of slope come from the same summit may be called, for want of a better name, hills.

Hence the whole earth may be naturally divided into Basins or Dales, and also, by an independent division, into hills, each point of the surface belonging to a certain dale and also to a certain hill.

On Watersheds and Watercourses

Dales are divided from each other by Watersheds, and Hills by Watercourses. To draw these lines, begin at a pass or a bar. Here the ground is level, so that we cannot begin to draw a line of slope; but if we draw a very small closed curve round this point, it will have highest and lowest points, the number of maxima being equal to the number of minima, and each one more than the index number of the pass or bar. From each maximum point draw a line of slope upwards till it reaches a summit. This will be a line of Watershed. From each minimum point draw a line of slope downwards till it reaches a bottom. This will be a line of Watercourse. Lines of Watershed are the only lines of slope which do not reach a bottom, and lines of Watercourse are the only lines of slope which do not reach a summit. All other lines of slope diverge from some summit and converge to some bottom, remaining throughout their course in the district belonging to that summit and that bottom, which is bounded by two watersheds and two watercourses.

In the pure theory of surfaces there is no method of determining a line of watershed or a line of watercourse, except by first finding a pass or a bar and drawing the line of slope from that point. /.../ "

The watercourse is, in our frame, the IRC of Fukui. We believe there is no further

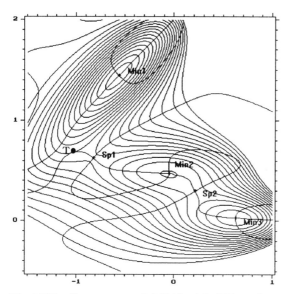

Figure 15. Müller-Brown potential [28] with GEs and a turning point T of a VF-GE characterizing the closing valley.

need for a discussion of Maxwells words. However, there is now a second type of curves: VF-GE. They are a supplement to Maxwell's procedure giving us a more accurate means of detecting an ending valley or a bifurcation of a valley.

A gradient extremal (GE) of the Müller-Brown potential (see **Figure 15** [28]) is a suitable example showing how the definition of these curves works: The bowl of Min_1 is a deep, long, and relatively straight valley. The col of SP_1 opens to this main valley and a steepest descent path goes downhill perpendicular to the contour lines of the floor. At the valley floor, it joins in the floor line. However, we cannot decide at which point the lines cross, because this is an asymptotic junction.

The GE curve, an the other hand, shows a totally different behaviour. It also runs along the col of SP_1. But then it turns sideways and goes uphill! Very strange? No! The behaviour is a consequence of the GE definition to indicate a valley floor line: The col of SP_1 ends at the slope to the broad valley of Min_1. Thus, the GE has to end, too. The final point is a turning point (T). The curve continues as a so-called flank line of the potential, i.e., the GE line changes from a minimum GE (VF-GE) to a maximum GE.

GEs (cf. [24] and Refs. therein) form a new tool that, in comparison to SDP treatments only, significantly broadens the possibilities to explore a PES. However, even GEs cannot answer the ultimate question: "What is the 'true' valley floor line", in the general case? There is an example [15,16] of a model PES where the GE curves do not describe the structure of two branches of a bifurcating valley (shown in **Figure 16**). The model PES is linear in y and quartic in x

$$E(x,y) = y + x^2\left(x^2 + y\right). \tag{21}$$

The 2D gradient extremal condition gives:

$$2x\left(-\left(1+x^2\right)^2 + 4x^2\left(2x^2+y\right)^2 - 2\left(1+x^2\right)\left(2x^2+y\right)\left(6x^2+y\right)\right) = 0. \quad (22)$$

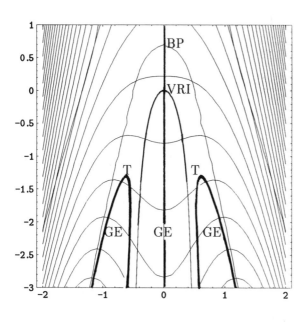
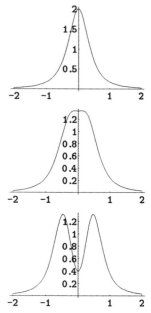

Figure 16. GEs (bold-faced). They have turning points (T) in a valley which does not end in T. Inner parabolic curve through VRI: Line connecting points of contour lines with curvature zero indicating the flank line. Outer curve through BP: Line of points with maximum curvature of the contour line indicating a valley path. Right column: Curvature of the contour lines with E=[1, 2/3 (BP), 0.2, -1].

Expanded, the gradient extremal is the implicit function:

$$GE(x,y) = 2x\left(\left(1+2x^2+25x^4+8x^6\right)+16x^2y+2\left(1-x^2\right)y^2\right) = 0. \quad (23)$$

Its solutions (with a seeming singularity at $x = \pm 1$) are given in Figure 16.

$$\{y_1(x) = \frac{8x^2 + \sqrt{2}\left(1+x^2\right)\sqrt{-1+x^2+8x^4}}{2\left(-1+x\right)\left(1+x\right)}, \quad (24)$$

$$y_2(x) = \frac{8x^2 - \sqrt{2}\left(1+x^2\right)\sqrt{-1+x^2+8x^4}}{2\left(-1+x\right)\left(1+x\right)},$$

$$x \equiv 0, \quad \text{thus} \quad y_3(x) = (y - axis)\},$$

The three gradient extremal curves (bold) are combinations of curves with different GE character. $y_3(x)$ is composed of a valley floor path (VF-GE) and a ridge GE. The two parts are divided by the valley ridge inflection point (VRI). The other two branches, $y_1(x)$ and $y_2(x)$, are also composed of two different curves, the valley floor lines of the both side valleys each ending at a turning point. The curves change into a flank line, dividing the corresponding valleys from the middle ridge. In our example, there is no connection between the three VF-GE curves, which should represent the reaction path. The valley floor curves are continued by GE curves, but, thereby, change the GE type (i) into a ridge, characterized by passing the valley ridge inflection point, and (ii) into two flank lines.

In order to better analyse the PES, we have calculated a new kind of valley floor line: We connect the points of the contour lines with maximal line curvature, termed curvature line of the PES. The contour lines are $y(x) = (E - x^4)/(1 + x^2)$ where E is any constant contour of the PES $E(x,y)$. The curvature of a contour line in a contour plane is

$$\kappa(x) = y''/(1 + y'^2)^{3/2} = \frac{2\,(-6\,x^2 - 3\,x^4 - x^6 - E + 3\,x^2\,E)\,(1 + x^2)^3}{[4\,x^2\,(2\,x^2 + x^4 + E)^2) + (1 + x^2)^4]^{3/2}}. \tag{25}$$

In a point (x,y) the curvature of the contour line at height $E(x,y)$ is

$$\kappa(x,y) = \frac{2\,(-6\,x^2 + 2\,x^4 - y + 3\,x^2\,y)\,(1 + x^2)}{\left[4\,x^2\,(2\,x^2 + y)^2 + (1 + x^2)^2\right]^{3/2}}. \tag{26}$$

An extremum of $\kappa(x)$ in a fixed contour plane is calculable by setting its first derivative to zero. Thus, we arrive at third derivatives of the PES! In the case of the simple test potential, the calculation leads to an implicitly given curvature line

$$CL(x,y) = \frac{24\,x\,(2\,x^2 + y)\,(-6\,x^2 + 2\,x^4 - y + 3\,x^2\,y)^2}{\left[4\,x^2\,(2\,x^2 + y)^2 + (1 + x^2)^2\right]^{5/2}} \tag{27}$$

$$+ \frac{24\,x\,(1 - x^2)\,(-1 + x^2 + y)}{\left[4\,x^2\,(2\,x^2 + y)^2 + (1 + x^2)^2\right]^{3/2}} = 0.$$

If we set $\kappa(x,y) = 0$, we obtain a flank line starting in the valley-ridge-inflection point (VRI) at (0,0) where, per definition, the curvature of the contour line is zero. Its equation is

$$FL(x) = 2x^2(3 - x^2)/(3x^2 - 1). \tag{28}$$

In contrast to this flank line, we may define the branching point (BP) of the valley in **Figure 16** at the point $(0,\frac{2}{3})$ where the trivial curvature line, the y axis, bifurcates. This BP is situated in front of the VRI point. If we assume the curvature line to be the "true" valley line in this case, the GE curves in the example of a bifurcating path are curves between the "true" valley floor lines and the "true" flank lines. For the symmetrical example of test potential (21), the curvature lines must be assumed to follow the "idealized lines". However, GEs are easier to calculate than curvature lines. Note, the SDPs are much easier to calculate because only first derivatives of the PES are needed. But, the SDPs are too coarse tools to be used in the analysis of complex valley structures that include bifurcations.

Hence, VF-GE tracing is a new procedure that can characterize a valley floor line to a good approximation under an accessible computational effort, with the possibility to indicate and detect branching and other interesting points.

Acknowledgment

This work was supported in part by the Deutsche Forschungsgemeinschaft. We thank J. Fischer for the preparation of Figure 7.

References

1. D. Heidrich, W. Kliesch, and W. Quapp, *Properties of Chemically Interesting Potential Energy Surfaces,* Lecture Notes in Chemistry **56**, Springer, Berlin, 1991.
2. A. Tachibana, K. Fukui, Theor. Chim. Acta **49**, 321 (1978).
3. C. J. Cerjan and W. H. Miller, J. Chem. Phys. **75**, 2800 (1981).
4. J. Nichols, H. Taylor, P. Schmidt, and J. Simons J. Chem. Phys. **92**, 340 (1990).
5. C. J. Tsai and K. D. Jordan, J. Phys. Chem. **97**, 11227 (1993).
6. R. A. Marcus, J. Chem. Phys. **45**, 4493 and **49**, 2610 (1966), K. Fukui, J. Phys. Chem. **74**, 4161 (1970).
7. D. G. Truhlar and A. J. Kuppermann, J. Am. Chem. Soc **93**, 1840 (1971).
8. K. Fukui, in R. Daudel and B. Pullman (eds.) The World of Quantum Chemistry, Dordrecht, Reidel, 1974.
9. W. Quapp and D. Heidrich, Theor. Chim. Acta **66**, 245 (1984).
10. J.-Q. Sun and K. Ruedenberg, J. Chem. Phys. **98**, 9707 (1993).
11. D. J. Rowe and A. Ryman, J. Math. Phys. **23**, 732 (1982).
12. B. L. Garret, R. Steckler, D. G. Truhlar, K. K. Baldrige, D. Bartol, M. V. Schmidt, and M. S. Gordon, J. Phys. Chem. **92**, 1476 (1988).
13. V. S. Melissas, D. G. Truhlar, and B. L. Garret, J. Chem. Phys. **96**, 5758 (1992).
14. D. A. Liotard, Int. J. Quant. Chem. **44** 723 (1992).
15. H. B. Schlegel, J. Chem. Soc.,Faraday Disc. **90**, 1569 (1994).
16. W. Quapp, J. Chem. Soc.,Faraday Disc. **90**, 1607 (1994).
17. S. Pancíř, Coll. Czech. Chem. Commun. **40**, 1112 (1975).
18. M. V. Basilevsky and A. G. Shamov, Chem. Phys. **60**, 347 (1981).
19. D. K. Hoffman, R. S. Nord, and K. Ruedenberg, Theor. Chim. Acta **69**, 265 (1986).
20. H. Primas and U. Müller-Herold, *Elementare Quantenchemie,* Teubner, Stuttgart, 1984.
21. W. Quapp, Theor. Chim. Acta **75**, 447 (1989).
22. J. W. McIver, A. Komornicki J. Am. Chem. Soc. **94**, 2625 (1972).
23. A. Niño, C. Munoz-Caro, and D. C. Moule, J. Mol. Struct. **318**, 237 (1994).
24. O. Imig, D. Heidrich, and W. Quapp, in preparation (1995).
25. L. F. Colegrove, J. C. Wells, and J. Laane, J. Chem. Phys. **93**, 6299 (1990).
26. M. V. Basilevsky, Chem. Phys. **67**, 337 (1982).
27. J. C. Maxwell, Philosophical Magazine **40**, 421 (1870), reprinted in: *The Scientific Papers,* Vol.II, 233 (1890).
28. K. Müller and L. D. Brown, Theor. Chim. Acta **53**, 75 (1979).

DENSITY FUNCTIONAL THEORY, CALCULATIONS OF POTENTIAL ENERGY SURFACES AND REACTION PATHS

GOTTHARD SEIFERT [a] and KERSTIN KRÜGER [b]
Institut für Theoretische Physik [a]
and Institut für Analytische Chemie [b],
Technische Universität Dresden,
D-01062 Dresden, Germany

1. Introduction

The Density-Functional-Theory (DFT) [1] is an alternative to the traditional wave function based methods, such as Hartree-Fock plus Configuration Interaction or many-body perturbation theory, for calculations of the potential energy surfaces of molecules. In principle, this formalism allows to solve the many body problem of the electron-electron interaction through a set of quasi one particle Schrödinger-like equations - the Kohn-Sham equations [2]. In practice, approximations for the consideration of exchange and correlation energy are necessary. The so-called local density approximation (LDA) [2] was used mostly for an approximate treatment of exchange and correlation. The LDA seems to be a useful tool for handling of many electron systems like molecules. Originally, a domain of solid state physics since the pioneering investigations on molecules by Gunnarsson, Harris, Jones, Baerends and Ellis in the 1970's [3, 4, 5] the DFT-LDA has become increasingly attractive for chemical systems, and in this way, also for studying chemical reactions. This holds essentially true due to the success of the corrections to LDA by the gradient corrections [6, 7]. One advantage of the DFT is a much lower computational effort required for the calculations of rather large molecules, compared with the "post-Hartree Fock methods" like Configuration Interaction or many body perturbation theory. Several computer codes for DFT calculations of molecules have been developed- as to review, see e.g. [8].

However, there is still much less experience in DFT calculations of molecules, than with more traditional methods. Investigations of chemical reactions on the basis of DFT have only just begun, and this article presents a first overview of the present stage of these investigations - pointing out success and problems.

Eventually, the DFT opens the possibility of direct dynamical simulations of chemical reactions via the combination of DFT with molecular dynamics [9], and moreover on the base of a real quantum molecular dynamics combined with DFT. But these concepts are beyond the scope of this book.

We will discuss only the treatments concerning the "reaction path". We will not repeat all the problems of defining and determining the "reaction path", since these are outlined in other contributions to this book. Especially, transition structures on

D. Heidrich (ed.), The Reaction Path in Chemistry:
Current Approaches and Perspectives, 161–189.
© 1995 *Kluwer Academic Publishers. Printed in the Netherlands.*

saddle points between two energy minima on the calculated potential energy surface will be considered. For convenience, we will use the term transition state (TS) for these structures, although - as pointed out in [10] - this term is not identical with the conception of Eyring [11] about the Transition State.

Concerning the Density Functional Theory and its application in general, meanwhile there exist many review articles or books - among them we want to mention, for example [12, 8, 13, 14].

First, a summarized description and discussion of DFT within LDA and gradient corrections will be given. After that, the present stage of the calculations of energy derivatives (gradients - forces, Hessians etc.) is outlined. Finally, we discuss a series of examples of molecular chemical reactions.

2. The Density Functional Theory - DFT

2.1. GENERAL CONCEPTS

Hohenberg and Kohn [1] proved that the ground state energy of a many particle system can be written in a unique way as a functional of the electron density $\rho(\vec{r})$:

$$E[\rho(\vec{r})] = \int d^3r V_{ext}(\vec{r})\rho(\vec{r}) + F[\rho(\vec{r})] \tag{2.1.}$$

$V_{ext}(\vec{r})$ is the Coulomb potential of the nuclei.
The functional $F[\rho(\vec{r})]$ may be divided [2] into:

  i the mean field interaction energy of the electrons (Hartree energy - E_H)

$$2E_H = \int d^3r V_H(\vec{r})\rho(\vec{r}) = \int d^3r \int d^3r' \frac{\rho(\vec{r})\rho(\vec{r'})}{|\vec{r} - \vec{r'}|} \tag{2.2.}$$

  ii the kinetic energy of a free electron system with the density $\rho(\vec{r})$ - $T_0[\rho(\vec{r})]$

  iii the exchange and correlation energy $E_{XC}[\rho(\vec{r})]$.

$$F[\rho(\vec{r})] = T_0[\rho(\vec{r})] + E_H[\rho(\vec{r})] + E_{XC}[\rho(\vec{r})] \tag{2.3.}$$

(Concerning a detailed discussion of the mathematical properties see, e.g., [15, 16].)
Kohn and Sham [2] proposed the following functional ansatz for $\rho(\vec{r})$:

$$\rho(\vec{r}) = \sum_i^N |\psi_i(\vec{r})|^2 \tag{2.4.}$$

with N, the number of electrons. The functions $\psi_i(\vec{r})$ are the variation functions, and the variation gives a set of one particle like equations, the Kohn-Sham (KS) equations:

$$[-\frac{1}{2}\Delta + V_{ext}(\vec{r}) + V_H(\vec{r}) + V_{XC}(\vec{r})]\psi_i(\vec{r}) = \varepsilon_i\psi_i(\vec{r}) \qquad (2.5.)$$

$V_{XC}(\vec{r})$ is the exchange-correlation potential, defined as the variational derivative of the exchange-correlation energy E_{XC}: $V_{XC} \equiv \delta E_{XC}/\delta\rho$. The density ρ is calculated from the wave functions ψ_i, where the summation is to be done over the N lowest lying one particle states. Then the total energy E can be calculated from the density ρ.

The exchange-correlation energy E_{XC} may be interpreted as the interaction energy of an electron with its surrounding exchange-correlation hole [17]:

$$E_{XC} = \frac{1}{2} \int d^3r\rho(\vec{r}) \int d^3r' \frac{\rho_{XC}(\vec{r},\vec{r}')}{|\vec{r} - \vec{r}'|} \qquad (2.6.)$$

The exchange-correlation density ρ_{XC} describes this exchange-correlation hole:

$$\rho_{XC}(\vec{r},\vec{r}') = \rho(\vec{r}') \int_0^1 [g(\vec{r},\vec{r}',\lambda) - 1]d\lambda \qquad (2.7.)$$

$g(\vec{r},\vec{r}',\lambda)$ is the pair correlation function of the system with the electron density $\rho(\vec{r})$ and a strength λ of the electron-electron interaction. (The "physical" value of λ is equal to one.)

In the local density approximation (LDA) the pair correlation function of a homogeneous electron gas is used to calculate ρ_{XC}. In this case, the exchange-correlation energy may be written in the following form:

$$\rho_{XC}(\vec{r},\vec{r}') \approx \rho(\vec{r}') \int_0^1 [g^h(\vec{r},\vec{r}',\lambda) - 1]d\lambda \qquad (2.8.)$$

Within this approximation for ρ_{XC}, Hedin and Lundqvist [18] derived the current expression for V_{XC}:

$$V_{XC}^{HL} = -\alpha(r_S)3(\frac{3}{8\pi}\rho(\vec{r}))^{1/3} \qquad (2.9.)$$

with

$$\alpha(r_S) = 2/3 + 0.0368r_S ln(1 + 21/r_S) \qquad (2.10.)$$

(r_S is the density parameter - $r_S = (3/\rho)^{1/3}$.)

Somewhat different expressions for $\alpha(r_S)$ were given for example by Vosko et al. [19]. Neglecting correlation, i.e., considering exchange only, then V_X is given by the same expression as above, but $\alpha(r_S)$ simply becomes a constant equal to 2/3 [2, 20] ($V_X(Kohn - Sham - Gaspar)$. The averaged exchange potential in the Hartree-Fock approximation (Slater) [21] differs only from $V_X(Kohn - Sham - Gaspar)$ by the factor α: $\alpha_{Slater}(r_S) = 1$. Fixed values for α are used in the so-called X_α method, where the α parameters are fixed by fitting the X_α total energy of an atom to the corresponding HF total energy [22]. The X_α parameters obtained in this way range from 0.7 - 0.8 for the light atoms, and converge to 2/3 for heavy atoms. An extensive discussion on this traditional X_α method is given in [23].

Von Barth and Hedin extended the Kohn-Sham theory to spin-polarized systems [24], where the Kohn-Sham equations become spin dependent:

$$\sum_{\sigma'}\{[-\frac{1}{2}\Delta + V_{ext}(\vec{r}) + V_H(\vec{r})]\delta_{\sigma\sigma'} + V_{XC}^{\sigma\sigma'}(\vec{r})\}\psi_{i\sigma'}(\vec{r}) = \varepsilon_i\psi_{i\sigma}(\vec{r}) \qquad (2.11.)$$

(The spin is indicated by σ.) Von Barth and Hedin [24] as well as Gunnarsson and Lundqvist [25] derived similar expressions as in the spin-free case for the spin-dependent exchange and correlation potential $V_{XC}^{\sigma\sigma'}(\vec{r})$ under consideration of the pair correlation function of a spin-polarized homogeneous electron gas. This approximation is called the local spin density approximation - LSDA. A relativistic generalization of the Hohenberg-Kohn-Sham theory may be obtained replacing the charge density by the four-current density j_μ [26, 27].

2.2. THE LOCAL DENSITY APPROXIMATION AND ITS EXTENSIONS

Most of the calculations of atoms, molecules, clusters and solids within the density functional theory have been done using the local density or local spin density approximation. The empirical material accumulated up to now as well as theoretical considerations allow to estimate the power and the limits of the L(S)DA.

The approximation of the exchange and correlation density $\rho_{XC}(\vec{r}, \vec{r}')$ (see eq. 2.7.) by the pair correlation function of the homogeneous electron gas for inhomogeneous electron systems, as done in the LDA, means that the exchange and correlation energy per electron $(\epsilon_{XC}(\vec{r}))$ at $\vec{r}$ is determined by the density $\rho(\vec{r})$ at $\vec{r}$ only (thus the name local density approximation):

$$E_{XC}(\rho) \approx \int \rho(\vec{r})\epsilon_{XC}[\rho(\vec{r})]d^3r \qquad (2.12.)$$

It has been found, for example, that by this approximation the ionization energies of atoms are achieved in a reasonable accuracy (see, e.g., [25]). The calculated binding energies of some diatomic molecules are given together with the experimental data and results from Hartree-Fock calculations in Figure 1a.

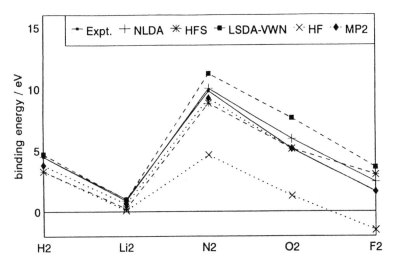

Figure 1a: Binding energies of some diatomic molecules.
HFS - Hartree-Fock-Slater, LSDA - Local Spin Density Appr., NLDA - gradient corrected DFT-LDA, HF - Hartree-Fock, MP2 - HF plus MP2, data taken from [28]

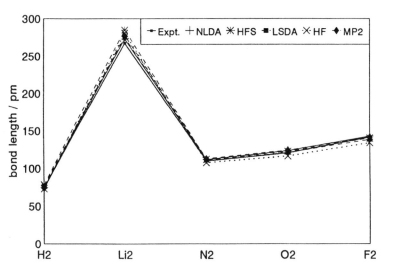

Figure 1b: Bond lengths of some diatomic molecules.
(For explanation see Figure 1a.)

As it is seen in this Figure, there is an overall qualitative agreement between LDA results and the experimental values. In particular, the binding energies from LDA calculations are generally closer to the experiment than those from the Hartree-Fock calculations. However, the quantitative deviations cannot be neglected. For example, the calculated binding energy for the O_2 molecule is about 2 eV above the experimental one ($\Delta E/E \sim 0.4!$). Generally, LDA gives an overbinding - on an average of about 1 eV for the diatomic molecules considered in Figure 1. The calculated bond lengths (see Figure 1b) agree very well with the experimental data. The data given in Figure 1 suggest also that the accuracy of the results seems to be influenced only weakly by the "LDA-variant"(formulas for E_{XC} after [25], after VWN (Vosko, Wilk, Nusair) [19] or after $X\alpha$). Even the results from the simple $X\alpha$ model may yield better results in some cases than the more refined schemes.

Success of LDA for rather inhomogeneous systems, like molecules, was substantiated e.g. by Gunnarsson et al. [25]. Their arguments are: while the exchange-correlation density $\rho_{XC}(\vec{r}, \vec{r'})$ is described rather poorly by the LDA, the evaluation of the exchange-correlation energy E_{XC} is given by an integral average over the XC-hole (see equations 2.6. and 2.7.). This "average" is obtained with a reasonable accuracy within LDA.

One qualitative defect in LDA for example is the imperfect cancellation of the Coulomb self-interaction in the mean field Coulomb energy (Hartree energy E_H - see eq. 2.2.) and the corresponding potential V_H (eq. 2.5.), due to the approximate nature of $E_{X(C)}$. There are hints that this defect might have a significant influence on reaction barriers [29] - see also chapter 3.3. The self-interaction may be corrected in DFT by a self-interaction correction (SIC) [29, 30, 31]. However, these corrections are rather cumbersome and therefore they have been applied up to now only very rarely.

In general, several trials to develope calculation methods for atoms and molecules within the DFT beyond LDA have been made (see e.g. [17, 32, 33] etc.).

Nearly from the beginning of the DFT it has been tried to improve LDA by a gradient expansion (see, e.g., [34]):

$$E_{XC} = \int d^3r\rho(\vec{r})\epsilon_{XC}(\rho) + \int b(\rho)\left(\frac{\partial\rho(\vec{r})}{\partial\vec{r}}\right)^2 d^3r + ... \qquad (2.13.)$$

Only the latest developments by Becke [6, 35] and Perdew [7, 36] of generalized gradient expansions [6, 7] led to a real success for molecular calculations - especially for binding energies - see Figure 1a. The significant improvement of calculated binding energies by consideration of the gradient corrections is illustrated for hydrocarbons in Figure 2. As a short-hand, the results obtained with the several gradient corrected XC-functionals, will be denoted as N-LDA.

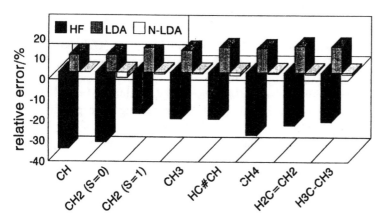

Figure 2: Atomization energies of some hydrocarbons.
(For explanation see Fig. 1, data from [28].)

2.3. SOLUTION OF THE KOHN-SHAM EQUATIONS

The non-linear coupling of the Kohn-Sham wave function (ψ) with the effective one particle potential in the Kohn-Sham equation requires an iterative solution of the KS equations (SCF treatment similar to the traditional Hartree-Fock scheme). As in the solution of the HF equations for molecules mainly the linear combination of atomic orbitals is used (HF-LCAO):

$$\psi = \sum_{\nu}^{M} c_\nu \phi_\nu \qquad (2.14.)$$

But, due to the rather involved dependence of the XC potential (V_{XC}) on the charge density (ρ) the LCAO treatment cannot be applied simply for the DFT-KS method. Sambe and Felton [37] first suggested the use of auxiliary functions (f_j) for the representation of the KS potential:

$$V(\vec{r}) = \sum_{j}^{N} a_j f_j(\vec{r}) \qquad (2.15.)$$

The coefficients a_j may be determined by a least square fit for the expansion. Despite of the fit of the potential $V(\vec{r})$ the XC-potential (V_{XC}) only, and the density (ρ) may be expanded [4]. Such auxiliary basis for the potential enables an analytical calculation of the Hamiltonian matrix elements ($h_{\nu\mu} \equiv \langle \phi_\nu \hat{h} \phi_\mu \rangle$). Furthermore, the use of auxiliary basis scales down the M^4 dependence in the complexity of the Coulomb integral evaluation to a M^3 dependence. On the other hand, the introduction of the additional auxiliary basis leads to an additional basis set size dependence (for a discussion, see Ref. [4]).

Implementations have been realized using Gaussian functions (GTO's) ([38, 39] and Slater-type orbitals (STO's) [5, 40, 41], and numerical basis sets [42, 43, 44]. The auxiliary basis may be avoided by the use of a purely numerical representation of the potential on a grid (usually called DVM - Discrete Variational Method [45, 5]), by certain approximations for the potential (Multiple Scattering concept within the so-called "muffin-tin approximation" - [46]), the linear combination of "muffin-tin" orbitals [47, 3], and in connection with the pseudopotential concept the application of plane-wave basis expansions - see, e.g., Ref.[112].

2.4.   CALCULATION OF ENERGY GRADIENTS

Following the Hellmann-Feynman theorem [49, 50] the gradient of energy is simply determined by the electron density $\rho(\vec{r})$:

$$-\frac{\partial E}{\partial \vec{R}_j} = \int \rho(\vec{r}) \frac{\partial}{\partial \vec{R}_j} \left[\frac{Z_j}{|\vec{r} - \vec{R}_j|} \right] d^3r$$

$$-\sum_{j \neq k} \frac{\partial}{\partial \vec{R}_j} \left[\frac{Z_j Z_k}{|\vec{R}_j - \vec{R}_k|} \right] d^3r \qquad (2.16.)$$

It describes the gradient, or the force on the atom j ($\vec{F}_j = -\frac{\partial E}{\partial \vec{R}_j}$), as an electronic contribution (the first term) and the nuclear repulsion (the second term). The Hellmann-Feynman theorem is rigorously satisfied in the DFT (see [51, 52]), as well as for the KS-LDA approximation [53]. Thus, the electronic part of the force ($\vec{F}_j^e$) may be written as a sum of orbital contributions:

$$\vec{F}_j^e = \sum_i^N \vec{F}_j^i = -\sum_i^N \langle \psi_i | \frac{\partial}{\partial \vec{R}_j} \frac{Z_j}{|\vec{r} - \vec{R}_j|} | \psi_i \rangle \qquad (2.17.)$$

However, this relation holds only if the variationally exact solution of the KS equation (eq. 2.5.) has been achieved. Otherwise, the Hellmann-Feynman term for each orbital has to be completed by the following contribution [54] - Pulay-correction:

$$\vec{F}_j^i(Pulay) = -Re \int \frac{\partial \psi_i^*}{\partial \vec{R}_j} \hat{h} \psi_i d^3r \qquad (2.18.)$$

($\hat{h}$ is the KS-Hamiltonian). These corrections have been discussed often in the literature (see, e.g., [52, 55, 56]). The corrections due to finite basis sets in DFT-LDA-LCAO calculations are of special importance, and can be expressed in terms of derivatives of the basis functions [56]:

$$\vec{F}_j^i(Pulay - BS) = -Re \sum_{\mu\nu} C_\mu^i C_\nu^{i*} \int \frac{\partial \phi_\mu^*}{\partial \vec{R}_j} [\hat{h} - \epsilon_i] \phi_\nu d^3r \qquad (2.19.)$$

Summarizing, the forces in KS-DFT may be written as a sum of the following contributions:

$$\vec{\mathbf{F}} = \vec{\mathbf{F}}(HF) + \vec{\mathbf{F}}(BS) + \vec{\mathbf{F}}(SCF) + \vec{\mathbf{F}}(FS) \qquad (2.20)$$

The first term ($\vec{\mathbf{F}}(HF)$ - Hellmann-Feynman term) is sufficient only if basis set free methods would be applied or if plane-wave basis sets are used [52]. For basis set methods - like LCAO - the basis set term ($\vec{\mathbf{F}}(BS)$) is usually an important correction. The correction due to an incomplete self-consistency in solving the KS-equations [52] ($\vec{\mathbf{F}}(SCF)$) can be estimated most easily. For methods with an additional auxiliary basis for fitting potential or density (see eq. 2.15., [57, 58]) the last term ($\vec{\mathbf{F}}(FS)$) must also be considered. These corrections are, however, much smaller than the orbital basis term [58].

As to higher derivatives, especially the second derivative of energy, which is important for the characterization of special points on the potential surface, the above considerations may be extended in a similar way. However, especially in the case of basis set methods, the corresponding expressions becomes rather complicated. Detailed discussions were given quite recently by Komornicki and Fitzgerald [59], Handy et al. [60] as well as by Johnson et al. [61].

Usually, the first derivatives of the energy, which are necessary for the geometry optimization procedures are calculated analytically with expressions derived by Fournier et al. [57] or techniques from Versluis and Ziegler [62]. However, second derivatives of the energy were calculated numerically. Analytical treatments have been implemented into DFT computer codes [61] only quite recently .

3. Results of Calculations for Molecular reactions

3.1. TRANSITION-STATES AND BARRIER HEIGHTS

3.1.1. Some random examples

Transition states (TS) of molecular reactions were first studied by Ziegler and Fan [63] with density functional schemes. At the same time Seminario, Grodzicki and Politzer [48] applied LDA to study symmetry allowed isomerization reactions and the ring dissociation of 1,3-diazocyclobutane. More recently, Bickelhaupt et al. [64] investigated some base-induced elimination reactions, whereas Stanton and Merz [65] calculated density functional transition structures of organic and organometallic reactions. In both cases the algorithm of Simons et al. [66] in the implementation of Baker [67] was used for the location of the transition state. Abashkin and Russo [68] used a constrained optimization technique for finding the transition state structure in their DFT investigations.

Fan and Ziegler [69] studied the applicability of LDA and the influence of the gradient corrections (N-LDA) in calculations of transition state structures and activation energies. They studied elementary reaction steps in reactions such as

I. the hydrogen abstraction reactions

$$\cdot CH_3 + CH_4 \longrightarrow CH_4 + \cdot CH_3 \qquad\qquad 3.1.1a)$$

$$\cdot CH_3 + CH_3Cl \longrightarrow CH_4 + \cdot CH_2Cl \qquad\qquad 3.1.1b)$$

II. the dissociation reaction

$$H_2CO \longrightarrow H_2 + CO \qquad\qquad 3.1.1c)$$

III. the migration reactions

$$RCN \longrightarrow CNR \qquad (R = H, CH_3) \qquad\qquad 3.1.1d)$$

$$HOC^+ \longrightarrow HCO^+ \qquad\qquad 3.1.1e)$$

The reactions 3.1.1a - 3.1.1e are characterized by open shell systems (a,b,e), with high barriers (c) and without symmetry restrictions. The results for structures and energies of reactants, transition states and products, obtained by Fan and Ziegler can be compared to those of HF calculations and to experimental data - see Table 3.1.

For the hydrogen abstraction reactions 3.1.1a and b, the energy barriers from various HF investigations are significantly higher than the experimental data, as shown in Table 3.1. For reaction 3.1.1a the experimental value for the activation barrier has been determined, 14.1 kcal/mol [71], whereas the HF and the post HF calculations yield 29.7 kcal/mol [70] and 19.7 kcal/mol [70], respectively. The much too low LDA activation barrier (1.9 kcal/mol) was significantly improved to 11.7 kcal/mol by including gradient corrections [69]. Similar results were obtained by Pederson [80]. On the other hand, the calculated structures of the reactants, TS and products of the reactions 3.1.1a - e are quite similar for the different methods [69].

For reaction 3.1.1b LDA predicts no energy barrier (-2.1kcal/ mol) in contrast to the experimental observed activation energy of 9.4 kcal/mol [71]. Gradient corrections gave a barrier of 9.3 kcal/mol. Including the zero point correction the barrier is 8.5 kcal/mol.

Fan and Ziegler concluded [69] that non-local corrections according to Becke and Perdew are essential for an accurate description of hydrogen abstraction reactions 3.1.1a and 3.1.1b. The reaction 3.1.1c will be discussed separately below.

Table 3.1. Energetics of the species involved in hydrogen abstraction reactions and the isomerization reactions, relative to the energies of reactants (in kcal/mol), calculated with different methods.

A) $\cdot CH_3 + CH_3X \rightarrow CH_4 + \cdot CH_2X$ [for $X = H$ and Cl]

Reaction	LDA[a]	N-LDA[a]	UHF	CIS	exp.
TS 3.1.1a)	2.8	12.6	29.7[b]	19.7[b]	
barrier [c]	1.9	11.7			14.1[d]

Reaction	LDA[a]	N-LDA[a]	UHF	MP2	exp.
$CH_4 + CH_2Cl$	-7.6	-4.8			
TS 3.1.1b)	-1.3	9.3	28.5[e]	18.7[e]	
barrier [c]	-2.1	8.5			9.4[d]

B) $C \equiv N\text{-}R \rightarrow R\text{-}C \equiv N$ [for $R = H$ and CH_3]

Reaction	LDA[a]	N-LDA[a]	HF[f]	MP3[f]	CISD[g]	MP4[h]	exp.[i]
$H\text{-}C \equiv N$	-14.7	-15.5	-11.0	-16.0	-14.6	-16	-14.8
TS 3.1.1d)	30.5	30.7	39.5	35.4	34.9	33	

Reaction	LDA	N-LDA	HF[k]	MBPT[k]			exp..
$CH_3\text{-}C \equiv N$	-23.1	-23.0	-19.2	-22.7			-23.7[l]
TS 3.1.1e)	42.1	39.5	48.1	45.3			38.4[m]

[a] Ref. [69], [b] with a 6-31G basis set, Ref. [70], [c] zero point corrected value, [d] Ref. [71], [e] with a 6-31G* basis set, Ref. [72], [f] with a 6-31G** basis set, Ref. [73], [g] with a 6-31 G** basis set, Ref. [74], [h] Ref. [75], [i] Ref. [76], [k] a DZP calculation, Ref. [77], [l] Ref. [78], [m] Ref. [79a]

LDA - Local Density Approximation, N-LDA - gradient corrected LDA, (U)HF - (Unrestricted) Hartree-Fock, CIS(D) - Configuration Interaction including Single (and Double) excitations, MP2(3 or 4) - second (third or fourth) order Moller-Plesset perturbation theory, MBPT - Many Body Perturbation Theory

For the isomerization reactions in 3.1.1d and e a similar picture for the activation barrier is obtained: whereas the HF calculations overestimate the barriers, the calculated with LDA and N-LDA are lower than the post HF results, but in a fairly good agreement with the experimental data. The difference between LDA and N-LDA results is small (at most 3 kcal/mol, see Table 3.1). As shown in Table 3.2., the structures obtained by the different methods differ slightly for the migration reactions 3.1.1d and e. Whereas N-LDA brings about a nearly equidistant arrangement of the migratory group $R = H$ or $R = CH_3$ between C and N, the group R is placed closer to the carbon atom in the HF calculations [73, 79] and the distance C...N is smaller than in the N-LDA calculations.

Table 3.2. Geometrical parameters of the Transition State involved in the $RNC \rightarrow NCR$ Isomerization reactions 3.1.1d), for different methods (TS are constrained in C_S symmetry, bond length r in Å, bond angles α in degree).

A) Transition State for $HN \equiv C \rightarrow N \equiv CH$

parameter	N-LDA [a]	MP3 [b]	CISD [c]
$r(C \equiv N)$	1.190	1.187	1.181
r (N-H)	1.365	1.430	1.430
α (HNC)	56.5	51.8	52.2

B) Transition State for $CH_3N \equiv C \rightarrow N \equiv CCH_3$

parameter	N-LDA [a]	HF [d]
$r(C \equiv N)$	1.197	1.178
$r(C-N)$	1.825	1.898
$r(C-C)$	1.818	1.740
$r(C-H)$	1.102	1.078
	1.096	1.076
α (HCH)	115.3	113.0
	110.8	110.1

[a] Ref. [69], [b] with a 6-31G** basis set, Ref. [73], [c] Ref. [76], [d] with a DZP basis set, Ref. [77]

N-LDA - gradient corrected LDA, MP3 - third order Moller-Plesset perturbation theory, CISD - Configuration Interaction including Single and Double excitations, HF - Hartree-Fock

Beside Stanton and Merz [65] also Carpenter and Sosa [81] have studied some representative pericyclic reactions as the [1,5] sigmatropic hydrogen shift in (Z)-1,3-pentadiene

$$
\begin{array}{ccc}
& CH & \\
H_2C & \diagdown & CH_2 \\
| & & \\
HC & & CH_3 \\
& \diagdown CH \diagup &
\end{array}
\longrightarrow
\begin{array}{ccc}
& CH & \\
HC & \diagdown & CH_3 \\
| & & \\
H_2C & & CH_2 \\
& \diagdown CH \diagup &
\end{array}
\tag{3.1.1f}
$$

the electrocyclic ring openning of cyclobutene (only in [81])

$$
\begin{array}{c}
HC-CH_2 \\
\| \quad | \\
HC-CH_2
\end{array}
\longrightarrow
\begin{array}{c}
\quad CH_2 \\
HC \diagup \\
| \\
HC \diagdown \\
\quad CH_2
\end{array}
\tag{3.1.1g}
$$

and the Diels-Alder reaction between ethylene and butadiene

$$
\begin{array}{c}
\quad CH_2 \\
HC \diagup \\
| \\
HC \diagdown \\
\quad CH_2
\end{array}
+
\begin{array}{c}
CH_2 \\
\| \\
CH_2
\end{array}
\longrightarrow
\begin{array}{cc}
CH_2 & \\
HC \diagup \diagdown & CH_2 \\
\| & | \\
HC \diagdown \diagup & CH_2 \\
CH_2 &
\end{array}
\tag{3.1.1h}
$$

using LDA and N-LDA schemes. In general, the N-LDA method brings about results that are comparable to those obtained at the MP2 level of theory and is significantly more efficient with regard to computations required. A rather strong sensitivity on the exchange-correlation functional within gradient corrections was found by Sosa et al. [82] for the calculated barrier heights of the silylene insertion into the hydrogen molecule

$$
SiH_{2-n}F_n + H_2 \longrightarrow SiH_{4-n}F_n
\tag{3.1.1i}
$$

with n = 0, 1 or 2.
The decomposition of the quantumchemically large 1,3,3-trinitroazetidine

$$
\begin{array}{c}
O_2N \\
\quad \diagdown \\
\quad\quad N- \\
\quad\quad | \quad\quad |-NO_2 \\
\quad\quad NO_2
\end{array}
\tag{3.1.1j}
$$

was investigated by Politzer et al. [83] using an N-LDA density functional approach. The authors were mainly interested in the energy changes for the individual steps of three possible pathways of decomposition reaction.

3.1.2. *Molecules with transition metal atoms*

DFT is well suited for the calculation of systems containing transition metals. Thus, investigations of reactions of molecules or complexes with transition metal atoms with DFT methods are quite obvious. Ziegler et al. investigated some organometallic reactions involving transition metals such as scandium [84], zirconium [85] or molybdenum [86]. Ziegler, Folga and Berces [84] have studied the activation of hydrogen-hydrogen and hydrogen-carbon bonds by $Cp_2Sc - H$ and $Cp_2Sc - CH_3$ (Cp is the cyclopentadienyl anion). Their interest was directed towards the ability of electron-poor metal centers to activate, break and form H-H and H-C bonds in the σ-bond metathesis reaction via the four-center transition state I.

$$Cp_2Sc\text{ - }R' + H\text{ - }R \longrightarrow \begin{bmatrix} Cp_2\,Sc\,....\,R' \\ \vdots \qquad \vdots \\ R\,....\,H \end{bmatrix}^{\ddagger} \longrightarrow Cp_2Sc\text{ - }R + H\text{ - }R' \qquad 3.1.2a)$$

Transition State I

The optimized structures for various hydrocarbonyl derivatives of Scandocene, $Cp_2Sc - R'$ with $R' = H, CH_3, CH_2CH_3, CH_2CH_2CH_3, C_2H_3$ and C_2H, show that the hydrocarbonyl group is bonded exclusively to the scandium atom by a single carbon atom. The only exception is the ethyl derivative, which forms a structure comparable to TS I by an interaction between the metal and the β-hydrogen. For the $Cp_2Sc - R'$ compounds the bond dissociation energies of the $Sc - R'$ bond were also calculated. The errors, compared with the experimental values, are less than 10 percent.

The reaction 3.1.2a proceeds from an adduct **A** between $Cp_2Sc - R'$ and $H - R$ over a kite-shaped four-center TS I with a $Sc - R' - H - R$ core to an adduct **C** between $Cp_2Sc - R$ and $H - R'$. The energies to form an adduct **A** or **C** and the activation barriers for 3.1.2a are listed in Table 3.3. The reactions with $R = H$ and $R' = H$ (**a1**), $R = C_2H_2$, $R' = H$ (**a6**) and $R = C_2H_2$, $R' = CH_3$ (**a7**) have "negative activation energies". If only one alkyl (**a2**) or alkenyl (**a4**) group participate in the TS complex the activation barrier is increased to 8 kcal/mol and 20 kcal/mol, respectively. The highest barriers were calculated in the cases where either $R, R' = CH_3$ (**a3**) or $R = C_2H_4$ and $R' = CH_3$ (**a5**).

The calculated trends in activation energies follow closely the order of the rates obtained in experiments. The reason for the increasing of the activation energy on going from alkynyl over hydrogen to alkenyl and alkyl, is the geometrical arrangement of the σ orbitals on these groups which makes it impossible to maintain optimal interactions with both neighbours in the $Sc - R' - H - R$ core.

Instead of the σ-metathesis, other reactions may appear for $R = C_2H_n$ (n=2 or 4), such as the insertion into the $Sc - R'$ bond

$$Cp_2Sc - R' + C_2H_2 \longrightarrow Cp_2Sc - CH = CHR' \qquad 3.1.2b)$$

$$Cp_2Sc - R' + C_2H_4 \longrightarrow Cp_2Sc - CH_2 - CH_2R' \qquad 3.1.2c)$$

The insertion of C_2H_4 or C_2H_2 into the $Sc-R'$ bonds should be preferred energetically over the alternative alkynylic or alkenylic $C - H$ bond activation (see Table 3.3), whereas for the alkynylic group the opposite behaviour was observed experimentally.

Table 3.3. Energetics of the reactions 3.1.2a - 3.1.2c, relative to the energies of the reactants (in kcal/mol), as calculated in Ref. [84].

No.:	Reactants	Adduct A	TS I	Adduct C	Products
a1	Cp_2Sc-H + H-H	-16	-7	-16	0
a2	Cp_2Sc-CH_3 + H-H	-17	8	-62	-42
a3	Cp_2Sc-CH_3 + CH_4	-25	45	-25	0
a4	Cp_2Sc-H + C_2H_4	-12	20	3	16
a5	Cp_2Sc-CH_3 + C_2H_4	-12	39	-48	-26
a6	Cp_2Sc-H + C_2H_2	-37	-31	-108	-86
a7	Cp_2Sc-CH_3 + C_2H_2	-18	-4	-157	-128

No.:	Reaction	Products
b	Cp_2Sc-CH_3 + C_2H_2 $\rightarrow$ Cp_2Sc-CH=CH-CH_3	-191
c1	Cp_2Sc-H + C_2H_4 $\rightarrow$ Cp_2Sc-CH_2-CH_3	-115
c2	Cp_2Sc-CH_3 + C_2H_4 $\rightarrow$ Cp_2Sc-CH_2-CH_2-CH_3	-65

Stanton and Merz [65] studied the reaction $ZnOH^+ + CO_2$, as an example for a simple organometallic reaction. They found a four-center TS too:

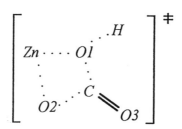

N-LDA, HF 6-31G** as well as MP2 calculations [65] give very similar activation energies, whereas LDA predict much lower activation barriers. As to the geometry of the TS from the N-LDA calculations follows a shorter $Zn - O1$ distance than from MP2 calculations, but for the $Zn - O2$ and $O1 - C$ distances the situation is reversed.

3.1.3. *Base-induced elimination reactions*

In a series of studies Bickelhaupt et al. [64, 87, 88] investigated base-induced elimination reactions experimentally with Fourier Transform Ion Cyclotron Resonance (FT-ICR) mass spectroscopy and theoretically with DF calculations. They applied the computational scheme of Baerends et al. [5], using the gradient corrections of Becke [89] for the exchange potential. They studied the fundamental class of reactions

$$B^- + H\text{-}CR_2\text{-}CR_2\text{-}L \longrightarrow B\text{-}H + CR_2\text{=}CR_2 + L^- \qquad \text{3.1.3a)}$$

especially

$$F^- + CH_3CH_2F \begin{cases} \xrightarrow{E2} FHF^- + CH_2\text{=}CH_2 \\ \xrightarrow{S_N2} CH_3CH_2F + F^- \end{cases} \qquad \text{3.1.3b)}$$

and furthermore the following base-induced elimination and nucleophilic substitution reactions:

$$F^- + CH_3 - O - NH_2 \quad \begin{array}{l} \xrightarrow{\quad S_N2 \text{ on } C \quad} CH_3F + NH_2O^- \\[2em] \xrightarrow{\quad S_N2 \text{ on } N \quad} NH_2F + CH_3O^- \end{array} \qquad 3.1.3c)$$

$$ClCH_2^- + CH_3Cl \quad \begin{array}{l} \xrightarrow{\quad E2 \quad} HCl + CH_2{=}CH_2 + Cl^- \\[2em] \xrightarrow{\quad S_N2 \quad} ClCH_2CH_3 + Cl^- \end{array} \qquad 3.1.3d)$$

$$ClCH_2^- + NH_3 \longrightarrow [ClCH_2^- ... NH_3]^{\neq} \longrightarrow CH_3NH_2 + Cl \qquad 3.1.3e)$$

A theoretical investigation has been performed by Bickelhaupt et al. [88] on the gas phase reaction of $F^- + C_2H_5F$. For this reaction the geometries of the reactants, reactant complexes, transition states, product complexes and the products for anti-E2 and syn-E2 as well as for S_N2 pathways were optimized with the $X\alpha$ potential [23]. The energetically favoured products $FHF^- + C_2H_4$ are formed in an anti as well as in a syn E2 mechanism. In their study Bickelhaupt et al. presented a qualitative MO theoretical analysis enabling to understand and to predict which reaction - the elimination E2 or the substitution S_N2 - dominates for a given substrate C_2H_5L. From their results they concluded that the base plays a key role as a catalyst which strongly influences the competition between syn and anti elimination. Its catalytic effect consists in a considerable stabilization of the transition state.

An interesting question for the reaction 3.1.3c is where the nucleophilic substitution does take place:

- on the nitrogen atom, forming NH_2F and $CH_3 - O^-$ anions

- on the carbon atom, forming CH_3F and $H_2N - O^-$ ions.

The DF calculations show that the reaction of $CH_3 - O - NH_2$ with F^- leads preferentially to the formation of NH_2F and the methylate anion, since the corresponding reaction complex is 2.55 kcal/mol more stable than the alternative complex.

Bickelhaupt et al. also studied the multi-step processes in the gas-phase reactions of the chloromethyl anion $ClCH_2^-$ with CH_3Cl and NH_3 [87]. Three competing reaction mechanism are possible

i A one step S_N2 substitution of $ClCH_2^-$ on CH_3Cl (direct S_N2), which is the dominant process, as shown experimentally.

ii The product complex $[ClCH_2CH_3 \cdots Cl^-]^*$ of the S_N2 substitution undergoes a Cl^- induced 1,2-elimination to form the products HCl, C_2H_4 and Cl^-, as postulated in reaction 3.1.3d)

iii A consecutive endothermic proton transfer (PT) from NH_3 to $ClCH_2^-$ and a very exothermic S_N2 substitution of the resulting amide on $ClCH_3$, forming the products Cl^- and CH_3NH_2.

Theoretical calculations show that for iii) the PT/S_N2 reaction proceeds in two distinct phases, but does not have stable intermediate. Therefore, this reaction is not a two-step, but a one-step process.

3.1.4. *A case study - formaldehyde and its isomers*

The potential energy surface of formaldehyde was extensively studied by Deng, Ziegler and Fan [90] and by Stanton and Merz [65]. They have investigated, in particular, the following elementary steps:

3.1.4a) the H_2-elimination

$$\underset{H_a}{\overset{H_b}{\diagdown}}C{=}O \quad \longrightarrow \quad \left[\begin{array}{c} H_b \\ \vdots \\ H_a{-}C{\equiv}O \end{array}\right]^{\ddagger} \quad \longrightarrow \quad \overset{H_b}{\underset{H_a}{|}} \; + \; C{\equiv}O$$

3.1.4b) the 1,2 hydrogen migration process

$$\underset{H_a}{\overset{H_b}{\diagdown}}C{=}O \quad \longrightarrow \quad \left[\underset{H_a}{\overset{H_b}{\diagdown}}C{=}O\right]^{\ddagger} \quad \longrightarrow \quad \underset{H_a}{\diagup}C{-}O^{\diagup H_b}$$

3.1.4c) the trans-cis isomerization of hydroxymethylene

$$\underset{H_a}{\diagup}C{-}O^{\diagup H_b} \quad \longrightarrow \quad \left[\underset{H_a}{\diagup}C{-}O^{\vartriangle H_b}\right]^{\ddagger} \quad \longrightarrow \quad \underset{H_a}{\diagup}C{-}O^{\diagdown}H_b$$

3.1.4d) the 1,2 hydrogen elimination from cis-hydroxymethylene

$$\underset{H_a}{\diagup}C{-}O^{\diagdown}H_b \quad \longrightarrow \quad \left[\begin{array}{c} C{\cdots}O \\ H_a{\cdots} \\ H_b \end{array}\right]^{\ddagger} \quad \longrightarrow \quad \begin{array}{c} C{\equiv}O \\ + \\ H_a{-}H_b \end{array}$$

Deng et al. evaluated the structures and harmonic vibrational frequencies for all of the reactants, products and transition states. Stanton and Merz present only the structures and the energy relations of the reactants. The results have been compared with those from HF and post HF calculations and with available experimental data. To illustrate the quality of the results we have compiled the data of the reaction 3.1.4a and 3.1.4c in Table 3.4.

Table 3.4.   Results for the Transition states of elementary processes of H_2CO, calculated with different
methods : geometrical parameters in Å (for distances r) and degrees (for angles α); the activation barriere ΔE_e and its zero point corrected value ΔE_0 in kcal/mol.

A) 1,1-hydrogen elimination 3.1.4a)

Parameter	LDA		N-LDA		HF	post-HF			exp.
	a	b	a	b		MP2	CISD	MP4	
$r(C=O)$	1.164	1.173	1.169	1.180	1.141	1.180	1.179	1.180	
$r(C-H_b)$	1.676	1.660	1.706	1.721	1.663	1.726	1.572	1.719	
$r(H_a-H_b)$	1.312	1.320	1.341	1.349	1.267	1.356	1.213	1.356	
$\alpha\,(OCH_b)$	111.5	111.9	112.5	114.1	114.3	112.3	110.8		
$\alpha\,(H_aH_bC)$	41.3	41.7	40.2	39.7	41.3	39.2	44.0		
ΔE_e	78.0	79.3	78.8	79.9	102.0	95.2	98.1	85.8	
ΔE_0	73.4		74.0		96.0	90.3	92.8		79.2±0.8
basis set	TZ+2P	6-31G**	TZ+2P	6-31G**	TZ+2P	6-31G**	DZP	6-11G**	
Ref.	[90]	[65]	[90]	[65]	[91]	[92]	[93]	[94]	[95]

B) cis-trans Isomerization 3.1.4c)

Parameter	LDA		N-LDA		HF	post-HF		
	a	b	a	b		MP2	MP4	MP2
$r(C=O)$	1.324	1.325	1.353	1.358	1.347	1.360		1.347
$r(C-H_a)$	1.144	1.163	1.142	1.160	1.109	1.127		1.124
$r(O-H_b)$	0.969	0.962	0.969	0.973	0.951	0.971		0.946
$\alpha\,(H_aCO)$	104.7	104.3	104.9	104.1	104.9	104.2		104.1
$\alpha\,(H_bOC)$	128.4	129.2	121.3	120.3	115.2	117.4		117.6
$\alpha\,(H_bOCH_a)$	91.9	93.0	90.1	92.1	90.1	90.8		91.1
ΔE_e	30.6	31.6	30.0	31.8	27.7		33.2	30.5
ΔE_0	27.9		27.8		24.9		30.3	
basis set	TZ+2P	6-31G**	TZ+2P	6-31G**	6-31G*	6-31G*	6-311G**	6-31G**
Ref.	[90]	[65]	[90]	[65]	[92]	[92]	[94]	[65]

[a] calculations from Deng, Ziegler and Fan [90], [b] calculations from Stanton and Merz [65]

LDA - Local Density Approximation, N-LDA - gradient corrected LDA, HF - Hartree-Fock, CISD - Configuration Interaction including Single and Double excitations, MP2(4) - second (fourth) order Moller-Plesset perturbation theory

For the first elementary step a comparison with experimental results [95] and with a large number of HF and high level post HF calculations [91-94] is possible, for the following steps only a few HF and post HF calculations are available. Concerning the geometries of the TS for the reactions 3.1.4a and 3.1.4c there are small differences in the results of Deng et al. and those of Stanton and Merz. The calculated energy barriers are very similar (the barriers of Deng et al. are 1 -1.3 kcal/mol smaller than those of Stanton). High level post HF calculations, such as MP4(SDTQ)/ 6-311G**//MP2/6-31G* of Kamaiya et al. [94] give a barrier for reaction 3.1.4a of 80.2 kcal/mol, which is in very good agreement with the experimental value of 79.2±0.8 kcal/mol [95]. The zero point energy corrected value of Deng et al. of 73.4 kcal/mol is also quite close to the experimental estimate. But HF and "low level" post HF calculations - such as MP2 or CISD - underestimate the barrier for the reaction 3.1.4a with 11...16 kcal/mol. The situation is quite similar in the reaction 3.1.4c.

Deng et al. [90] were the first to investigate a DFT generated potential energy surface to find the intrinsic reaction pathway from the transition state to the reactants and products for the elementary reaction steps 3.1.4a - d. They have implemented the Intrinsic Reaction Coordinate (IRC) method by Fukui [96] into the AMOL program using mass-weighted internal coordinates and the constrained search method of Gonzalez and Schlegel [97]. Deng et al. [90] have also traced selected reaction coordinates, such as lengths and angles in the IRC evalution . In reaction 3.1.4-a), for example, the distance $R(H_a - H_b)$ decreases strongly from approximately 1.9 Å (in H_2CO) to 0.75 Å in the H_2 molecule. Simultaneously the distance $R(C - H_b)$ increases from 1.1 Å to about 3 Å for the products, whereas the $R(C - H_a)$ distance increases only after reaching the TS. The $C - O$ bond length decreases from 1.209 Å in H_2CO to 1.130 Å in carbon monoxide. In the case of the trans-cis isomerization (reaction 3.1.4c the simplest reaction coordinate, the dihedral angle $H_a - C - O - H_b$, increases linearly from $0°$ in the trans-conformer, to approximately $90°$ in the TS to $180°$ in the cis-hydroxymethylene. In this reaction the other bond lengths and angles are relatively constant, except the $C - O$ bond length, which reaches a maximum in the TS, due to the breaking of the $C = O$ double bond.

3.2.   CALCULATIONS OF THE REACTION PATHWAY

In addition to the study of the IRC in formaldehyde, Deng and Ziegler [98] studied the IRC for the following chemical reactions:

I. the isomerization of hydrocyanic acid:

$$HCN \longrightarrow CNH \qquad\qquad 3.2a)$$

II. the S_N2 reaction

$$H^- + CH_4 \longrightarrow CH_4 + H^- \qquad\qquad 3.2b)$$

III. the exchange process

$$\cdot H + HX \longrightarrow HX + H \cdot \qquad with \quad X = F, Cl \qquad\qquad 3.2c)$$

IV. the elimination reaction:

$$C_2H_5Cl \longrightarrow C_2H_4 + HCl \qquad\qquad 3.2d)$$

A complete characterization of the reactants, the TS and products are presented, with optimized structures and calculated harmonic vibrational frequencies. Deng and Ziegler concluded that with the N-LDA scheme they obtained the same accuracy as the MP4 level, but their IRC method has the advantage of providing more information about the energy minimum path connecting the stationary points on the potential energy surface. In contrast, IRC calculations based on post HF methods are very time consuming and thus restricted to small molecules.

The isomerization reaction 3.2a is characterized by a nearly elliptical motion of the hydrogen atom from HCN to CNH. The angle $\angle C - N - H$ increases monotonically from $0°$ in HCN to $180°$ in CNH, whereas the distance $R(C \equiv N)$ changes only very little (0.03 Å).

In contrast to the unimolecular reaction 3.2a, where the energy profiles obtained by LDA and N-LDA calculations are nearly identical, the LDA and the N-LDA energy profiles in the bimolecular S_N2 reaction 3.2b are rather different. During the S_N2 reaction the total bond order is increased. With the LDA method, which favors bond formation and overestimates the bonding energy, the weakly bonded intermediate $[H^-, CH_4]$ is much more stable than with the N-LDA scheme, whereas the activation barrier is smaller for the LDA method.

For the exchange process 3.2c the geometry of the transition state $H...X...H$ is linear for HF. The authors expect a slightly bent transition state for HCl, whereas they obtained a second order saddle point for HCl with D_∞ symmetry constrain. The activation barriers are rather different for the DFT and post level HF methods: the N-L(S)DA barriers are too small (23.4 kcal/mol for $H + HF$ and 5.7 kcal/mol for $H + HCl$[98]). The MP4 values (38.2 kcal/mol and 18.4 kcal/mol [99], respectively) are higher than the experimental estimates (34 kcal/mol [100] and as an upper bound 52 kcal/mol from Bartoszek et al. [101] for $H + HF$, for $H + HCl$ the experimental data range from 1.3 to 8.0 kcal/mol - see, e.g., [102, 103]).

In agreement with previous HF investigations [104], Deng and Ziegler found for the 1,2 elimination of HCl from C_2H_5Cl (reaction 3.2d) at first an isomerization of the staggered equilibrium structure of C_2H_5Cl to an eclipsed structure, followed by an elimination of HCl via the four-centred transition state structure

Before HCl is separated from C_2H_4 an adduct $[C_2H_4 + HCl]$ is formed, which is a very shallow minimum on the energy surface, 0.7 kcal/mol below the energy of the final products C_2H_4 and HCl.

Deng and Ziegler have concluded [98] that the IRC paths generated by the DFT schemes are in qualitative agreement with similar paths from HF or post HF calculations. The combination of approximate density functional theory with the IRC formalism of Fukui makes it possible to scan the essential points of the potential energy surface generated by DFT and to trace the pathways, connecting the transition states with the products and the reactants by an efficient method for large systems, such as organometallic complexes.

3.3.   STUDIES TO IMPROVE DFT BARRIERS

Summarizing the results presented in the chapters above, it has become clear that density functional calculations within the LDA significantly underestimate the reaction barriers. The gradient corrections improve this situation, but the quality of the results is still not perfect. Johnson et al. [61], Sosa et al. [105] and Baker et al. [106] studied in detail the barrier heights of some simple model reactions with a variety of density functionals.

Johnson et al. investigated the model reaction

$$H + H_2 \longrightarrow H_2 + H \qquad\qquad 3.3a)$$

as the simplest hydrogen abstraction reaction (a reaction where radical species are involved such as in reactions 3.1.1a, 3.1.1b, and 3.2c), whereas Baker et al. [106] performed similar calculations of another simple radical reaction:

$$OH + H_2 \longrightarrow H_2O + H \qquad\qquad 3.3b)$$

Sosa et al. studied the F_3^- anion dissociation

$$F_3^- \longrightarrow F^- + F_2 \qquad\qquad 3.3c)$$

These investigations have included several exchange-correlation functionals, such as those of Lee, Yang and Parr [107] and the generalized gradient correction for the exchange-correlation energy of Perdew and Wang [108] and the gradient corrected exchange functionals of Becke [35]. Furthermore, Baker et al. also applied Becke's Adiabatic Connection Method (**ACM**) [109]. This method represents a hybrid of several exchange and correlation functionals and a term describing the exact exchange. Sosa et al. pointed out in [105] that the effect of non-local corrections to the exchange functionals is to elongate the bond and to underestimate the dissociation energy. The opposite effect is observed for the correlation energy functionals: the bond length is shortened and the dissociation energy becomes larger. To obtain accurate results, both corrections must be well balanced.

Johnson et al. has also considered the influence of the self-interaction correction on the barrier height. Their results show that even for gradient corrected functionals the barrier height is substantially too low, by several kcal/mol without the self-interaction correction. When this correction is added, the barrier height is reproduced far better, i.e., the addition of the self-interaction correction as implemented in [29] gives an large energy change in the correct direction. Similar results were obtained for the reaction 3.3b $(OH + H_2)$. In that case the "mixture" of different XC-functionals (**ACM**) gives reasonable results. One essential feature of the ACM is an improved treatment of exchange, and exact handling of the exchange also includes self-interaction. Thus, the correct description of the exchange or considering the self-interaction seems to be essential for a proper description of reaction barriers in many cases.

4.   Summary and outlook

The description of chemical reactions on the basis of DFT is only at the beginning. The techniques for calculations of "reaction paths" are now more or less developed, i.e., there exist efficient schemes for solving the Kohn-Sham equations and computation of gradients and methods for the second derivatives of the energy are available. The last of these still require improvement.

The survey above of a rather wide range of chemical reactions shows that due to its tendency of overbinding, usually activation barriers are too low. In extreme cases this may lead to zero or "negative" barriers, whereas more sophisticated methods or experiments clearly show a barrier. Artificial metastable transition state complexes (local minima on the potential surface) may also occur, as in the case of the simple hydrogen abstraction reaction $(\cdot CH_3 + CH_4$ - reaction 3.1.1a). The consideration of the gradient correction improves the accuracy of the calculated barrier heights significantly. However, as shown for some simple model reactions, even the gradient correction may fail qualitatively. This is not surprising, since it is well known that the gradient correction does not remove the qualitative defects of the LDA (e.g., the imperfect self-interaction correction), and one has to consider also that the gradient expansion for the exchange-correlation energy is divergent [110].

Reaction barriers clearly remain challenge for improved density functionals. Nevertheless, even LDA calculations are already quite helpful for qualitative discussions f reaction paths, for example in cases where transition metals are involved. Finally, it is worthwhile to mention the possibility for direct simulations of the dynamics of chemical reactions within DFT, as recently shown for the protolysis of ,3,5-trioxane and 1,3-dioxolane [111]. And at least in an approximate way such imulations can be done including the statistics of collision processes in chemical eactions [9].

5. Acknowledgement

The authors thank J. Fabian, R.O. Jones and T. Ziegler for valuable discussions nd critical comments. K.K. gratefully acknowledges for the kind support of G. Grossmann.

References

[1] P. Hohenberg and W. Kohn, *Phys. Rev.* **136**, B864 (1964).

[2] W. Kohn and L.J. Sham, *Phys. Rev.* **140**, A1133 (1965).

[3] O. Gunnarsson, J. Harris, and R.O. Jones, *J. Phys.* **C9**, 739 (1976).

[4] B.I. Dunlap, J.W.D. Connolly, and J.R. Sabin, *J. Chem. Phys.* **71**, 3396, 4993 (1979).

[5] E.J. Baerends, D.E. Ellis, and P. Ros, *Chem. Phys.* **2**, 41 (1973).

[6] A.D. Becke, *J. Chem. Phys.* **84**, 4524 (1986).

[7] J.P. Perdew and Y. Wang, *Phys. Rev.* **B33**, 8800 (1986).

[8] T. Ziegler, *Chem. Rev.* **91**, 651 (1991), *J. Chem. Phys.* **95**, 7401 (1991).

[9] J. Schulte and G. Seifert, *Chem. Phys. Lett.* **221**, 230 (1994).

[10] K.N. Houk, Y. Li, and J.D. Evanseck *Angew. Chem. Int. Ed. Engl.* **31**, 682 (1992).

[11] S. Gladstone, K.J. Laidler, and H.M. Eyring, *The Theory of Reactions*, Mc Graw Hill, New York, 1941.

[12] R.O. Jones, *Angew. Chem.* **103**, 647 (1991).

[13] E.S. Kryachko and E.V. Ludena, *Energy Density Functionals. Theory of Many Electron Systems*, Kluwer Academic Publishers, Dordrecht, Boston, London, 1990.

[14] R.G. Parr and Y. Weitao, *Density-Functional Theory of Atoms and Molecules*, Oxford University Press, Oxford 1989.

[15] E.H. Lieb, *Int. J. Quant. Chem.* **24**, 243 (1983).

[16] H. Englisch and R. Englisch, *phys. stat. sol.* **b123**, 711 (1984), **b124**, 373, (1984).

[17] O. Gunnarsson, M. Jonson, and B.I. Lundqvist, *Phys. Rev.* **B20**, 3136 (1979).

[18] L. Hedin and B.I. Lundqvist, *J. Phys.* **C4**, 2064 (1971).

[19] S.H. Vosko, L. Wilk, and M. Nusair, *Can. J. Phys.* **58**, 1200 (1980).

[20] R. Gaspar, *Acta Phys. Akad. Sci. Hung.* **3**, 263 (1954).

[21] J.C. Slater, *Phys. Rev.* **81**, 385 (1951).

[22] K.H. Schwarz, *Phys. Rev.* **B5**, 2466 (1972).

[23] J.C. Slater, *Quantum Theory of Atoms and Molecules*, Vol. 4, Mc Graw-Hill, New York, 1974.

[24] U. von Barth and L. Hedin, *J. Phys.* **C5**, 1629 (1972).

[25] O. Gunnarsson and B.I. Lundqvist, *Phys. Rev.* **B13**, 4274 (1976).

[26] A.K. Rajagopal and J. Callaway, *Phys. Rev.* **B7**, 1912 (1973).

[27] H. Eschrig, G. Seifert, and P. Ziesche, *Sol. State Comm.* **56**, 777 (1985).

[28] B.G. Johnson, P.M.W. Gill, and J.A. Pople, *J. Chem. Phys.* **98**, 5612 (1993).

[29] E.S. Fois, J.I. Penman, and P.A. Madden, *J. Chem. Phys.* **98**, 6352 (1993).

[30] J.P. Perdew and A. Zunger, *Phys. Rev.* **B23**, 5048 (1981).

[31] M.R. Pederson and C.C. Lin, *J. Chem. Phys.* **88**, 1807 (1988).

[32] L. Fritsche and H. Gollisch, *Z. Phys.* **B48**, 209 (1982).

[33] D.C. Langreth and D.J. Mehl, *Phys. Rev.* **B28**, 1809 (1983).

[34] W. Kohn, L.J. Sham, F. Herman, J.P. van Dyke, and I.B. Ortenburger, *Phys. Rev. Lett.* **22**, 807 (1969).

[35] A.D. Becke, *Phys. Rev.* **A38**, 3098 (1988).

[36] J.P. Perdew, *Phys. Rev.* **B33**, 8822 (1986).

[37] H. Sambe and R.H. Felton, *J. Chem. Phys.* **62**, 1122 (1975).

[38] D. Salahub, *Adv. Chem. Phys.* **69**, 447 (1987).

[39] J. Andzelm and E. Wimmer, *J. Chem. Phys.* **96**, 1280 (1992).

[40] A. Rosen, D.E. Ellis, H. Adachi, and F.W. Averill, *J. Chem. Phys.* **65**, 3629 (1976).

[41] W. Bieger, G. Seifert, H. Eschrig, and G. Grossmann, *Z. Phys. Chem.* **266**, 751 (1985).

[42] B. Delley and D.E. Ellis, *J. Chem. Phys.* **76**, 1949 (1982).

[43] F.W. Averill and D.E. Ellis, *J. Chem. Phys.* **59**, 6412 (1973).

[44] B. Delley, *J. Chem. Phys.* **90**, 508 (1990).

[45] D.E. Ellis and G.S. Painter, *Phys. Rev.* **B2**, 2887 (1970).

[46] K.H. Johnson, *Adv. Quant. Chem.* **7**, 143 (1973).

[47] O.K. Andersen and R.G. Woolley, *Mol. Phys.* **26**, 905 (1973).

[48] J.M. Seminario, M. Grodzicki, and P. Politzer, in J.K. Labanowski and J.W. Andzelm (eds.) *Density Functional Methods in Chemistry*, Springer-Verlag, New York, 1991, p 419.

[49] H. Hellmann, *Einführung in die Quantenchemie*, Deuticke, Leipzig, 1937.

[50] R.P. Feynman, *Phys. Rev.* **56**, 340 (1939).

[51] R.A. Harris and D.F. Heller, *J. Chem. Phys.* **62**, 3601 (1975).

[52] M. Scheffler, J.P. Vigneron, and G.B. Bachelet, *Phys. Rev.* **B31**, 6541 (1985).

[53] J.C. Slater, *J. Chem. Phys.* **57**, 2389 (1972).

[54] P. Pulay, *Mol. Phys.* **17**, 197 (1969).

[55] C. Satoko, *Phys. Rev.* **B30**, 1754 (1984).

[56] F.W. Averill and G.S. Painter, *Phys. Rev.* **B32**, 2141 (1985).

[57] R. Fournier, J. Andzelm, and D.R. Salahub, *J. Chem. Phys.* **90**, 6371 (1989).

[58] B. Delley, *J. Chem. Phys.* **94**, 7245 (1991).

[59] A. Komornicki and G. Fitzgerald, *J. Chem. Phys.* **98**, 1398 (1993).

[60] N.C. Handy, T.J. Tozer, G.J. Laming, C.W. Murray, and R.D. Amos, *Israel J. of Chem.* **33**, 331 (1993).

[61] B.G. Johnson, C.A. Gonzales, P.M.W. Gill, and J.A. Pople, *Chem. Phys. Lett.* **221**, 100 (1994).

[62] L. Versluis and T. Ziegler, *J. Chem. Phys.* **88**, 322 (1988).

[63] L. Fan and T. Ziegler, *J. Chem. Phys.* **92**, 3645 (1990).

[64] F.M. Bickelhaupt, Ph. D. Thesis, Amsterdam, 1993.

[65] R.V. Stanton and K.M. Merz Jr., *J. Chem. Phys.* **100**, 434 (1994).

[66] J. Simons, P. Jorgenson, H. Taylor, and J. Ozment, *J. Phys. Chem.* **87**, 2745 (1983).

[67] J. Baker, *J. Comp. Chem.* **7**, 385 (1986).

[68] Y. Abashkin and N. Russo, *J. Chem. Phys.* **100**, 4477 (1994).

[69] L. Fan and T. Ziegler, *J. Am. Chem. Soc.* **114**, 10890 (1992).

[70] M. Sana, G. Leroy, and J.L. Villaveces, *Theoret. Chim. Acta* **65**, 109 (1984).

[71] CRC Handbook of Bimolecular and Trimolecular Gas Reactions; J.A. Kerr and S.J. Moss (eds.) vol.1 CRC Press Boca Raton, Fl, 1981.

[72] Y. Chen, A. Rauk, and E.T Roux, cited in [69] as private communication.

[73] R.H. Nobes and L. Radom, *Chem. Phys.* **60**, 1 (1981).

[74] P.K. Pearson and H.F. Schaefer III, *J. Chem. Phys.* **62**, 350 (1975).

[75] W.J. Hehre, L.Radom, P.v.R. Schleyer, and J.A. Pople, in *Ab-initio Molecular Orbital Theory*, John Wiley & Sons, New York, 1986.

[76] C.F. Pan and W.J. Hehre, *J. Phys. Chem.* **86**, 321 (1982).

[77] L.T. Redmon, G.D. Purvis, and R.J. Barlett, *J. Chem. Phys.* **69**, 5386 (1978).

[78] M.H. Baghal-Vayjooee, J.L. Collister, and H.O. Pritchard, *Can. J. Chem.* **55**, 2634 (1977).

[79] a) F.W. Schneider and B.S. Rabinovitch, *J. Am. Chem. Soc.* **84**, 4215 (1962). b) F.J. Fletcher, B.S. Rabinovitch, K.W. Watkins, and D.J. Locker, *J. Phys. Chem.* **70**, 2823 (1966). c) F.M. Wang and B.S. Rabinovitch, *J. Phys. Chem.* **78**, 863 (1974).

[80] M.R. Pederson, *Chem. Phys. Lett. in press*

[81] J.E. Carpenter and C.P. Sosa, *J. Mol. Struct.* **311**, 325 (1994).

[82] C. Sosa and C.Lee, *J. Chem. Phys.* **98**, 8004 (1993).

[83] P. Politzer and J.M. Seminario, *Chem. Phys. Lett.* **207**, 27 (1993).

[84] a) T. Ziegler, E. Folga, and A. Berces, *J. Am. Chem. Soc.* **115**, 636 (1993). b) T. Ziegler and E. Folga, *J. Organomet. Chem.* **478**, 57 (1994).

[85] T.K. Woo, L. Fan, and T. Ziegler, *Organometallics* **13**, 432, 2252 (1994).

[86] a) E. Folga and T. Ziegler, *Organometallics* **12**, 325 (1993). b) T.K. Woo, E. Folga, and T. Ziegler, *Organometallics* **12**, 1289 (1993).

[87] F.M. Bickelhaupt, L.J. de Koning, N.M.M. Nibbering, and E.J. Baerends, *J. Phys. Org. Chem.* **5**, 179 (1992).

[88] F.M. Bickelhaupt, E.J. Baerends, N.M.M. Nibbering, and T. Ziegler, *J. Am. Chem. Soc.* **115**, 9160 (1993).

[89] a) A.D. Becke, *Int. J. Quantum Chem.* **23**, 1915 (1983). b) A.D. Becke, *J. Chem. Phys.* **85**, 7184 (1986).

[90] L. Deng, T. Ziegler, and L. Fan, *J. Chem. Phys.* **99**, 3823 (1993).

[91] G.E. Scuseria and H.F. Schaefer III, *J. Chem. Phys.* **90**, 3629 (1989).

[92] L.B. Harding, H.B. Schlegel, R. Krishnan, and J.A. Pople, *J. Chem. Phys.* **84**, 3394 (1980).

[93] J.D. Goddard, Y. Yamaguchi, and H.F. Schaefer III, *J. Chem. Phys.* **75**, 3459 (1981).

[94] K. Kamaiya and K. Morokuma, *J. Chem. Phys.* **94**, 7287 (1991).

[95] W.F. Polik, D.R. Guyer, and C.B. Moore, *J. Chem. Phys.* **92**, 3453 (1990).

[96] K. Fukui, *Acc. Chem. Res.* **14**, 363 (1981).

[97] C. Gonzalez and H.B. Schlegel: *J. Chem. Phys.* **94**, 5523 (1990).

[98] L. Deng and T. Ziegler, *Int. J. Quantum Chem.* **52**, 731 (1994).

[99] K.D. Dobbs and D.A. Diox, *J. Phys. Chem.* **97**, 2085 (1993).

[100] J.F. Bott, *J. Chem. Phys.* **65**, 1976 (1976).

[101] F.E. Bartoszek, D.M. Manos, and J.C. Polannyi, *J. Chem. Phys.* **69**, 933 (1978).

[102] D.L. Thompson, H.H. Suzukava,Jr., and L.M. Raff, *J. Phys. Chem.* **75**, 1844 (1971).

[103] H. Endo and G.P Glass, *Chem. Phys. Lett.* **44**, 180 (1976); *J. Phys. Chem.* **80**, 1519 (1976).

[104] a) J.A. Pople, M.J. Frisch, and J.E. Del Bene, *Chem. Phys. Lett.* **91**, 185 (1982). b) S. Nagase and T. Kudo, *J. Chem. Soc. Chem. Comm.* 363 (1983). c) C. Clavero, M. Duran, A. Lledos, N. Ventura, and J. Bertran *J. Comput. Chem.* **8**, 481 (1987).

[105] C. Sosa, C.Lee, G. Fitzgerald, and R.A. Eades, *Chem. Phys. Lett.* **211**, 265 (1994).

[106] J. Baker, J. Andzelm, and M. Muir, *J. Chem. Phys. in press.*

[107] C. Lee, W. Yang, and R.G. Parr, *Phys. Rev.* **B37**, 785 (1988).

[108] J.P. Perdew, J.A. Chevary, S.H. Vosko, K.A. Johnson, M.R. Pederson, D.J. Singh, and C. Fiolhais, *Phys. Rev.* **B46**, 6671 (1992).

[109] A.D. Becke, *J. Chem. Phys.* **98**, 5648 (1993).

[110] L. Kleinman, *Phys. Rev.* **B10**, 2221 (1974).

[111] A. Curioni, W. Andreoni, J. Hutter, H. Schiffer, and M. Parrinello, *J. Amer. Chem. Soc. in press.*

[112] R. Car and M. Parrinello, in *Simple Molecular Systems at Very High Density*, NATO Advanced Study Institute, Series **B186**, p. 455.

USING THE REACTION PATH CONCEPT TO OBTAIN RATE CONSTANTS FROM *AB INITIO* CALCULATIONS

ALAN D. ISAACSON
Department of Chemistry
Miami University
Oxford, Ohio 45056

1. Introduction

Even after so many years of progress in theoretical chemistry, the accurate first-principles description of chemical reactions still poses a major challenge. Under the Born-Oppenheimer approximation, the dynamics of an elementary chemical reaction are determined by the sum of the electronic energy plus the internuclear coulomb repulsion energy as a function of the nuclear geometry, i.e., the potential energy surface (PES). (Although we shall restrict the present section to bimolecular elementary gas-phase reactions occurring on a single PES, the concepts and methodologies discussed below are applicable to unimolecular reactions as well as to processes occurring in condensed phases or at phase interfaces; for some examples of such applications, see [1].) For example, given the entire PES for a particular reaction, the thermal rate constant can be obtained through a Boltzmann average of the reaction cross section [2,3], which can be approximated with classical trajectory [3-5] or quantum mechanical coupled-channel calculations [6,7]. More recently, a discrete variable representation approach [8] to the calculation of the cumulative reaction probability has also shown great promise. However, since such methods require a great deal of information about the PES, and, in addition, are generally not practical for three-dimensional studies of reactions involving more than three (or, perhaps, four) atoms, the most common approach for calculating thermal rate constants is transition state theory. In the conventional formulation of transition state theory (TST) [9-13], one assumes [14] that the net rate of forward reaction at equilibrium is given by the flux of reaction complexes in the product direction across a coordinate-space hyperplane that divides reactants from products and passes through the saddle point of the PES, which is the point at which the matrix of second derivatives of the potential energy with respect to atomic coordinates (hessian) has exactly one negative eigenvalue. This point is also the

191

D. Heidrich (ed.), The Reaction Path in Chemistry:
Current Approaches and Perspectives, 191–228.
© 1995 *Kluwer Academic Publishers. Printed in the Netherlands.*

highest energy point on the minimum energy path (MEP) from reactants to products. (We take the MEP to mean the path of steepest descents from the saddle point to reactants and products in an isoinertial set of coordinates, i.e., a set of coordinates in which the same arbitrary mass μ is associated with each degree of freedom, so that the motion of the many-atom system is equivalent to the motion of a point of mass μ on the PES. In practice, we use mass-scaled coordinates [15,16], which are obtained by multiplying each atomic Cartesian coordinate (measured relative to the system center of mass) by $(m_i/\mu)^{1/2}$, where m_i is the mass of the corresponding atom. For example, choosing $\mu = 1$ amu gives coordinates that are numerically equivalent to the conventional mass-weighted coordinates [17] used in vibrational spectroscopy. Thus, the MEP discussed here is the intrinsic reaction coordinate [18-21], and we define the reaction coordinate s as the signed distance from the saddle point along the MEP such that negative and positive values of s refer to reactant and product sides, respectively.)

1.1.   CONVENTIONAL TRANSITION STATE THEORY

The fundamental TST assumption stated above leads to the following expression for the rate constant k at temperature T:

$$k^{TST}(T) = \frac{\sigma}{\beta h} \frac{Q^{\ddagger}(T)}{Q^R(T)} \exp(-\beta V^{\ddagger}), \tag{1}$$

where σ is the symmetry factor [13,22] (number of equivalent reaction paths), β is $(k_B T)^{-1}$, where k_B is Boltzmann's constant, h is Planck's constant, $V^{\ddagger}$ is the classical potential energy at the saddle point (i.e., not including zero point energy) with the zero of energy at the reactants' classical equilibrium position, $Q^R(T)$ is the reactants' partition function per unit volume with the zero of energy at the reactants' classical equilibrium position, and $Q^{\ddagger}(T)$ is the transition state partition function with the zero of energy at $V^{\ddagger}$. The reactants' partition function per unit volume factors as

$$Q^R(T) = Q_{rel}(T)Q^A(T)Q^B(T), \tag{2}$$

where the relative translational factor is

$$Q_{rel} = (2\pi\mu_{AB} / \beta h^2)^{3/2}, \tag{3}$$

Q^A and Q^B are the partition functions for the internal degrees of freedom of reactants A and B, and $\mu_{AB} = m_A m_B/(m_A + m_B)$ is the reduced mass of reactant relative translational motion. For molecules, the usual separable approximation allows the reactant and transition state partition functions to be written as products of electronic, vibrational, and rotational partition functions, which are computed from quantized energy levels in each of these degrees of freedom. (In practice, the electronic partition function is often well approximated by direct summation over the lowest multiplet of electronic states, while the rotational partition function is usually approximated accurately enough by the standard classical formula involving the determinant of the moment of inertia tensor, which is easily calculated from the Cartesian geometry [15,23]; the symmetry numbers in the rotational partition functions are set to unity as they are already included in σ. The vibrational partition function can be evaluated, in principle, by a direct summation over the bound vibrational energy levels of the ground electronic state; this is further discussed below.) Thus, the calculation of a TST rate constant is not only simple, but also requires a minimum amount of information on the PES, namely, the energy, determinant of the moment of inertia tensor, and bound vibrational frequencies (and, optionally, higher derivatives of the PES if one includes anharmonic vibrational effects) at the saddle point and at the reactants. Because they avoid the errors inherent in numerical differentiation and do not require the calculation of the potential energy at a grid of points in the $(3N_{atom}-6)$-dimensional space of internal degrees of freedom, recent advances in *ab initio* electronic structure theory that provide an analytic gradient (the vector of first derivatives of the potential energy with respect to the atomic Cartesians) [24-30], hessian [31], and even higher derivatives of the potential energy [32] have allowed accurate calculations of these stationary points and their properties to be carried out routinely for systems of many atoms [see, e.g., 33-41].

1.2. CANONICAL VARIATIONAL TRANSITION STATE THEORY

The two major drawbacks of the TST approach are that it is applicable only to reactions that have a barrier and that it can be inaccurate [42,43]. This inaccuracy arises (within a classical picture for the reaction-coordinate motion) from trajectories that cross the transition state dividing surface more than once [15,42,43], and, therefore, predict too large a rate constant, as well as from the neglect of quantum mechanical effects on this degree of freedom. One very practical approach to reducing these errors that has been shown to provide reliable thermal rate constants for a given PES even for systems of many atoms is to employ canonical variational transition state theory (CVT) with multidimensional semiclassical transmission coefficients [15,16,30,42-77]. In this approach, a generalized transition state (GTS) dividing surface (which, in general, can be any hypersurface in phase space, i.e., the space of all positions and momenta, that divides reactants from products) is taken to be a hyperplane in mass-scaled Cartesian coordinates that intersects and is perpendicular to the MEP at s, but, more globally, is bent if necessary to truly separate reactants from products. The generalized TST rate constant $k^{GT}(T,s)$ at temperature T for the GTS that intersects the MEP at s is then given by [15,16,44,69]

$$k^{GT}(T,s) = \frac{\sigma}{\beta h} \frac{Q^{GT}(T,s)}{Q^R(T)} \exp[-\beta V_{MEP}(s)]. \tag{4}$$

Here $V_{MEP}(s)$ is the classical potential energy along the MEP with the zero of energy at the reactants' classical equilibrium position and $Q^{GT}(T,s)$ is the quantum mechanical partition function for the GTS at s with the zero of energy at $V_{MEP}(s)$ and rotational symmetry numbers set to unity. To reduce the error due to trajectories that cross the transition state dividing surface more than once, the CVT rate constant $k^{CVT}(T)$ with quantized internal degrees of freedom but classical reaction-coordinate motion is obtained by minimizing the generalized TST rate constant $k^{GT}(T,s)$ with respect to the position of the GTS along the MEP, i.e., with respect to s [15,16,44,69]:

$$k^{CVT}(T) = \min_s k^{GT}(T,s). \tag{5}$$

The location of the GTS corresponding to this minimum flux can be interpreted as a dynamical bottleneck to the reaction, in that it includes both entropic effects (associated with the ratio of the partition functions for the GTS and the reactants) and zero-point energy effects as well as the energetic effects in $V_{MEP}(s)$. In addition, since this minimum corresponds to a maximum in the generalized standard-state free energy change for the formation of the GTS at s from the reactants, $\Delta G^{GT,0}(T,s)$ [42,44,47,78,79], the CVT approach can also be applied to reactions in which $V_{MEP}(s)$ does not exhibit a barrier between the reactants and products.

1.2.1. *Tunneling corrections*

To incorporate multidimensional quantum effects arising from the motion of the system along the reaction coordinate, the CVT rate constant is multiplied by a ground state semiclassical transmission coefficient $\kappa^{CVT/G}$ that primarily accounts for reaction-path tunneling and nonclassical reflection [15,16,43,47]. The final semiclassical CVT rate constant is thus given by

$$k^{CVT/G}(T) = \kappa^{CVT/G}(T)k^{CVT}(T). \tag{6}$$

(In TST, the transmission coefficient is often approximated by the Wigner correction, which only requires the imaginary frequency at the saddle point. However, this correction is rarely justifiable [80,81], and is not recommended.) For reaction paths exhibiting very small curvature, tunneling (i.e., motion through the classically forbidden region) occurs along the MEP through an effective potential given in an adiabatic approximation (in which all GTS modes are assumed to adjust adiabatically to changes in s) by the vibrationally adiabatic ground state potential curve [18,47],

$$V_a^G(s) = V_{MEP}(s) + \varepsilon_{int}^G(s), \tag{7}$$

where $\varepsilon_{int}^G(s)$ is the total zero point energy for the bound vibrational degrees of freedom transverse to the MEP at s. Thus, in the minimum-energy-path semiclassical adiabatic ground-state (MEPSAG) method [47,82], the $\kappa^{CVT/MEPSAG}(T)$ transmission coefficient is computed from the Boltzmann average of the semiclassical ground-state probability $P^G(E)$ of tunneling through $V_a^G(s)$ at total energy E, where $P^G(E)$ is obtained by integrating the multidimensional imaginary action along the one-dimensional tunneling path by numerical quadrature [82]. (Although the tunneling probability $P^G(E)$ could be determined by solving the effective one-dimensional Schrödinger equation numerically rather than by semiclassical methods, the small improvement in accuracy obtained with the numerical quantal solution is generally not worth the extra effort [47,82,83].) The advantage of this method is that no additional information beyond that required for computing $k^{CVT}(T)$ is needed [assuming that vibrational energies, and hence $V_a^G(s)$, have been obtained over the entire MEP, rather than just over the region near the saddle point needed to find the minimum in eq. (5)]. However, when the reaction path has a significant degree of curvature, such as in reactions involving the transfer of a light atom between two heavy moieties, tunneling occurs on the concave side of the MEP [15,49,58,63,64,69,84,85], since such "corner-cutting" shortens the tunneling path and leads to a larger tunneling contribution. For cases of small to moderate reaction-path curvature in collinear atom-diatom reactions, Marcus and Coltrin [85] showed that the optimum tunneling path corresponds to the path of concave-side turning points for the ground state of the bound vibrational mode transverse to the reaction coordinate (i.e., the symmetric stretch mode) in the classically forbidden region. A straightforward multidimensional generalization of the Marcus-Coltrin path in which the one-dimensional tunneling path is distorted from the MEP out to the concave-side ground-state vibrational turning point of each bound mode transverse to the MEP in the classically forbidden region leads to the small-curvature semiclassical adiabatic ground-state (SCSAG) approximation to the transmission coefficient, $\kappa^{CVT/SCSAG}(T)$ [15,49], while a better generalization in which the tunneling path is distorted from the MEP in the direction of the internal centrifugal force in the classically forbidden region (thereby providing a more reliable treatment when the curvature is significant in more than one bound mode transverse to the MEP) leads to the centrifugal-dominant small-curvature semiclassical adiabatic ground-state (CD-SCSAG) approximation to the transmission coefficient, $\kappa^{CVT/CD-SCSAG}(T)$ [74,77]. Both methods require a knowledge over the whole reaction path of the turning point $t_m(s)$ and its derivative $dt_m(s)/ds$, the eigenvector $L_m(s)$, and the curvature component $B_{mF}(s)$ for each of the $3N_{atom}-7$ (or, for a linear transition state, $3N_{atom}-6$) generalized normal modes m transverse to the MEP at s. [The curvature component $B_{mF}(s)$, where F signifies the reaction coordinate, can be computed from the eigenvector $L_m(s)$ and the derivative of the gradient (which represents reaction-path motion) with respect to s.]

When the curvature of the MEP is large, the optimum tunneling path often lies outside the region of the PES that can be represented by a set of quadratic (or even higher-order) force fields along the MEP. In fact, in the limit of very large curvature, the optimum tunneling paths are straight lines from the reactant valley to the product valley [15,56,57,77]. The large-curvature ground-state approximation, version 3 (LCG3) [15,86,87] provides an efficient method for obtaining the tunneling probability $P^G(E)$ in just such a case. However, it requires more global information about the PES than can be obtained just from the properties along the MEP. Specifically, it requires information on the reaction "swath" that contains the MEP and the concave region it borders [63].

1.2.2. PES information needed for CVT calculations

The above discussion shows that while the calculation of a semiclassical CVT rate constant does not require the entire PES and is quite practical even for many-atom systems, it does require more information about the PES than is needed for calculating a TST rate constant. In particular, one must determine the MEP from the reactants to the products and the properties (geometry, energy, gradient, hessian, etc.) along it. To accomplish this, one first determines the geometries and energies of the reactants, products, and saddle point, and performs normal-mode analyses at these points. The normal-mode analysis can be carried out conveniently by diagonalizing the hessian in mass-scaled Cartesians. (To avoid excessive numerical error, the hessian should be computed either analytically or by numerical first derivatives of an analytic gradient.) One then computes the steepest-descents path downhill from the saddle point to the reactants and the products by following the direction of the negative of the gradient in mass-scaled Cartesians. [If there is more than one saddle point, the portion of the MEP originating from each of them must be determined, and the individual MEP segments joined together; if there is no saddle point, the MEP can be obtained by starting from an arbitrary geometry that corresponds to separated (but not infinitely separated) reactants when the reaction is written in the exoergic direction.] Since the gradient is zero at the saddle point, an initial finite step in mass-scaled Cartesians is taken from the saddle point toward the reactants or products in the imaginary-frequency normal mode direction, as this direction is tangent to the MEP at the saddle point. Alternatively, one can base this initial step on a higher-order representation of the PES at the saddle point [88,89]. After this initial step, the simplest approach to determining the MEP, called the Euler single-step method [90], is to take steps of fixed length, δs, in the direction of the negative of the gradient in mass-scaled Cartesians. However, since such steps tend to "zig-zag" back and forth across the true MEP, a very small value of δs (generally 0.001 a_0 or less) is required for an accurate approximation to the MEP [63]. More sophisticated algorithms involving higher-order derivatives of the PES and/or a reduced-dimensionality search for a corrected point on the MEP have been developed to allow for larger step sizes [29,30,88,89,91-95]. Since these methods require additional energy, gradient, or even higher-derivative calculations at each point on the MEP, it is not clear whether any of these approaches will produce converged properties along the MEP (especially the curvature components, B_{mF}, which are very sensitive to the accuracy of the MEP) more efficiently than the Euler single-step method for a particular problem.

In order to calculate $Q^{GT}(T,s)$ as well as $V_a^G(s)$ and the other quantities needed for computing the semiclassical CVT rate constant, a generalized normal mode analysis must be performed at regular intervals Δs (typically 0.01 to 0.1 a_0) along the MEP. Since the gradient is not zero except at the saddle point, contributions to the hessian arising from motion along the reaction coordinate as well as from overall translations and rotations must first be projected out [15,16,96]. The generalized normal modes transverse to the MEP at s are then determined by diagonalizing the projected hessian in mass-scaled Cartesians. Higher potential energy derivatives along the generalized normal mode directions can then be calculated if one wishes to include anharmonic vibrational effects. Once $Q^{GT}(T,s)$, $V_a^G(s)$, and the other reaction-path data have been calculated on the Δs grid, they can be computed for arbitrary values of s within the span of the grid by, for example, Lagrangian interpolation or cubic spline fits. When such data are needed outside the span of the Δs grid, such as when the reaction path becomes too difficult to follow accurately or a minimum in $V_{MEP}(s)$ is encountered before $V_a^G(s)$ is low enough for computing converged tunneling probabilities, one may be able to extrapolate to the reactant or product limits; simple exponential extrapolations [16] are often sufficiently accurate, although a variety of functional forms has been used [75].

Within the harmonic approximation, one can easily obtain the total vibrational zero point energy, $\varepsilon_{int}^G(s)$, and the vibrational partition function factor, $Q_{vib}^{GT}(T,s)$, contained in $Q^{GT}(T,s)$ from the generalized normal mode frequencies discussed above. Specifically, $\varepsilon_{int}^G(s)$ can be written as the sum of the harmonic zero point energies in each generalized bound mode m, $\varepsilon_{int,m}^G(s)$, while $Q_{vib}^{GT}(T,s)$ can be written as the product of the harmonic vibrational partition functions for each generalized bound mode m, $Q_{vib,m}^{GT}(T,s)$, where $\varepsilon_{int,m}^G(s)$ and $Q_{vib,m}^{GT}$ are given by standard formulas found in almost any physical chemistry text. However, the vibrational degrees of freedom are, in general, bound by an anharmonic PES, and the harmonic approximation can be inaccurate [45,79,97-101]. Due to the large number of coupled-mode energy levels required for convergence at each GTS location along the MEP, computing the vibrational partition function by direct summation over multidimensional energy levels obtained from quantal or even semiclassical methods is not practical for polyatomic systems, both in terms of the computational effort and the amount of required information about the PES. If the mode-mode coupling of the normal modes can be neglected, one can employ the independent normal mode (INM) approximation [15,16], in which the motion in each generalized bound normal mode is treated as an independent anharmonic oscillator by one of several models, e.g., the Morse model [15,16], the quadratic-quartic model [15,16], or the WKB approximation [15,102-104]. Since the total INM vibrational energy is the sum of the vibrational energies for each mode, the vibrational partition function $Q_{vib}^{GT}(T,s)$ can still be written as the product of the vibrational partition functions $Q_{vib,m}^{GT}(T,s)$ for each generalized bound mode m, where $Q_{vib,m}^{GT}(T,s)$ is obtained by direct summation over the levels in that mode. Alternatively, one could treat internal rotational motions by the hindered rotor model [77,105]. When the anharmonic effects due to mode-mode coupling are large, however, the INM approach will not be accurate [98,100]. In such cases, $Q_{vib}^{GT}(T,s)$ can be computed by direct summation over the coupled-mode energy levels obtained from perturbation theory (PT) through second order in the cubic anharmonic terms and first order in the quartic anharmonic terms of the PES [63,98,100,101,106-112]. In larger systems, a more practical approach, called simple perturbation theory (SPT) [106], can be used to obtain $Q_{vib}^{GT}(T,s)$ using only the perturbation theory zero point energy and the fundamental excitation frequencies for each mode. Both the PT and SPT approaches require information on the third and fourth derivatives of the PES along the MEP; these can be computed directly or modeled through the use of a local quadratic expansion of the PES in internal coordinates [63].

1.3.   OBTAINING PES INFORMATION

For a first-principles calculation of a semiclassical CVT rate constant, one would like to extract all of the PES information discussed above from *ab initio* calculations. Traditionally, this has involved constructing an analytic function that represents the PES for all values of the nuclear coordinates [37,48,63,113]. For example, by fitting a London-Eyring-Polanyi-Sato (LEPS) potential function [114-116] to experimental data, semiempirical PESs were obtained for the H + H_2 and Cl + H_2 reactions [117]. More recently, advances in electronic structure calculations, especially in locating and determining accurate properties for reactants, products, and saddle points [33-41,118,119], have allowed analytic PESs to be constructed for a variety of systems by fitting to *ab initio* properties at these geometries or to grids of *ab initio* points in the vicinities of these geometries, sometimes together with experimental data [see, e.g., 37,56,57,67,120-130]. The obvious advantage of constructing such a PES is that analytic values of the energy and its first (and even higher) derivatives can be computed very cheaply, so that one can easily obtain an accurate MEP by using a very small δs, one can study the effects of including vibrational anharmonicity and reaction-path curvature, and, since the PES is available for all nuclear geometries, one can determine whether the small-curvature or large-curvature description of tunneling is more appropriate. Furthermore, one can ensure that the PES reproduces certain reaction properties, such as the *ab initio* saddle point geometry or the experimental thermal rate constant at a particular temperature. In fact, by allowing the parameters in a simple functional form having the proper global topology (such as the LEPS potential) to vary with geometry, one can conveniently make localized changes in a PES so that a given set of data on the reactant, product, and interaction regions is reproduced [67,131-135]. This is readily accomplished by the use of switching functions (such as exponential, Gaussian, or hyperbolic forms) that smoothly change the parameters between their asymptotic limits. In Section 2, we present an application of this strategy to the construction of improved PESs for the OH + H_2 system [135] that accurately reproduce a large amount of *ab initio* data.

There are two major difficulties with the traditional approach of constructing an analytic PES. First, the predicted dynamical results can be very sensitive to both the functional form assumed for the PES and how well it reproduces the data to which it is fit [135]. Second, since the PES is a function of $3N_{atom}-6$ coordinates, the effort required to obtain the necessary high-quality *ab initio* information, devise a reasonable functional form for the PES, and carry out the fitting becomes prohibitive as we go to larger systems. In addition, from the point of view of calculating the semiclassical CVT rate constant, this procedure is inefficient, in that only the portion of the PES in the vicinity of the MEP is used. An approach that avoids these problems is called direct dynamics [1,30,136, and references therein], in which one obtains the potential energy information needed for a semiclassical CVT calculation directly from electronic structure calculations, rather than from an analytic PES. That is, the MEP is determined by taking finite steps in the negative of the direction of the electronic structure gradient, with further electronic structure calculations used to extract potential energy information at various points along the MEP. One can imagine two basic ways to accomplish this. One way is to make the electronic structure calculation a "subroutine" inside the semiclassical CVT program, so that it can be called for each geometry at which the energy or its derivatives are needed. For larger systems, such a scheme is not currently practical with *ab initio* molecular orbital methods, but it is practical with various semiempirical molecular orbital methods that are based on the neglect of diatomic differential overlap (NDDO) [137,138]. In fact, the MORATE program [139] employs this approach with the MINDO/3 [140], MNDO [141], AM1 [142], or PM3 [143] methods, using either standard parameter sets or specific reaction parameters optimized for the particular reaction(s) of interest (NDDO-SRP) [144,145]. For example, Truhlar and co-workers [74,136] obtained good agreement between the experimental kinetic isotope effects for the [1,5] sigmatropic rearrangement reaction of *cis*-1,3-pentadiene and for the $CF_3 + CD_3H \rightarrow CF_3H + CD_3$ reaction and those computed from direct dynamics calculations of semiclassical CVT rate constants using the AM1 parametrization and the NDDO-SRP approach based on the standard AM1 parametrization, respectively.

The other basic way of carrying out a direct dynamics calculation is to first perform as high quality *ab initio* calculations along the MEP as one can afford, and then use this information as input data for the semiclassical CVT program [75-77]. Thus, the PES in this case is the implicit one defined by the geometries, energies, gradients, and hessians (and higher derivatives, if desired) at a set of points along the MEP. Examples of processes to which this approach has been applied include the $CH_3 + H_2 \rightarrow CH_4 + H$ reaction [30], the geometrical isomerization of cyclopropane [89], hydrogen atom transfers in small hydrogen-bonded clusters of water molecules and of formaldehyde with water molecules

[146], and proton transfer in the $CH_4 + CH_3^-$ system [147]. This last example is described in detail in Section 3. It is important to note that since the electronic structure information used in a direct dynamics calculation must be computed along the MEP in isoinertial coordinates (e.g., mass-scaled Cartesians), a different set of electronic structure data is required for each choice of isotopes.

2. Analytic PES for a Four-Atom System

As an example of the construction of an analytic PES for a bimolecular reaction by fitting to *ab initio* data, this section presents our development of a PES for the reaction

$$OH + H_2 \rightarrow H_2O + H . \tag{R1}$$

This simple four-atom abstraction reaction, which is exoergic by about 16 kcal/mol [148], is central to an understanding of combustion [149,150]; for example, it is believed to be the major source of water production in high-temperature hydrogen-oxygen flames [151] and in hydrocarbon-air flames at atmospheric pressure [152]. Several experimental studies of the kinetics of this reaction and its isotopic analogues have been carried out. Brown and co-workers [151] determined a rate constant for reaction (R1) in low pressure hydrogen/oxygen flame studies, while Ravishankara *et al.* [153] measured thermal rates of reaction (R1) and its isotopic analogue,

$$OH + D_2 \rightarrow HOD + D , \tag{R2}$$

from 250 to 1050 K by a flash photolysis-resonance fluorescence method. Their results for reaction (R1) generally agree with earlier thermal rate measurements by Dixon-Lewis and Williams [154] over the range 300 to 1600 K, and are quite close to the recommended values of Cohen and Westberg [155]. In addition, the effects of vibrational excitation in either OH [156,157] or H_2 [158-160] on the rate of reaction (R1) have been measured. The latter has a much greater effect on enhancing the reaction rate than the former, although the effect of exciting H_2 from v = 0 to v = 1 at 298 K is not as large as when an equivalent amount of extra energy is put into the relative translation of the reactants [159]. These observations are consistent with a direct abstraction process (i.e., without energy randomization) [161] on a PES that has an early saddle point [162], with the OH bond acting as a spectator.

2.1. ORIGINAL PES STUDIES OF THE OH + H_2 REACTION

This view of reaction (R1) is supported by various early theoretical studies. Walch and Dunning [126] performed accurate extended basis set configuration-interaction (CI) calculations for the energies at a limited number of geometries in the reactant, product, and saddle point regions. As those authors point out, the saddle point barrier derived from these points (referred to herein as the WD data) is coplanar and is indeed in the entrance channel, with a classical height (i.e., without considering the zero point energy contributions) of 6.19 kcal/mol with respect to the reactants. Furthermore, the equilibrium properties of the reactants and products obtained from the WD data differ from experiment by ~2% in the distances, by $\leq$5% in the frequencies, and by ~13% in the dissociation energies, which are typical errors for this level of calculation [163]. Due to a cancellation of errors, though, their predicted reaction exoergicity differs from experiment by only 0.6 kcal/mol, suggesting that their calculated barrier height is reasonably accurate. The characteristics of the *ab initio* PES were further discussed by Dunning, Walch, and Wagner [164], who pointed out that there is reasonable agreement between the vibrationally adiabatic threshold at the saddle point computed under the harmonic approximation (5.9 kcal/mol) and the available experimental activation energies (4-6 kcal/mol). In addition, TST calculations [165] employing the *ab initio* properties at the saddle point and the reactants and the Wigner transmission coefficient lead to thermal rate constants for reaction (R1) that are in good agreement with those of Dixon-Lewis and Williams [154] over the range 300 - 1600 K.

The first analytic PES for reaction (R1) (herein called No. 1 or SE) was obtained by Schatz and Elgersma [127] by an approximate fit to the WD data. Those authors then used classical trajectory calculations to investigate the product vibrational state distributions for four different translational energies, and found that most of the energy available to the products goes into H_2O vibrations. Further classical trajectory calculations employed the same PES to examine the product rotational distributions [166] and the effects of reactant vibrational excitation on reactivity [167]; at 300 K, the latter study predicts that exciting H_2 to v = 1 enhances the rate constant of reaction (R1) by a factor of 390. Rashed and Brown [168] modified the SE surface to remove what they called "spurious wells" in the asymptotic reactant region, and used classical trajectory calculations with their new surface (herein called No. 2) to investigate how the reactivity of reaction (R1) changes with initial translational temperature between 300 and 4000 K and with vibrational or rotational excitation of each reactant. (However, the asymptotic reactant well may indeed exist; see below.) They found that rotational excitation of either reactant lowers the reactivity, that vibrational excitation of OH has no effect on reactivity, and that vibrational excitation of H_2 has little effect on reactivity at a translational temperature of 1200 K, but leads to a mild increase in reactivity at a translational temperature of 2000 K (e.g., the cross section increases by about 20% for excitation to v = 1). In contrast, calculations with the SE surface by Truhlar and Isaacson [161] using a statistical-diabatic version of variational transition state theory (in which the OH and H_2 vibrations are treated diabatically with fixed vibrational quantum numbers while the other degrees of freedom are treated adiabatically and thermally) show that exciting H_2 to v = 1 enhances the rate constant for reaction (R1) at 298 K by a factor of 27-108, depending on the treatment of the quantal effects on the reaction-path motion, but that exciting OH to v = 1 enhances the rate constant at 298 K by about 50%. These results are in much better agreement with the measured v = 1 enhancement factors at 298 K of 120±40 [159] and 155±38 [160] for H_2 and <1.6 for OH [156].

Isaacson and Truhlar [16] also used the SE surface to compute CVT rate constants and semiclassical adiabatic transmission coefficients for reactions (R1) and (R2). Their results show that locating the GTS at the variational bottleneck, which is $\leq$0.1 a_0 toward the reactants from the saddle point, decreases the rate constant for reaction (R1) by a factor of 1.9 at 298 K. This effect is mainly due to a stronger H_2 bond at the variational bottleneck than at the saddle point; specifically, at 300 K the frequency of the vibrational mode that is primarily the H_2 stretch is 738 cm^{-1} higher at the CVT bottleneck than at the saddle point. That work also demonstrates the importance of reaction-path tunneling and the large contribution from the curvature of the MEP for reaction (R1) below about 600 K. For example, at 298 K k$^{CVT/MEPSAG}$ is a factor of 4.6 larger than k^{CVT} [135], while k$^{CVT/SCSAG}$ is a factor of 21 larger than k^{CVT} [16], when the generalized bound normal modes are treated harmonically. However, that work additionally shows that incorporating the effects of anharmonicity within each of the generalized bound modes but neglecting mode-mode coupling (i.e., the INM approach) leads to significantly different rate constant predictions than the harmonic approximation; at 298 K, k$^{CVT/SCSAG}$ for reaction (R1) is a factor of 2.3 smaller when INM anharmonicity is included. Finally, while the anharmonic k$^{CVT/SCSAG}$ value for reaction (R1) computed from the SE surface at 298 K and 1000 K is a factor of 1.7 larger and 1.7 smaller than the corresponding experimental value [16,153], respectively, the calculated kinetic isotope effect (KIE), i.e., the ratio of the rate constant for reaction (R1) to that for reaction (R2), is larger than experiment [153] by a factor of 1.7 and 1.2 at 298 K and 1000 K, respectively. This quantitative disagreement is partly due to the fact that the SCSAG method overestimates the effects of reaction-path tunneling here (see below), but may also be due somewhat to deficiencies in the SE surface.

When the CVT bottleneck is a strong function of temperature, the variationally best dividing surface should be located for each total energy [42,43], and the resulting microcanonical rate constants thermally averaged to obtain the rate at a given temperature. For the SE surface, the CVT bottleneck varies quite little with temperature, never by more 0.04 a_0 from the maximum in the vibrationally adiabatic ground state potential curve [16]. Thus, it is not surprising that the much simpler CVT method predicts rate constants that are within 2% of the microcanonical variational transition state theory values over the range 200 to 2400 K [169].

2.2.   IMPROVED PESs BASED ON THE WD DATA

Although the SE surface was used for the various theoretical studies described above, the actual fit of this surface to the saddle point properties derived from the WD data [126] is not very accurate. In fact, harmonic TST rate constants for reaction (R1) computed from the *ab initio* saddle point properties differ somewhat from the corresponding values obtained with the SE surface. These differences, which range from 29% to 46% over the temperature range 200 to 4000 K, are primarily due to errors in the saddle point geometry for the SE surface [16]. To obtain a PES that more closely fits the WD data as well as one that yields rate constants that are more quantitatively consistent with experiment, we constructed two improved PESs [135] (called Nos. 3 and 4) for reaction (R1) by fitting to the *ab initio* reactant, product, and saddle point properties derived from the WD points. The saddle point and reactant properties were taken directly from [126] (except for the saddle point frequencies, which were taken from [165]), while the H_2O properties were obtained by a least-squares fit of the equilibrium geometry (R_e, θ_e) and the three force constants k_{RR}, $k_{R\theta}$, and $k_{\theta\theta}$ in the potential function

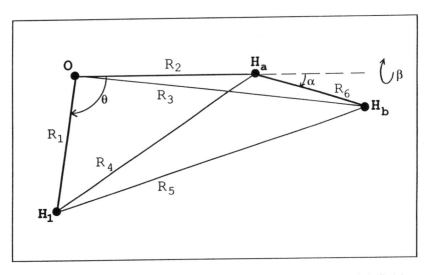

Figure 1. OH + H$_2$ saddle point geometry [126] (drawn to scale) and definitions of the internal coordinates (reproduced with permission from [176]).

$$V(R_1, R_2, \theta) = \tfrac{1}{2} k_{RR}[(R_1 - R_e)^2 + (R_2 - R_e)^2] + \tfrac{1}{2} k_{\theta\theta}(\theta - \theta_e)^2 +$$

$$\tfrac{1}{2} k_{R\theta}(\theta - \theta_e)[(R_1 - R_e) + (R_2 - R_e)] \qquad (8)$$

to the six H$_2$O energies given in Table V of [126], followed by a harmonic analysis. The resulting values are listed in the first column of data in Table 1; Figure 1 shows the saddle point geometry and defines the internal coordinates of the OH + H$_2$ system. (Although experimental as well as more accurate *ab initio* properties for the reactants and products are available [148,170,171], the present fits are based on the WD data for these quantities. In this way, the same basis set is used consistently for the data in all regions of the PES. We note that this is in contrast to the SE surface, which was fit to more accurate *ab initio* information on the H$_2$O properties [171].)

The idea behind these improved PESs is to use a simple form consisting of some three-body terms together with some additional bending contributions, but to allow the various parameters in these terms to vary with geometry along the reaction path. This approach allows one to specify a global form for the PES, while incorporating local changes [63,67,131-134]. In this way, even if the fit is not accurate for all nuclear configurations, one can probably obtain a reliable fit in a usefully large region (e.g., along the MEP). Thus, by analogy to the PES for the CH$_3$ + H$_2$ system [66,67,122], we chose the form of the PES for reaction (R1) to be [135]

$$V = V_3(R_1, R_3, R_5) + V_3(R_2, R_3, R_6) + V_{bend} + V_0. \qquad (9)$$

Table 1. Reactant, product, and saddle point properties of *ab initio* [126], SE [127], and improved [135] OH + H_2 potential energy surfaces [frequencies in cm^{-1}, distances in bohr, angles in degrees, and energies in kcal/mol relative to the reactants of reaction (R1)] (adapted from [135]).

Surface	*Ab Initio*	SE	No. 3	No. 4
		H_2		
R_e	1.428	1.429	1.428	1.428
ω_e	4277	4261	4277	4277
D_e	95.1	95.1	95.1	95.1
		OH		
R_e	1.864	1.863	1.864	1.864
ω_e	3637	3624	3637	3637
D_e	93.0	93.0	93.0	93.0
		H_2O		
R_e	1.8426	1.8081	1.8426	1.8426
θ_e	104.16	104.62	104.16	104.16
ω_1 (s str)	4032	3864	4032	4032
ω_2 (bend)	1612	1687	1612	1612
ω_3 (a str)	4020	3975	4020	4020
Endoergicity	-16.73	-15.20	-16.73	-16.73
		Saddle Point		
$R_1^{\ddagger}$	1.863	1.856	1.857	1.852
$R_2^{\ddagger}$	2.522	2.308	2.523	2.521
$R_6^{\ddagger}$	1.619	1.640	1.631	1.631
$\theta^{\ddagger}$	97.6	116.7	97.8	97.9
$\alpha^{\ddagger}$	15.0	-15.8	14.9	15.0
$\beta^{\ddagger}$	0.0	0.0	0.0	0.0
$\omega_1(a'')$ (β bend)	440	830	440	440
$\omega_2(a')$ (α bend)	686	572	686	686
$\omega_3(a')$ (θ bend)	1248	857	1248	1248
$\omega_4(a')$ (HH str)	1945	1921	1945	1945
$\omega_5(a')$ (OH str)	3368	3545	3368	3368
$\omega^{\ddagger}$	1655i	1526i	1655i	1655i
Energy	6.19	6.09	6.19	6.19

The first two terms in eq. (9) are three-body potentials obtained from a London equation [114,117]:

$$V_3(R_k, R_m, R_n) = Q_k + Q_m + Q_n - \{[(J_k - J_m)^2 + (J_m - J_n)^2 + (J_k - J_n)^2]/2\}^{1/2}, \quad (10)$$

where Q_i and J_i are the Coulomb and exchange integrals, respectively, for the pair of atoms that define R_i. These are approximated by

$$Q_i(R_i) = [^1E_i(R_i) + {}^3E_i(R_i)] / 2, \tag{11}$$

$$J_i(R_i) = [^1E_i(R_i) - {}^3E_i(R_i)] / 2, \tag{12}$$

where ^1E_i and ^3E_i are attractive and repulsive interactions, respectively, between the pair of atoms that define R_i. These interactions are in turn approximated by Morse and anti-Morse functions:

$$^1E_i(R_i) = D_i^{(1)}\{\exp[-2\alpha_i(R_i - R_i^e)] - 2\exp[-\alpha_i(R_i - R_i^e)]\}, \tag{13}$$

$$^3E_i(R_i) = D_i^{(3)}\{\exp[-2\alpha_i(R_i - R_i^e)] + 2\exp[-\alpha_i(R_i - R_i^e)]\}, \tag{14}$$

where $D_i^{(1)}$, $D_i^{(3)}$, α_i, and R_i^e are adjustable parameters. The V_{bend} term in eq. (9) is given by

$$V_{bend} = \tfrac{1}{2}k_\theta(\theta - \theta_0)^2 + \tfrac{1}{2}k_\alpha(\alpha - \alpha_0)^2 + \tfrac{1}{2}k_\beta\beta^2, \tag{15}$$

where θ_0, α_0, k_θ, k_α, and k_β are adjustable parameters. (Due to the limited nature of the *ab initio* data, the present surfaces contain simple harmonic terms for the three internal angles; for consistency, the dynamical calculations discussed below are based on the harmonic approximation for the bound vibrational degrees of freedom.) Finally, the constant term (V_0) is chosen so that the zero of energy corresponds to infinitely separated reactants each at its own equilibrium geometry:

$$V_0 = D_1^{(1)}(R_2 = \infty) + D_6^{(1)}. \tag{16}$$

2.2.1. *Variation of the PES parameters along the MEP*

Depending on the type and amount of information available for determining the adjustable parameters noted above, they were either chosen to be constant or to vary with geometry along the MEP. Thus, the H_2 Morse parameters ($D_i^{(1)}$, α_i, and R_i^e for i = 5, 6), the OH_b Morse parameter ($D_i^{(1)}$), and the OH and H_2 anti-Morse parameters ($D_1^{(3)} = D_2^{(3)}$ and $D_5^{(3)} = D_6^{(3)}$, respectively) are assumed to be constant, while the remaining parameters in eqs. (13) through (15) are given by functions of one or more of the internal coordinates. Specifically, the remaining OH parameters are given by

$$\alpha_i = a_1 + [b_1 - t_1(T - R_p)]\exp[-c_1(T - R_p)^2], \quad i = 1, 2, 3, \tag{17}$$

$$R_i^e = a_2 + b_2\{\tanh[c_2(T - R_p)] + 1\} / 2, \quad i = 1, 2, 3, \tag{18}$$

$$D_i^{(1)} = a_3 + b_3\exp[-c_3(T - R_p)^2], \quad i = 1, 2, \tag{19}$$

$$D_3^{(3)} = a_4 + b_4 (\exp\{-[4(R_2 - R_t)/c_4]^4\} - 1),\tag{20}$$

where $T = (R_1 + R_2)/2$ and R_p is the WD value for the OH distance in H_2O. In eqs. (17) through (19), using T as the independent variable ensures that the H_2O vibrations have the correct symmetries. Furthermore, the form of eq. (17) allows for a smooth transition from the WD stretching frequency in OH to those in H_2O, with a somewhat lower OH stretching frequency in the saddle point region, as suggested by the WD values in Table 1. Similarly, eqs. (18) and (19) allow for smooth transitions between the WD equilibrium distance and dissociation energy in OH to the equilibrium OH distance in H_2O and the exoergicity of reaction (R1), respectively, while eq. (20) allows the barrier along the MEP to be made narrower or broader.

The bending parameters in eq. (15) are given by

$$\theta_0 = a_\theta + b_\theta \{\tanh[c_\theta(T - R_p)] + 1\}/2,\tag{21}$$

$$\alpha_0 = \alpha^* \exp[-p_\alpha(R_3 - R^\ddagger)^2],\tag{22}$$

$$k_\theta = f_\theta \exp[-g_\theta(T - R_p)^2],\tag{23}$$

$$k_\alpha = f_\alpha \exp[-p_\alpha(R_3 - R^\ddagger)^2] S_\alpha,\tag{24}$$

$$k_\beta = f_\beta \exp[-p_\beta(R_3 - R_\beta)^2] S_\beta,\tag{25}$$

where $R^\ddagger$ is the WD value of R_3 at the saddle point. The form of eq. (21) allows for a smooth transition from the WD HOH angle in the reactants to that in the products, while the form of eq. (23) requires k_θ to be a maximum at products and go to zero at reactants, as expected for this system. Similarly, eqs. (22), (24), and (25) require that α_0, k_α, and k_β go to zero at both reactants and products, but maximize in the interaction region, as expected. The switching functions S_α and S_β are included to force the lowest-frequency bending vibrations to behave in a reasonable manner along the MEP; specifically, they prevent the corresponding frequencies from becoming large but imaginary on the reactant side of the saddle point. They are given by

$$S_\alpha = \begin{cases} 1, & R_2 \geq R' \\ w \exp[-y(R_2 - R')^2] + (1 - w), & R_2 < R' \end{cases},\tag{26}$$

$$S_\beta = \begin{cases} 1, & R_2 \geq R* \\ h \exp[-z(R_2 - R*)^2] + (1 - h), & R_2 < R* \end{cases},\tag{27}$$

Eqs. (21) through (27) thus ensure that the bending frequencies change smoothly from the reactant limits to the product limits, with realistic values in the saddle point region; such behavior is important for obtaining physically reasonable semiclassical CVT rate constants, such as those discussed below.

2.2.2. *Determination of the fit constants*

To complete the specification of the improved PESs for reaction (R1), we now discuss the strategy used to determine the thirty-seven independent constants (not counting R_p and $R^{\ddagger}$) referred to in the preceding subsection, which are listed in Table 2. For H_2 (i = 5, 6), the $D_i^{(1)}$ and R_i^e Morse parameters were set to the WD $D_e(H_2)$ and $R_e(H_2)$ values, while α_i was obtained from the WD H_2 frequency ω_e:

$$\alpha_i = 2\pi c \omega_e (\mu / 2D_e)^{1/2}, \tag{28}$$

where c is the speed of light, μ is the reduced mass of the diatomic molecule, and ω_e is in cm^{-1}. For OH, the value of a_1 in eq. (17) was obtained from the WD OH frequency by eq. (28), a_2, b_2, and c_2 in eq. (18) were obtained from the WD OH distances in the reactants $[R_e(OH)]$, in the products (R_p), and at the saddle point $(R_1^{\ddagger}$ and $R_2^{\ddagger})$ by the formulas

$$a_2 = 2R_p - R_e(OH), \qquad b_2 = 2[R_e(OH) - R_p], \tag{29}$$

$$c_2 = \ln[(1+u)/(1-u)]/(R_1^{\ddagger} + R_2^{\ddagger} - 2R_p), \tag{30}$$

$$u = [2(R_1^{\ddagger} - a_2)/b_2] - 1, \tag{31}$$

a_3 in eq. (19) was set to the WD value for $D_e(OH)$, and b_3 in eq. (19) was determined from the WD exoergicity (ΔE_{Rx}) of reaction (R1):

$$b_3 = \tfrac{1}{2}[D_e(OH) + D_e(H_2) - \Delta E_{Rx}] - a_3, \tag{32}$$

where we have used the fact that at products, eq. (9) reduces to $-2D_1^{(1)}$. For θ_0, the values of a_θ and b_θ in eq. (21) were obtained from the WD value of θ in H_2O (θ_e) and at reactants $(\theta_r = 90°)$ [126],

$$a_\theta = 2\theta_e - \theta_r, \qquad b_\theta = 2(\theta_r - \theta_e), \tag{33}$$

while c_θ was determined by eq. (30) with

$$u = [2(\theta^{\ddagger} - a_\theta)/b_\theta] - 1, \tag{34}$$

where $\theta^{\ddagger}$ is the WD value of θ at the saddle point. Next, the values of b_1, c_3, and f_θ were chosen by trial and error such that eq. (9) yields the WD frequencies for H_2O at products.

The remaining parameters were determined by the following approach: (1) The values of b_4, c_4, R_t, p_α, w, y, and R' were adjusted as a group by trial and error so that the lowest a' frequency and the classical energy, $V_{MEP}(s)$, behave as reasonably as possible along the MEP and to satisfy an additional criterion. The additional criterion for surface 3 was that $V_{MEP}(s)$ fit the shape of the energy barrier on the SE surface as closely as possible for positive energies with respect to the reactants, while for surface 4 it was that harmonic

Table 2. Fit constants for the improved surfaces in atomic units, except angles in radians (reproduced with permission from [135]).

Surface	No. 3	No. 4
$R_5^e = R_6^e$	1.428	1.428
$\alpha_5 = \alpha_6$	1.07280	1.07280
$D_5^{(1)} = D_6^{(1)}$	0.151548	0.151548
$D_3^{(1)}$	0.102083017	0.064207223
$D_1^{(3)} = D_2^{(3)}$	0.073339295	0.077891618
$D_5^{(3)} = D_6^{(3)}$	0.045852978	0.050227465
a_1	1.26537	1.26537
b_1	0.05793	0.05793
c_1	14.4987057	15.3588191
t_1	3.67379268	4.06152210
R_p	1.84263	1.84263
a_2	1.82126	1.82126
b_2	0.04274	0.04274
c_2	5.33263769	5.33263769
a_3	0.148201	0.148201
b_3	0.015003	0.015003
c_3	4.505	4.505
a_4	0.058929417	0.046381036
b_4	0.053625769	0.039423881
c_4	3.450	3.450
R_t	2.22	2.14
a_θ	2.0649137	2.0649137
b_θ	-0.4941174	-0.4941174
c_θ	1.43268829	1.43268829
f_θ	0.1587	0.1587
g_θ	3.50543837	3.58677818
α^*	1.32170694	0.838779182
f_α	0.018000611	0.028826226
p_α	0.20	0.20
$R^\ddagger$	4.10726472	4.10726472
w	0.7709	0.7709
y	40.0	40.0
R'	3.50	3.45
f_β	0.0045	0.004831638
p_β	0.10	0.40
R_β	4.107265	4.107265
h	1.00	1.00
z	5.46256	5.46246
R^*	3.20	3.20

$k^{CVT/SCSAG}$ values for reaction (R1) (see below) agree with experiment [153]. The rationale for the additional criterion for surface 3 is that more extensive high-quality *ab initio* calculations [29,172] along the reaction path for reaction (R1) yield a barrier shape that is quite close to that of the SE surface at positive energies [173], and, in addition, exhibit a reactant-side well that is even deeper than that on the SE surface, in contrast to surface 2 [168]. (2) For each choice of the seven constants in (1), the values of the five independent constants $D_3^{(1)}$, $D_1^{(3)} = D_2^{(3)}$, $D_5^{(3)} = D_6^{(3)}$, a_4, and α^* were adjusted by a weighted partial Newton search to reproduce the WD saddle point geometry, barrier height, and imaginary frequency as closely as possible. Specifically, a Newton search for simultaneous zeros for the relative errors in $V^{\ddagger}$, $\omega^{\ddagger}$, $\alpha^{\ddagger}$, $\theta^{\ddagger}$, and $R_2^{\ddagger}$ weighted by factors of 50, 50, 10, 1, and 10, respectively, was run for several (typically 10) iterations. (3) For each choice of the five constants in (2), the values of the five constants c_1, t_1, g_θ, f_α, and z were determined by a converged Newton search so that the WD frequencies for the five bound vibrational modes at the saddle point are reproduced. (4) As a final independent step, the values of the five constants f_β, p_β, R_β, h, and R^* were chosen as a group by trial and error to ensure that the a'' frequency behaves as reasonably as possible along the reaction path. (The value of z was then readjusted to yield the WD value for the a'' frequency at the saddle point.)

Table 1 shows that the surfaces obtained by this procedure have reactant, product, and saddle point properties that are in very good agreement with the WD data. In particular, both surfaces 3 and 4 provide a much better fit to the WD data at the saddle point than does the SE surface. This demonstrates that varying the surface parameters with geometry provides a relatively simple and efficient method for obtaining an accurate fit of a PES to a set of *ab initio* data. Further comparisons between surfaces 3 and 4 and the SE surface can be seen in Figure 2, which shows the classical potential energy curves [$V_{MEP}(s)$] and the ground-state vibrationally adiabatic potential energy curves [$V_a^G(s)$] along the corresponding reaction paths. As noted above, the constants for surface 3 were chosen so that the positive energy portion of $V_{MEP}(s)$ would be as close as possible to that of the SE surface, and Figure 2 shows that the fit is quite good. The well in $V_{MEP}(s)$ on the reactant side of the saddle point is even deeper for surface 3 (1.6 kcal/mol) than for the SE surface (1.3 kcal/mol), in better agreement with the *ab initio* results [29,173]. In contrast, the $V_{MEP}(s)$ and $V_a^G(s)$ curves for surface 4, which was constrained to yield harmonic CVT/SCSAG rate constants for reaction (R1) that agree with experiment, are much broader on the reactant side of the saddle point than those for surface 3, and there is no reactant-side minimum in $V_{MEP}(s)$.

2.2.3. *Rate constants for the improved PESs*

Converged harmonic rate constants for reaction (R1) computed from the SE surface, surface 3, and surface 4 with various levels of approximation [135] are given in Table 3. These results were obtained from the POLYRATE program [75-77]; the MEP was determined by the Euler single-step method with step sizes between gradient and hessian calculations of 0.0001 a_0 and 0.01 a_0, respectively. [All coordinates are scaled to a reduced mass of $m_{OH}m_{H_2}/m_{OH_3}$, and spin-orbit coupling in OH($^2\Pi$) is included. In addition, the results for surface 3 at 200 K that include tunneling are only converged within 5%, as they depend on the range parameter for the exponential extrapolation [16] from the reactant-side local minimum in $V_{MEP}(s)$ to the reactants.] Several aspects of these results deserve

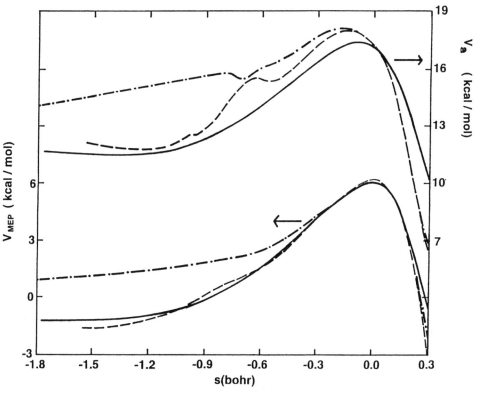

Figure 2. Classical potential energies, V_{MEP} (left scale), and ground-state vibrationally adiabatic potential energies, V_a^G (right scale), as functions of the corresponding reaction coordinate s for the SE surface (solid curves), surface 3 (dashed curves), and surface 4 (dot-dashed curves) (reproduced with permission from [135]).

comment. First, conventional transition state theory (TST) rate constants computed from surfaces 3 and 4 are very close to those using the *ab initio* data [16], while those obtained from the SE surface are somewhat lower, mainly due to the errors in its saddle point geometry [16]. Second, optimizing the location of the transition state results in a substantial decrease in the computed rate constants (CVT vs. TST), especially at lower temperatures. This effect increases on going from the SE surface to surface 3, and from surface 3 to surface 4, reflecting a more pronounced tightening of the H_2 bond (and, for surfaces 3 and 4, of the OH bond as well) between the saddle point and the CVT bottleneck on going from the SE surface to surface 3 and to surface 4 [135]. Third, the inclusion of reaction-coordinate tunneling along the MEP markedly increases the computed rate constants (CVT/MEPSAG vs. CVT) for T ≤ 600 K for all three surfaces; the similarity of the $V_a^G(s)$ curves for the SE surface and surface 3 (see Figure 2) leads to similar effects in this approximation, while the much broader $V_a^G(s)$ curve for surface 4 yields less pronounced increases. Fourth, incorporating the curvature of the reaction path in the

Table 3. Calculated and experimental rate constants for reaction (R1).[a]

T	TST	CVT	CVT/ MEPSAG	CVT/ CD-SCSAG	CVT/ SCSAG	CVT/ LCG3	Expt.[b]
200 K	2.08(-17)	9.98(-18)	4.91(-16)	2.80(-15)	6.87(-15)	1.36(-15)	---
	2.80(-17)	3.80(-18)	1.72(-16)	3.92(-15)	3.13(-14)	1.36(-15)	
	2.79(-17)	2.70(-18)	2.98(-17)	7.82(-17)	3.32(-16)	4.24(-17)	
298 K	1.76(-15)	1.11(-15)	5.10(-15)	1.38(-14)	2.38(-14)	7.21(-15)	6.08±0.37(-15)
	2.45(-15)	6.60(-16)	3.08(-15)	1.35(-14)	6.06(-14)	5.79(-15)	
	2.44(-15)	5.17(-16)	1.54(-15)	2.61(-15)	6.25(-15)	1.82(-15)	
400 K	1.76(-14)	1.30(-14)	2.86(-14)	5.15(-14)	7.26(-14)	3.32(-14)	3.7±1.1(-14)
	2.52(-14)	9.87(-15)	2.25(-14)	4.79(-14)	1.27(-13)	2.84(-14)	
	2.51(-14)	8.10(-15)	1.48(-14)	2.04(-14)	3.56(-14)	1.63(-14)	
600 K	1.82(-13)	1.57(-13)	2.16(-13)	2.83(-13)	3.35(-13)	2.28(-13)	2.9±0.9(-13)
	2.67(-13)	1.52(-13)	2.16(-13)	2.94(-13)	4.62(-13)	2.33(-13)	
	2.66(-13)	1.30(-13)	1.70(-13)	1.97(-13)	2.61(-13)	1.77(-13)	
1000 K	1.58(-12)	1.50(-12)	1.62(-12)	1.79(-12)	1.91(-12)	1.65(-12)	2.4±0.7(-12)
	2.32(-12)	1.75(-12)	1.94(-12)	2.16(-12)	2.54(-12)	1.99(-12)	
	2.31(-12)	1.56(-12)	1.70(-12)	1.80(-12)	2.00(-12)	1.73(-12)	
1500 K	6.40(-12)	6.26(-12)	6.30(-12)	6.59(-12)	6.78(-12)	6.34(-12)	---
	9.36(-12)	7.81(-12)	8.01(-12)	8.39(-12)	9.01(-12)	8.09(-12)	
	9.32(-12)	7.11(-12)	7.33(-12)	7.52(-12)	7.89(-12)	7.38(-12)	
2400 K	2.69(-11)	2.66(-11)	2.62(-11)	2.66(-11)	2.69(-11)	2.62(-11)	---
	3.92(-11)	3.42(-11)	3.40(-11)	3.47(-11)	3.56(-11)	3.42(-11)	
	3.91(-11)	3.14(-11)	3.16(-11)	3.19(-11)	3.26(-11)	3.17(-11)	
4000 K	1.09(-10)	1.08(-10)	1.06(-10)	1.07(-10)	1.08(-10)	1.07(-10)	---
	1.59(-10)	1.40(-10)	1.39(-10)	1.40(-10)	1.42(-10)	1.40(-10)	
	1.59(-10)	1.28(-10)	1.28(-10)	1.29(-10)	1.30(-10)	1.28(-10)	

[a]Units are cm^3 molecule^{-1} s^{-1}; power of 10 is given in parentheses. For each temperature, upper entry is for the SE surface, middle entry is for surface 3, and lower entry is for surface 4. [b]From [153]; for T = 298 K we quote the actual tabulated value and error bar, while for T > 298 K we employ those authors' three-parameter fit with 30% error bars.

tunneling calculation also substantially increases the computed rate constants (CVT/CD-SCSAG vs. CVT/MEPSAG) for T ≤ 600 K for all three surfaces, and the effects increase on going from surface 4 to the SE surface and to surface 3. Comparing the CVT/CD-SCSAG results of surface 4 to those of surface 3 and the SE surface shows that rate constants obtained from PESs fit to the same *ab initio* data (including the same barrier height) can still be profoundly different for low to intermediate temperatures, depending on the variational effect, the width of the barrier, and the effects of including reaction-path curvature. On the other hand, the similarity of the surface 3 and SE surface CVT/CD-SCSAG rate constants results from a cancellation of the larger variational effect and the larger effect of including reaction-path curvature in surface 3. Fifth, transmission coefficients computed in the large-curvature approximation (LCG3) are significantly lower than those computed in the small-curvature approximation (CD-SCSAG) for T ≤ 600 K. (The LCG3 calculation was performed with tunneling into all accessible product states. The excited-state contributions made no larger than 3%, 7%, and 13% differences in the SE surface, surface 3, and surface 4 results, respectively.) Thus, the small-curvature approximation provides a more accurate treatment of quantal effects on reaction-coordinate motion for the intermediate degree of curvature exhibited by reaction (R1).

Table 3 also shows that the CVT/SCSAG method, which leads to unrealistic results when curvature is non-negligible in several generalized normal modes [77,174], substantially overestimates the effects of including reaction-path curvature for these PESs. Since the width of the barrier and the curvature along the MEP for surface 4 were determined by fitting the CVT/SCSAG rate constants obtained from it to experiment, one or both of these may be incorrect. Indeed, since multiplying $k^{CVT}(298)$ for surface 4 by $\kappa^{CVT/MEPSAG}(298)$ for surface 3 and by the ratio of $\kappa^{CVT/CD\text{-}SCSAG}(298)$ to $\kappa^{CVT/MEPSAG}(298)$ for surface 4 still yields a rate constant that is less than the experimental value, both the true barrier width and the effect of including reaction-path curvature probably lie between those for surfaces 3 and 4. However, the fact that the CVT/MEPSAG rate constants for surface 3 and the SE surface are below the experimental ones, while the corresponding CVT/CD-SCSAG rate constants are higher than experiment for T ≤ 400 K, indicates that the effects of including reaction-path curvature are too large for these surfaces. Further indication that these effects are too large for surface 3 and the SE surface is presented in Table 4, which compares calculated and experimental values of the kinetic isotope effect, i.e., the ratio of the rate constant for reaction (R1) to that for reaction (R2). Although the TST values for all three surfaces are higher than experiment over the whole temperature range, both the CVT/MEPSAG and CVT/CD-SCSAG KIEs for surface 4 are in good agreement with experiment over this range. For surface 3, however, the CVT/MEPSAG KIEs are slightly higher than the upper ends of the experimental error bars at these temperatures, while the corresponding CVT/CD-SCSAG KIEs are much too large for T ≤ 600 K.

2.3. A NEW PES FOR THE OH + H₂ REACTION

To obtain a more reliable first-principles PES for reaction (R1) that also describes the anharmonic nature of the bound vibrational degrees of freedom transverse to the MEP, we recently began the development of a new PES based on the extensive high-quality *ab initio* data of Kraka and Dunning (KD) [29,172,175], which consist of several large grids of energy values at different internal coordinate geometries both around the saddle point and

Table 4. Calculated and experimental kinetic isotope effects.[a]

T	TST	CVT	CVT/ MEPSAG	CVT/ CD-SCSAG	Expt.[b]
200 K	5.53	3.42	17.2	21.5	
	6.52	2.61	12.7	41.2	----
	6.53	2.19	4.85	5.01	
298 K	4.09	2.98	5.72	7.62	
	4.40	2.40	4.63	9.96	3.3±0.9
	4.40	2.13	3.24	3.48	
400 K	3.34	2.69	3.78	4.59	
	3.47	2.24	3.22	4.64	2.5±0.7
	3.47	2.04	2.60	2.76	
600 K	2.58	2.31	2.66	2.93	
	2.62	2.02	2.38	2.74	1.7±0.5
	2.62	1.90	2.12	2.19	
1000 K	1.95	1.88	1.94	2.01	
	1.97	1.78	1.87	1.96	1.6±0.5
	1.97	1.70	1.77	1.79	
1500 K	1.66	1.64	1.63	1.66	
	1.68	1.61	1.62	1.65	----
	1.68	1.57	1.58	1.59	
2400 K	1.49	1.48	1.46	1.47	
	1.50	1.49	1.47	1.48	----
	1.50	1.47	1.47	1.47	
4000 K	1.41	1.40	1.39	1.39	
	1.42	1.43	1.42	1.42	----
	1.42	1.42	1.42	1.42	

[a]Ratio of the rate constant for reaction (R1) to that for reaction (R2). For each temperature, upper entry is for the SE surface, middle entry is for surface 3, and lower entry is for surface 4. [b]From Table V of [153].

along the reaction path. These energy calculations employed a polarized valence double zeta basis set and included all single and double excitations from the GVB wave function. As the first stage in the development of this new anharmonic global PES, we recently described the construction and investigation of surface 5SP [176], which is a fit of the

Table 5. Properties of the fits to the KD saddle point data (obtained with a precursor to the POLYRATE program [75-77]; frequencies in cm^{-1}, distances in bohr, and angles in degrees) (adapted from [176]).

	148-term	20-term	5SP
$R_1^{\ddagger}$	1.861	1.861	1.861
$R_2^{\ddagger}$	2.497	2.491	2.483
$R_6^{\ddagger}$	1.606	1.611	1.611
$\theta^{\ddagger}$	97.1	96.9	97.4
$\alpha^{\ddagger}$	17.2	16.1	16.5
$\beta^{\ddagger}$	0.0	0.0	0.0
$\omega_1(a'')$ (β bend)	620	684	624
$\omega_2(a')$ (α bend)	645	634	608
$\omega_3(a')$ (θ bend)	1114	1093	1137
$\omega_4(a')$ (HH str)	2399	2399	2396
$\omega_5(a')$ (OH str)	3721	3705	3706
$\omega^{\ddagger}$	1410i	1448i	1410i

same analytic form used for surfaces 3 and 4 [given by eqs. (9) through (15) above] to the grid of 148 KD data points in the saddle point region. The properties of the saddle point for this data set, which served as a guide to fitting surface 5SP (see below), were estimated by two preliminary fits around the saddle point geometry. We first represented the PES by the same 148-term (131 unique term) polynomial that had been used in the original analysis of this data set [29,175]. The terms in this polynomial contain up to the fourth power of the Simons-Parr-Finlan (SPF) [177] coordinates $(R_i-R_i^0)/R_i$ for $i = 1, 2,$ and 6, up to the fourth power of the angle displacements $(\theta-\theta^0)$ and $(\alpha-\alpha^0)$, and cos $n\beta$ for $n = 0, 1,$ and 2. For a particular guess $(R_1^0, R_2^0, R_6^0, \theta^0, \alpha^0, \beta^0 = 0)$ for the saddle point position, the polynomial coefficients were determined by a linear least-squares fit of the KD data with the SURVIB program [178]. Although the resulting polynomial leads to a reasonable saddle point geometry and set of normal mode frequencies (see Table 5), we were unable to converge on the saddle point position by repeatedly re-expanding the polynomial around the position predicted by the fit, suggesting that this representation of the PES is too oscillatory. We next used the SAS package [179] to perform a nonlinear simultaneous least-squares fit to the KD data for both the saddle point position, $(R_1^{\ddagger}, R_2^{\ddagger}, R_6^{\ddagger}, \theta^{\ddagger}, \alpha^{\ddagger}, \beta^{\ddagger} = 0)$, and the coefficients of a low-order polynomial in SPF coordinates $S_i = (R_i-R_i^{\ddagger})/R_i$ for $i = 1, 2,$ and 6, angle displacements $\phi_1 = (\theta-\theta^{\ddagger})$ and $\phi_2 = (\alpha-\alpha^{\ddagger})$, and cos β that is expanded around the saddle point. By examining the statistical uncertainties in the polynomial coefficients, the least-squares residual, and the overall behavior of the fit for a wide variety of trial terms, we obtained the following set of only 20 terms that each made an important contribution to the fit and whose coefficient could be determined from the KD data set with a reasonable degree of certainty:

$$V_{20} = c_1 + c_2 S_1^2 + c_3 S_2^2 + c_4 S_6^2 + c_5 S_1 S_2 + c_6 S_2 S_6 + c_7 S_1^3 + c_8 S_2^3 + c_9 S_6^3$$

$$+ c_{10} \phi_1^2 + c_{11} \phi_2^2 + c_{12} \phi_1 \phi_2 + c_{13} \phi_1^3 + c_{14} \phi_2^3 + c_{15} S_1 \phi_1 + c_{16} S_2 \phi_1$$

$$+ c_{17} S_2 \phi_2 + c_{18} S_6 \phi_1 + c_{19} S_6 \phi_2 + c_{20} \cos \beta. \tag{35}$$

This 20-term function also gives reasonable saddle point properties (see Table 5) that are quite similar to those for the 148-term function; we conclude that these sets of saddle point properties are consistent with the KD *ab initio* data, so that surface 5SP should be chosen to reproduce these properties as closely as possible.

For surface 5SP, the adjustable parameters in eqs. (9) through (15) are taken to be constant, and were chosen by a nonlinear least-squares fit of V in eq. (9) to the KD saddle point grid. Such an approach is similar to methods employed by Hase and co-workers [180-182]. [The constant term, V_0, in eq. (9) was also treated as an adjustable parameter in these fits.] Perhaps because the adjustable parameters do not vary with geometry, we found that the fitting procedure worked best when we restricted the point grid to the 112 KD points closest to the saddle point and employed the algorithm for weighting the points used in the SURVIB program [178]. Also, in order to obtain a PES having saddle point properties consistent with those of the preliminary fits described above, the values of some of the parameters were not varied during the least-squares fitting. Specifically, the $D_i^{(1)}$ values for i = 1, 2, 5, and 6 were set to their values at the surface 4 saddle point [135], and values for the fixed parameters $D_3^{(1)}$ and $D_i^{(3)}$ for i = 1, 2, and 3 that yield a surface having saddle point properties that best agree with those of the preliminary fits were then determined by trial and error. In addition, the k_β parameter was fixed at a value that provides an out-of-plane bending frequency consistent with those of the preliminary fits. The final fit parameters are listed in Table 6, and the resulting saddle point properties are given in Table 5. The excellent agreement between these properties and those of the preliminary fits demonstrates that the form of surface 5SP is sufficiently flexible to represent the information contained in the KD saddle point grid accurately.

Table 6. Fit constants for surface 5SP (all values in atomic units, except angles in radians) (reproduced with permission from [176]).

i	R_i^e	α_i	$D_i^{(1)}$	$D_i^{(3)}$
1,2	1.860894	1.253446	0.157007	0.089000
3	1.860894	1.253446	0.064000	0.039000
5,6	1.454002	1.099716	0.151548	0.053224

$k_\theta = 0.087398$ $\qquad$ $k_\alpha = 0.0061822$ $\qquad$ $k_\beta = 0.00087948$
$\theta_0 = 1.692656$ $\qquad$ $\alpha_0 = 0.750832$ $\qquad$ $V_0 = -76.434515$

2.3.1. *Anharmonic energy levels and partition functions for surface 5SP*

To study the effects of incorporating the anharmonic nature of the generalized normal modes transverse to the MEP on the vibrational partition function factor, $Q_{vib}^{GT}(T,s)$, in the generalized transition state partition function, $Q^{GT}(T,s)$, in eq. (4), we computed $Q_{vib}^{GT}(T,s)$ at the saddle point of surface 5SP from sets of either harmonic or anharmonic bound vibrational energy levels $E_{v_1 v_2 v_3 v_4 v_5}/hc$ (in wave numbers) [176], where $v_1,...,v_5$ are the vibrational quantum numbers and the energy is measured relative to the saddle point (i.e., from the bottom of the vibrational well). That is, we take

$$Q_{vib}^{GT}(T,s=0) = \sum_{v_1 v_2 v_3 v_4 v_5} \exp(-E_{v_1 v_2 v_3 v_4 v_5} / kT). \qquad (36)$$

For harmonic vibrational energy levels, eq. (36) reduces to the well-known closed-form result, which we write as

$$Q_{vib}^{GT}(T,s=0) = \exp(-hc\varepsilon_0 / kT)\prod_{i=1}^{5}[1 - \exp(-hc\Delta_i / kT)]^{-1}, \qquad (37)$$

where ε_0 is the ground-state energy and Δ_i is the lowest excitation energy of mode i (both in wave numbers); for harmonic levels, these are given by $\frac{1}{2}\Sigma_{i=1}^{5}\omega_i$ and ω_i, respectively. For anharmonic vibrational energy levels, we evaluated eq. (36) in two ways. One way is by direct summation over second order perturbation theory (PT) vibrational energy levels; for a system having no degenerate vibrations, these are given by [107,111]

$$E_{v_1 v_2 v_3 v_4 v_5} / hc = E_0 + \sum_i \omega_i(v_i + \tfrac{1}{2}) + \sum_i \sum_{j\geq i} x_{ij}(v_i + \tfrac{1}{2})(v_j + \tfrac{1}{2}), \qquad (38)$$

where the constant anharmonic term E_0 and the anharmonicity coefficients x_{ij} are expressed in terms of the third- and fourth-order dimensionless normal coordinate force constants k_{ijk}, k_{iiii}, and k_{iijj} by formulas summarized elsewhere [100,106]. In our calculations, these force constants were obtained from numerical second and third derivatives of the analytic gradient of the PES along the appropriate normal-mode directions, Coriolis contributions were included in the anharmonicity coefficients [106] and in E_0 [176], and resonant iij- and ijk-type interactions were removed from the perturbation theory treatment as recommended elsewhere [100,176]. The other way we approximated the anharmonic vibrational partition function is by simple perturbation theory (SPT) [106], in which one replaces ε_0 and Δ_i in eq. (37) by the PT ground-state energy and fundamental excitation energy of mode i, respectively, both of which are calculated from eq. (38). Thus, SPT avoids the direct multidimensional summation in the PT approach, and is much more practical.

Table 7 compares harmonic vibrational partition functions at the saddle point of surface 5SP for several temperatures between 200 K and 2400 K with those obtained from SPT and direct summation over the PT vibrational energy levels. For these last values, enough terms were included in the summation to achieve three-digit convergence; at 2400 K, we employed maximum quantum numbers of 38, 38, 36, 36, and 34 for v_1, v_2, v_3, v_4, and v_5, respectively. In addition, whenever a PT energy level predicted by eq. (38) decreased for

Table 7. Vibrational partition functions at the saddle point of surface 5SP (power of 10 in parentheses) (adapted from [176]).

T	Harmonic	PT	SPT
200 K	6.00(-14)	1.12(-13)	1.12(-13)
298 K	1.47(-9)	2.45(-9)	2.45(-9)
400 K	3.10(-7)	4.92(-7)	4.90(-7)
600 K	7.01(-5)	1.06(-4)	1.06(-4)
1000 K	8.43(-3)	1.24(-2)	1.28(-2)
1500 K	1.49(-1)	2.18(-1)	2.36(-1)
2000 K	8.69(-1)	1.28(0)	1.43(0)
2400 K	2.47(0)	3.70(0)	4.18(0)

increasing vibrational quantum number, it was replaced by the corresponding SPT energy level, which is given by $\varepsilon_0 + \Sigma_{i=1}^5 v_i \Delta_i$. This replacement serves as a compromise between an unphysically low PT value and an inaccurately high harmonic value. Table 7 shows that harmonic and anharmonic vibrational partition functions at the saddle point of surface 5SP are significantly different; the PT value of the partition function is 1.9 times larger than the harmonic value at 200 K and 1.5 times larger at 2400 K. At low temperature, this difference is mostly due to the fact that the PT ground-state energy is 75 cm^{-1} lower than the harmonic one, while at high temperature this difference is due predominantly to the differences in excited state energies. The PT and SPT values for the partition function are nearly identical below 1000 K, while at higher temperatures SPT leads to a slightly larger partition function (e.g., 13% larger at 2400 K). Thus, SPT yields partition function values that agree quite well with those obtained by direct summation over the PT energy levels, but without the effort required to obtain a converged result. Since SPT can also avoid the larger errors in the more highly excited PT energy levels [176], we recommend it in general for CVT applications that incorporate vibrational anharmonicity.

To obtain a global PES for the region around the MEP of reaction (R1), we are currently fitting a function similar to that given by eqs. (9) through (16) to several grids of *ab initio* points along the reaction path [175] by allowing the parameters of the function to vary with geometry in a similar manner as in surface 4 (see above). This surface will also be required to reproduce reactant and product properties that have been obtained from *ab initio* calculations that are consistent with those in the saddle point region, i.e., that employ the same basis set and list of configurations. Such a surface will then be used to compute first-principles anharmonic semiclassical CVT rate constants for reactions (R1) and (R2), which can then be compared with experiment.

3. Direct Dynamics Calculation of Proton Transfer

As discussed in Section 1, a useful approach for calculating thermal rate constants that does not require the intermediate step of fitting a PES is direct dynamics, in which one computes the potential energy information needed for a CVT calculation directly from an electronic structure program. As an example of an application of this approach in which *ab initio* electronic structure information was computed in advance at a set of points along the MEP and then used as input data for the dynamics calculation, this section summarizes a direct-dynamics investigation of the proton transfer step in the symmetric reaction $CH_4 + CH_3^- \rightarrow CH_3^- + CH_4$ [147]. That is, we consider the unimolecular proton transfer process that occurs in going from the reactant-side hydrogen-bonded complex shown in Figure 3 to the corresponding product-side complex.

Due to its simplicity and the importance of proton transfers in reactions of chemical and biological relevance, Latajka and Scheiner calculated optimized energies and geometries for the reactants, complexes, and saddle point of this reaction at the MP3 level with a 6-31G* basis set augmented by a set of diffuse p functions on each carbon and hydrogen [183]. Their calculations gave an energy barrier to proton transfer of 13 to 15 kcal/mol, depending on the exponent of the p functions on the hydrogens. Those authors then used the calculated properties of the H-bonded complex and the saddle point to compute proton transfer rate constants at various temperatures using microcanonical transition state theory; tunneling was incorporated at each energy below the adiabatic barrier by replacing the density of states with a transmission coefficient that was evaluated by approximating the adiabatic barrier with an Eckart function [184]. The resulting rate constants show a strong dependence on tunneling below 400 K, as might be expected for such a high classical energy barrier. Thus, a reliable treatment of the dynamics for this system clearly requires a knowledge of the properties along the MEP between the complex and the saddle point, rather than just at the two stationary points.

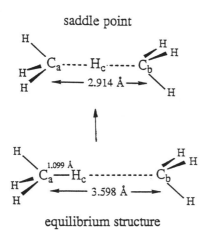

Figure 3. Optimized 4-31G geometries for the reactant H-bonded complex and saddle point structure in the $CH_4 + CH_3^-$ system (reproduced with permission from [147]).

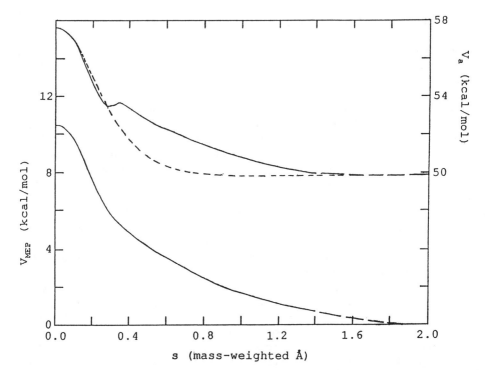

Figure 4. Classical potential energy, V_{MEP} (lower curve and left scale), and ground-state vibrationally adiabatic potential energy, V_a^G (upper solid curve and right scale), as functions of the reaction coordinate s along the MEP for the $CH_4 + CH_3^-$ system. The long-dashed portions are extrapolations (see text). The short-dashed curve is an Eckart potential fit to the adiabatic barrier (reproduced with permission from [147]).

To this end, the required *ab initio* PES information was computed at the Hartree-Fock level with a 4-31G basis set [185] using the GAUSSIAN-88 suite of programs [186]. First, the saddle point and complex structures and energies were determined. Optimized geometries are shown in Figure 3. Note that the C-C distance is 0.684 Å shorter at the saddle point than in the complex, while the C_a-H_c distance is 0.358 Å longer at the saddle point than in the complex, with H_c located at the midpoint of the C-C axis. The SCF/4-31G barrier height (10.6 kcal/mol) is only somewhat smaller than the range of MP3 values [183]. Thus, we expect the PES information obtained with a 4-31G basis set at the SCF level to give qualitatively correct dynamical results for the proton transfer process in this system.

3.1. PROTON TRANSFER REACTION PATH AND RATE CONSTANTS

With the origin defined as the center of mass of the saddle point geometry, the MEP from the saddle point to the product-side complex was determined by the Euler single-step method with step sizes between gradient and hessian calculations of 0.005 Å and 0.025 Å,

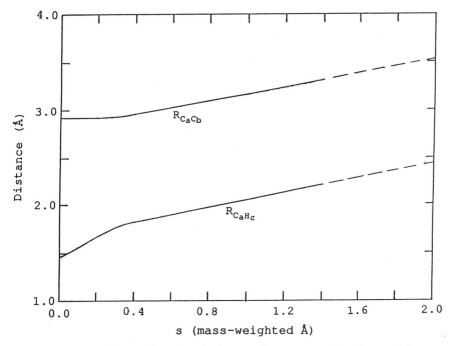

Figure 5. Carbon-carbon and carbon-hydrogen distances as functions of the reaction coordinate s along the MEP for the $CH_4 + CH_3^-$ system (reproduced with permission from [147]).

respectively; using the symmetry of the problem, the required information for the other half of the reaction path was obtained by reflection. (All coordinates are scaled to a reduced mass of 1 amu.) The solid curves in Figure 4 show the classical potential energy curve, $V_{MEP}(s)$, and the harmonic vibrationally adiabatic potential curve, $V_a^G(s)$, along the product side of the MEP. (The dashed portions of these curves for $s > 1.4$ Å indicate extrapolations of reaction path properties to the complex, as the computed path "zig-zagged" about the true MEP in that region.) Note in particular the sudden flattening of these curves for $s \approx 0.3$ Å, which is also where the reaction-path curvature is greatest. This behavior can be understood by considering the important geometric changes the system undergoes along the MEP. Figure 5 demonstrates that motion along the MEP occurs in two phases: near the saddle point, it corresponds to motion of the central hydrogen (H_c) toward the acceptor carbon (C_b) with a nearly fixed carbon-carbon distance, while for $s > 0.35$ Å it corresponds to fragment separation with a nearly fixed H_c-C_b distance. This second phase results in the broad tails for the $V_{MEP}(s)$ and $V_a^G(s)$ curves in Figure 4. To underscore this, we compare the *ab initio* $V_a^G(s)$ curve to a symmetric Eckart potential that has been fit to it at the saddle point (the short-dashed curve in Figure 4) [147]. This Eckart potential has the form

Table 8. Calculated proton transfer rate constants (in s^{-1}) for the $CH_4 + CH_3^-$ system (power of 10 in parentheses).

T	CVT	CVT/ MEPSAG	CVT/ CD-SCSAG	CVT/ MEPSAG (Eckart)
40 K	1.07(-32)	9.87(-5)	1.51(-2)	6.57(1)
60 K	3.08(-18)	4.29(-3)	3.12(-1)	2.16(2)
80 K	5.19(-11)	2.09(-1)	6.92(0)	5.88(2)
100 K	1.10(-6)	6.00(0)	8.85(1)	1.45(3)
150 K	5.86(-1)	1.70(3)	6.82(3)	1.11(4)
200 K	4.02(2)	4.50(4)	1.00(5)	7.43(4)
250 K	1.97(4)	4.08(5)	6.68(5)	4.12(5)
300 K	2.63(5)	2.09(6)	2.89(6)	1.97(6)
400 K	6.75(6)	2.08(7)	2.46(7)	1.93(7)
500 K	4.82(7)	9.75(7)	1.08(8)	9.23(7)
600 K	1.82(8)	2.95(8)	3.15(8)	2.83(8)
800 K	9.96(8)	1.30(9)	1.35(9)	1.27(9)
1000 K	2.85(9)	3.38(9)	3.46(9)	3.33(9)

$$V_E(s) = \frac{4E_0 A}{(1+A)^2}, \tag{39}$$

where E_0 is the adiabatic barrier height [i.e., the value of $V_a^G(s = 0)$ measured relative to the ground vibrational level of the reactant complex] and $A = \exp(as)$, where $a = (2F/E_0)^{1/2}$ and F is the magnitude of the second derivative of $V_a^G(s)$ at $s = 0$. [The local maximum in $V_a^G(s)$ for $s \approx 0.35$ Å arises from the sudden increase in the generalized vibrational frequency corresponding to an "asymmetric umbrella" motion of the system as the H_c-C_b bond forms. Higher level *ab initio* calculations are needed to determine if this effect is real or an artifact of the present calculations.]

Using four-point Lagrange interpolation to obtain the geometry, energy, generalized normal mode frequencies, and reaction-path curvature components at 0.02 a_0 intervals along the MEP, harmonic unimolecular CVT rate constants and semiclassical transmission coefficients for the proton transfer process were obtained with the POLYRATE program [75-77]. Rate constants for various levels of approximation are listed in Table 8 and plotted in Figure 6 as functions of temperature. At all temperatures considered in this study, the variational transition state (i.e., the highest maximum in the generalized free energy of activation curve) occurs at the saddle point, so that CVT rate constants are identical to those computed with conventional transition state theory. From the curvature of the three upper Arrhenius curves in Figure 6, it is clear that these results predict tunneling to be very important for this proton transfer process below 500 K. For example, the CVT/MEPSAG rate constants (which include tunneling along the MEP) are larger than the CVT results

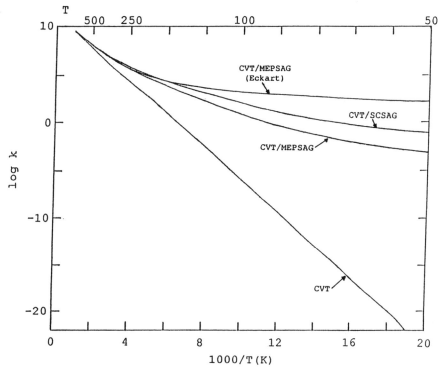

Figure 6. Arrhenius plot of calculated canonical rate constants for proton transfer in the $CH_4 + CH_3^-$ system (reproduced with permission from [147]).

(which ignore tunneling) by factors of 7.9, 1.1×10^2, and 5.5×10^6 at 300, 200, and 100 K, respectively. Incorporating the curvature of the reaction path in the calculation of the transmission coefficient (CVT/CD-SCSAG) further increases the results significantly for temperatures below 300 K; CVT/CD-SCSAG rate constants are larger than the CVT results by factors of 11, 2.5×10^2, and 8.1×10^7 at 300, 200, and 100 K, respectively. The major contribution to the enhancement of the CVT/CD-SCSAG rate constants over the CVT/MEPSAG results comes from the region of the MEP for $s \approx 0.3$ Å, where the character of the reaction-path motion changes from proton transfer to hydrocarbon fragment separation. Table 8 and Figure 6 also show that approximating the adiabatic barrier by a symmetric Eckart potential, which does not have the broad tail exhibited by the *ab initio* $V_a^G(s)$ curve, leads to substantial error for this system below 300 K. These results demonstrate the importance of determining dynamical quantities from properties of the system along the entire MEP, rather than just around the stationary points (reactants, products, and saddle point), as in conventional transition state theory or interpolated variational transition state theory [187,188]. Further studies involving better *ab initio* calculations (a larger basis set and the inclusion of electron correlation) are planned for the $CH_4 + CH_3^-$ system and its isotopic analogues in order to predict reliable proton transfer rate constants and kinetic isotope effects.

Acknowledgments

I am indebted to Donald Truhlar for my training in this field and for many helpful discussions and fruitful collaborations since that time. I am also grateful to Steve Scheiner for initiating our collaboration on the $CH_4 + CH_3^-$ system and to Lan Wang for carrying out the *ab initio* computations, with financial support provided by the National Institutes of Health (GM29391). I would also like to acknowledge Shu-Chin Hung Covick, who carried out the calculations needed for the development of surface 5SP for the $OH + H_2$ system. Some of the calculations reported here were carried out on the computers of Miami University, and the computer time is appreciated.

References

1. D. G. Truhlar and M. S. Gordon, Science **249**, 491 (1990).
2. M. A. Eliason and J. O. Hirschfelder, J. Chem. Phys. **30**, 1426 (1959).
3. D. G. Truhlar and J. T. Muckerman, Reactive scattering cross sections: quasiclassical and semiclassical methods, in R. B. Bernstein (ed.) Atom-Molecule Collision Theory: A Guide for the Experimentalist, Plenum, New York, 1979.
4. D. G. Truhlar, J. Phys. Chem. **83**, 188 (1979).
5. L. M. Raff and D. L. Thompson, The classical trajectory approach to reactive scattering, in M. Baer (ed.) The Theory of Chemical Reaction Dynamics, CRC Press, Boca Raton, FL, 1985, Vol. 3.
6. D. G. Truhlar and R. E. Wyatt, Annu. Rev. Phys. Chem. **27**, 1 (1976).
7. R. B. Walker and J. C. Light, Annu. Rev. Phys. Chem. **31**, 401 (1980).
8. T. Seideman and W. H. Miller, J. Chem. Phys. **96**, 4412 (1992); **97**, 2499 (1992).
9. S. Glasstone, K. J. Laidler, and H. Eyring, Theory of Reaction Rate Processes, McGraw-Hill, New York, 1941, p 10.
10. H. S. Johnston, Gas Phase Reaction Rate Theory, Ronald Press, New York, 1966, p 119.
11. D. L. Bunker, Theory of Gas Phase Reaction Rates, Pergammon Press, Oxford, 1966, p 9.
12. K. J. Laidler, Theories of Chemical Reaction Rates, McGraw-Hill, New York, 1969, p 42.
13. R. E. Weston and H. A. Schwartz, Chemical Kinetics, Prentice-Hall, Englewood Cliffs, N. J., 1972.
14. E. Wigner, Trans. Faraday Soc. **34**, 29 (1938).
15. D. G. Truhlar, A. D. Isaacson, and B. C. Garrett, Generalized transition state theory, in M. Baer (ed.) The Theory of Chemical Reaction Dynamics, CRC Press, Boca Raton, FL, 1985, Vol. 4, pp 65-137.
16. A. D. Isaacson and D. G. Truhlar, J. Chem. Phys. **76**, 1380 (1982).
17. E. B. Wilson, Jr., J. C. Decius, and P. C. Cross, Molecular Vibrations, Mc-Graw Hill, New York, 1955, p 19.
18. D. G. Truhlar and A. Kuppermann, J. Am. Chem. Soc. **93**, 1840 (1971).
19. K. Fukui, in R. Daudel and B. Pullman (eds.) The World of Quantum Chemistry, Reidel, Dordrecht, 1974, pp 113-141.
20. K. Fukui, S. Kato, and H. Fujimoto, J. Am. Chem. Soc. **97**, 1 (1975).

21. H. F. Schaefer III, Chem. Br. **11**, 227 (1975).
22. P. Pechukas, Annu. Rev. Phys. Chem. **32**, 159 (1981).
23. G. Herzberg, Molecular Spectra and Molecular Structure. II. Infrared and Raman Spectra of Polyatomic Molecules, Van Nostrand, Princeton, 1945, p 505.
24. S. K. Gray, W. H. Miller, Y. Yamaguchi, and H. F. Schaefer III, J. Chem. Phys. **73**, 2733 (1980).
25. K. Morokuma and S. Kato, in D. G. Truhlar (ed.) Potential Energy Surfaces and Dynamics Calculations, Plenum, New York, 1981, p 243.
26. A. Tachibana, I. Okazaki, M. Koizumi, K. Hori, and T. Yomabe, J. Am. Chem. Soc. **107**, 1190 (1985).
27. S. M. Colwell and N. C. Handy, J. Chem. Phys. **82**, 128 (1985).
28. G. Doubleday, J. McIver, M. Page, and T. Zielinski, J. Am. Chem. Soc. **107**, 5800 (1985).
29. T. H. Dunning, Jr., L. B. Harding, and E. Kraka, in A. Lagana (ed.) Supercomputer Algorithms for Reactivity, Dynamics, and Kinetics of Small Molecules, Kluwer, Dordrecht, 1989, p 57.
30. K. K. Baldridge, M. S. Gordon, R. Steckler, and D. G. Truhlar, J. Phys. Chem. **93**, 5107 (1989).
31. M. Page and J. W. McIver, Jr., J. Chem. Phys. **88**, 922 (1988).
32. J. F. Gaw, Y. Yamaguchi, and H. F. Schaefer III, J. Chem. Phys. **81**, 6395 (1984).
33. P. Pulay, in H. F. Schaefer (ed.) Applications of Electronic Structure Theory, Plenum, New York, 1977, p 153.
34. J. A. Pople, R. A. Krishnan, H. B. Schlegel, and J. B. Binkley, Int. J. Quantum Chem. Symp. **13**, 225 (1979).
35. T. H. Dunning, Jr., L. B. Harding, A. F. Wagner, G. C. Schatz, and J. M. Bowman, in R. J. Bartlett (ed.) Comparisons of *Ab Initio* Quantum Chemistry with Experiment for Small Molecules, Reidel, Dordrecht, 1985, p 67.
36. H. B. Schlegel, Adv. Chem. Phys. **67**, 249 (1987); H. B. Schlegel, in J. Bertran and I. G. Csizmadia (eds.) New Theoretical Concepts for Understanding Organic Reactions, Kluwer, Dordrecht, 1989, p 33.
37. D. G. Truhlar, R. Steckler, and M. S. Gordon, Chem. Rev. **87**, 217 (1987).
38. C. W. Bauschlicher, S. R. Langhoff, and P. R. Taylor, in A. Lagana (ed.) Supercomputer Algorithms for Reactivity, Dynamics, and Kinetics of Small Molecules, Kluwer, Dordrecht, 1989, p 1.
39. C. Gonzales, C. Sosa, and H. B. Schlegel, J. Phys. Chem. **93**, 2435 (1989).
40. Y. Li and K. Houk, J. Am. Chem. Soc. **111**, 1236 (1989).
41. D. A. Hrovat, W. T. Borden, R. L. Vance, N. G. Rondan, K. N. Houk, and K. Morokuma, J. Am. Chem. Soc. **112**, 2018 (1990).
42. B. C. Garrett and D. G. Truhlar, J. Phys. Chem. **83**, 1052, 1079, 3058 (1979); **84**, 682 (1980); **87**, 4553 (1983).
43. B. C. Garrett and D. G. Truhlar, Acc. Chem. Res. **13**, 440 (1980).
44. B. C. Garrett and D. G. Truhlar, J. Chem. Phys. **70**, 1593 (1979).
45. B. C. Garrett and D. G. Truhlar, J. Amer. Chem. Soc. **101**, 5207 (1979); **102**, 2559 (1980).
46. B. C. Garrett and D. G. Truhlar, J. Chem. Phys. **72**, 3460 (1980).

47. B. C. Garrett, D. G. Truhlar, R. S. Grev, and A. W. Magnuson, J. Phys. Chem. **84**, 1730 (1980); **87**, 4554E (1983).
48. B. C. Garrett, D. G. Truhlar, and R. S. Grev, in D. G. Truhlar (ed.) Potential Energy Surfaces and Dynamics Calculations, Plenum, New York, 1981, p 587.
49. R. T. Skodje, D. G. Truhlar, and B. C. Garrett, J. Phys. Chem. **85**, 3019 (1981); J. Chem. Phys. **77**, 5955 (1982).
50. D. G. Truhlar, A. D. Isaacson, R. T. Skodje, and B. C. Garrett, J. Phys. Chem. **86**, 2252 (1982).
51. D. K. Bondi, D. C. Clary, J. N. L. Connor, B. C. Garrett, and D. G. Truhlar, J. Chem. Phys. **76**, 4986 (1982).
52. N. C. Blais, D. G. Truhlar, and B. C. Garrett, J. Chem. Phys. **78**, 2363 (1983).
53. D. G. Truhlar, R. S. Grev, and B. C. Garrett, J. Phys. Chem. **87**, 3415 (1983).
54. D. G. Truhlar, W. L. Hase, and J. T. Hynes, J. Phys. Chem. **87**, 2664, 5523E (1983).
55. D. C. Clary, B. C. Garrett, and D. G. Truhlar, J. Chem. Phys. **78**, 777 (1983).
56. B. C. Garrett, D. G. Truhlar, A. F. Wagner, and T. H. Dunning, Jr., J. Chem. Phys. **78**, 4400 (1983).
57. D. K. Bondi, J. N. L. Connor, B. C. Garrett, and D. G. Truhlar, J. Chem. Phys. **78**, 5981 (1983).
58. D. G. Truhlar and B. C. Garrett, Annu. Rev. Phys. Chem. **35**, 159 (1984).
59. B. C. Garrett and D. G. Truhlar, J. Chem. Phys. **81**, 309 (1984).
60. B. C. Garrett and D. G. Truhlar, Int. J. Quantum Chem. **29**, 1463 (1986).
61. B. C. Garrett, D. G. Truhlar, J. M. Bowman, A. F. Wagner, D. Robie, S. Arepalli, N. Presser, and R. J. Gordon, J. Am. Chem. Soc. **108**, 3515 (1986).
62. B. C. Garrett, D. G. Truhlar, and G. C. Schatz, J. Am. Chem. Soc. **108**, 2876 (1986).
63. D. G. Truhlar, F. B. Brown, R. Steckler, and A. D. Isaacson, in D. C. Clary (ed.) The Theory of Chemical Reaction Dynamics, D. Reidel, Dordrecht, 1986, p 285.
64. D. G. Truhlar and B. C. Garrett, J. Chim. Phys. Phys.-Chim. Biol. **84**, 365 (1987).
65. B. C. Garrett and D. G. Truhlar, Int. J. Quantum Chem. **31**, 81 (1987).
66. R. Steckler, K. J. Dykema, F. B. Brown, G. C. Hancock, D. G. Truhlar, and T. Valencich, J. Chem. Phys. **87**, 7024 (1987).
67. T. Joseph, R. Steckler, and D. G. Truhlar, J. Chem. Phys. **87**, 7036 (1987).
68. T. Joseph, D. G. Truhlar, and B. C. Garrett, J. Chem. Phys. **88**, 6982 (1988).
69. S. C. Tucker and D. G. Truhlar, in J. Bertran and I. G. Csizmadia (eds.) New Theoretical Concepts for Understanding Organic Reactions, Kluwer, Dordrecht, 1989, p 291.
70. G. C. Lynch, P. Halvick, D. G. Truhlar, B. C. Garrett, D. W. Schwenke, and D. J. Kouri, Z. Naturf. **44a**, 427 (1989).
71. G. C. Lynch, D. G. Truhlar, and B. C. Garrett, J. Chem. Phys. **90**, 3102 (1989).
72. B. C. Garrett and D. G. Truhlar, J. Phys. Chem. **95**, 10374 (1991).
73. V. S. Melissas, D. G. Truhlar, and B. C. Garrett, J. Chem. Phys. **96**, 5758 (1992).
74. Y.-P. Liu, G. C. Lynch, T. N. Truong, D.-h. Lu, D. G. Truhlar, and B. C. Garrett, J. Am. Chem. Soc. **115**, 2408 (1993).
75. A. D. Isaacson, D. G. Truhlar, S. N. Rai, R. Steckler, G. C. Hancock, B. C. Garrett, and M. J. Redmon, Comput. Phys. Commun. **47**, 91 (1987).

76. D.-h. Lu, T. N. Truong, V. S. Melissas, G. C. Lynch, Y.-P. Liu, B. C. Garrett, R. Steckler, A. D. Isaacson, S. N. Rai, G. C. Hancock, J. C. Lauderdale, T. Joseph, and D. G. Truhlar, QCPE Bull. **12**, 35 (1992).

77. D.-h. Lu, T. N. Truong, V. S. Melissas, G. C. Lynch, Y.-P. Liu, B. C. Garrett, R. Steckler, A. D. Isaacson, S. N. Rai, G. C. Hancock, J. C. Lauderdale, T. Joseph, and D. G. Truhlar, Comput. Phys. Commun. **71**, 235 (1992).

78. A. Tweedale and K. J. Laidler, J. Chem. Phys. **53**, 2045 (1970).

79. B. C. Garrett and D. G. Truhlar, J. Am. Chem. Soc. **101**, 4534 (1979).

80. P. Pechukas, in W. H. Miller (ed.) Dynamics of Molecular Collisions, Part B, Plenum, New York, 1976, p 269.

81. B. C. Garrett and D. G. Truhlar, J. Phys. Chem. **83**, 200, 3058E (1979).

82. B. C. Garrett and D. G. Truhlar, J. Phys. Chem. **83**, 2921 (1979).

83. B. C. Garrett, D. G. Truhlar, and R. S. Grev, J. Phys. Chem. **84**, 1749 (1980).

84. R. A. Marcus, J. Chem. Phys. **45**, 4493 (1966).

85. R. A. Marcus and M. E. Coltrin, J. Chem. Phys. **67**, 2609 (1977).

86. B. C. Garrett and D. G. Truhlar, J. Chem. Phys. **79**, 4931 (1983).

87. B. C. Garrett, T. Joseph, T. N. Truong, and D. G. Truhlar, Chem. Phys. **136**, 271 (1989).

88. M. Page and J. W. McIver, Jr., J. Chem. Phys. **88**, 15 (1988).

89. C. Doubleday, Jr., J. W. McIver, Jr., and M. Page, J. Phys. Chem. **92**, 4367 (1988).

90. H. R. Schwarz, Numerical Analysis, Wiley, Chichester, 1989.

91. K. Ishida, K. Morokuma, and A. Komornicki, J. Chem. Phys. **66**, 2153 (1977); M. W. Schmidt, M. S. Gordon, and M. Dupuis, J. Am. Chem. Soc. **107**, 2585 (1985).

92. T. H. Dunning, Jr., E. Kraka, and R. A. Eades, Faraday Discuss. Chem. Soc. **84**, 427 (1987).

93. B. C. Garrett, M. J. Redmon. R. Steckler, D. G. Truhlar, K. K. Baldridge, D. Bartol, M. W. Schmidt, and M. S. Gordon, J. Phys. Chem. **92**, 1476 (1988).

94. J. Ischtwan and M. A. Collins, J. Chem. Phys. **89**, 2881 (1988).

95. C. Gonzalez and H. B. Schlegel, J. Chem. Phys. **90**, 2154 (1989); J. Phys. Chem. **94**, 5523 (1990).

96. W. H. Miller, N. C. Handy, and J. E. Adams, J. Chem. Phys. **72**, 99 (1980).

97. B. C. Garrett and D. G. Truhlar, J. Phys. Chem. **83**, 1915 (1979).

98. A. D. Isaacson, D. G. Truhlar, K. Scanlon, and J. Overend, J. Chem. Phys. **75**, 3017 (1981).

99. A. D. Isaacson and D. G. Truhlar, J. Chem. Phys. **75**, 4090 (1981); **80**, 2888 (1984).

100. A. D. Isaacson and X.-G. Zhang, Theor. Chim. Acta **74**, 493 (1988).

101. Q. Zhang, P. N. Day, and D. G. Truhlar, J. Chem. Phys. **98**, 4948 (1993).

102. J. N. L. Connor, Chem. Phys. Lett. **4**, 419 (1969).

103. G. C. Hancock, P. A. Rejto, R. Steckler, F. B. Brown, D. W. Schwenke, and D. G. Truhlar, J. Chem. Phys. **85**, 4997 (1986).

104. R. Steckler, K. Dykema, F. B. Brown, D. G. Truhlar, and T. Valencich, J. Chem. Phys. **87**, 7014 (1987).

105. D. G. Truhlar, J. Comp. Chem. **12**, 266 (1991).

106. D. G. Truhlar and A. D. Isaacson, J. Chem. Phys. **94**, 357 (1991).

107. H. H. Nielsen, Encycl. Phys. **37/1**, 173 (1959).

108. M. A. Pariseau, I. Suzuki, and J. Overend, J. Chem. Phys. **42**, 2335 (1965).
109. G. Amat, H. H. Nielsen, and G. Tarrago, Rotation-Vibration of Polyatomic Molecules, Marcel Dekker, New York, 1971.
110. D. G. Truhlar, R. W. Olson, A. C. Jeannotte, and J. Overend, J. Am. Chem. Soc. **98**, 2373 (1976).
111. S. Califano, Vibrational States, Wiley, London, 1976.
112. D. Papousek and M. R. Aliev, Molecular Vibrational-Rotational Spectra, Elsevier, New York, 1982.
113. J. M. Bowman and A. F. Wagner, in D. C. Clary (ed.) The Theory of Chemical Reaction Dynamics, Reidel, Dordrecht, 1986, p 47.
114. F. London, Z. Elektrochem. **35**, 552 (1929).
115. H. Eyring and J. Polanyi, Naturwissenschaften **18**, 914 (1930); Z. Phys. Chem. **B12**, 279 (1931).
116. S. Sato, J. Chem. Phys. **23**, 592 (1955).
117. C. A. Parr and D. G. Truhlar, J. Phys. Chem. **75**, 1844 (1971).
118. W. J. Hehre, L. Radom, P. v. R. Schleyer, and J. A. Pople, *Ab Initio* Molecular Orbital Theory, Wiley, New York, 1986.
119. C. W. Bauschlicher, S. R. Langhoff, and P. R. Taylor, Adv. Chem. Phys. **77**, 103 (1990).
120. D. L. Bunker and M. D. Pattengill, J. Chem. Phys. **53**, 3041 (1970).
121. D. R. McLaughlin and D. L. Thompson, J. Chem. Phys. **59**, 4393 (1973).
122. L. M. Raff, J. Chem. Phys. **60**, 2220 (1974).
123. N. Sathyamurthy and L. M. Raff, J. Chem. Phys. **63**, 464 (1975).
124. D. G. Truhlar and C. J. Horowitz, J. Chem. Phys. **68**, 2466 (1978); **71**, 1514E (1979).
125. R. Schinke and W. A. Lester, J. Chem. Phys. **70**, 4893 (1979); **72**, 3754 (1980).
126. S. P. Walch and T. H. Dunning, J. Chem. Phys. **72**, 1303 (1980).
127. G. C. Schatz and H. Elgersma, Chem. Phys. Lett. **73**, 21 (1980).
128. J. N. Murrell, S. Carter, S. C. Farantos, P. Huxley, A. J. C. Varandas, Molecular Potential Energy Functions, Wiley, New York, 1984.
129. A. J. C. Varandas, F. B. Brown, C. A. Mead, D. G. Truhlar, and N. C. Blais, J. Chem. Phys. **86**, 6258 (1987).
130. G. C. Schatz, Rev. Mod. Phys. **61**, 669 (1989).
131. N. C. Blais and D. G. Truhlar, J. Chem. Phys. **61**, 4186 (1974); **65**, 3803E (1976).
132. D. G. Truhlar, B. C. Garrett, and N. C. Blais, J. Chem. Phys. **80**, 232 (1984).
133. R. Steckler, D. G. Truhlar, and B. C. Garrett, J. Chem. Phys. **83**, 2870 (1985).
134. N. C. Blais and D. G. Truhlar, J. Chem. Phys. **83**, 5546 (1985).
135. A. D. Isaacson, J. Phys. Chem. **96**, 531 (1992).
136. Y.-P. Liu, D.-h. Lu, A. Gonzalez-Lafont, D. G. Truhlar, and B. C. Garrett, J. Am. Chem. Soc. **115**, 7806 (1993).
137. J. A. Pople and D. L. Beveridge, Approximate Molecular Orbital Theory, Mc-Graw Hill, New York, 1970.
138. M. J. S. Dewar and D. M. Storch, J. Am. Chem. Soc. **107**, 3898 (1985).
139. T. N. Truong, D.-h. Lu, G. C. Lynch, Y.-P.Liu, V. S. Melissas, J. J. P. Stewart, R. Steckler, B. C. Garrett, A. D. Isaacson, A. Gonzalez-Lafont, S. N. Rai, G. C. Hancock, T. Joseph, and D. G. Truhlar, Comput. Phys. Commun. **75**, 143 (1993).

140. R. C. Bingham, M. J. S. Dewar, and D. H. Lo, J. Am. Chem. Soc. **97**, 1294 (1975).
141. M. J. S. Dewar and W. Thiel, J. Am. Chem. Soc. **99**, 4899 (1977).
142. M. J. S. Dewar, E. G. Zoebisch, E. F. Healy, and J. J. P. Stewart, J. Am. Chem. Soc. **107**, 3902 (1985).
143. J. J. P. Stewart, J. Comput. Chem. **10**, 209, 221 (1989).
144. A. Gonzalez-Lafont, T. N. Truong, and D. G. Truhlar, J. Phys. Chem. **95**, 4618 (1991).
145. A. A. Viggiano, J. Paschkewitz, R. A. Morris, J. F. Paulson, A. Gonzalez-Lafont, and D. G. Truhlar, J. Am. Chem. Soc. **113**, 9404 (1991).
146. B. C. Garrett and C. F. Melius, in S. J. Formosinho, I. G. Csizmadia, and L. G. Arnaut (eds.) Theoretical and Computational Models for Organic Chemistry, Kluwer, Dordrecht, 1991, p 35.
147. A. D. Isaacson, L. Wang, and S. Scheiner, J. Phys. Chem. **97**, 1765 (1993).
148. JANAF Thermochemical Tables, Natl. Stand. Ref. Data Ser. Natl. Bur. Stand. **37**, (1970). (The exoergicity is obtained by removing zero point energy contributions from reactants and products.)
149. I. Glassman, Combustion, Academic Press, New York, 1977.
150. J. R. Creighton, J. Phys. Chem. **81**, 2520 (1977).
151. N. J. Brown, K. H. Eberius, R. M. Fristom, K. H. Hoyermann, and H. Gg. Wagner, Combust. Flame **33**, 151 (1978).
152. J. Warnatz, Sandia National Laboratories Report No. SAND83-8606, Livermore, CA (1983).
153. A. R. Ravishankara, J. M. Nicovich, R. L. Thompson, and F. P. Tully, J. Phys. Chem. **85**, 2498 (1981).
154. G. Dixon-Lewis and D. J. Williams, Comp. Chem. Kinet. **17**, 1 (1977).
155. N. Cohen and K. R. Westberg, Chemical Kinetic Data Sheets for High-Temperature Chemical Reactions, Aerospace Report No. ATR-82(7888)-3, El Segundo, CA, 1982.
156. J. E. Spencer, H. Endo, and G. P. Glass, Proc. 16th Symp. (Int.) Combustion, Combustion Institute, Pittsburgh, 1976, p 829.
157. G. C. Light and J. H. Matsumoto, Chem. Phys. Lett. **58**, 578 (1978).
158. W. Steinert, Ph.D. thesis, University of Göttingen, 1979.
159. R. Zellner and W. Steinert, Chem. Phys. Lett. **81**, 568 (1981).
160. G. P. Glass and B. K. Chaturvedi, J. Chem. Phys. **75**, 2749 (1981).
161. D. G. Truhlar and A. D. Isaacson, J. Chem. Phys. **77**, 3516 (1982).
162. M. H. Mok and J. C. Polanyi, J. Chem. Phys. **53**, 4588 (1970).
163. S. P. Walch, T. H. Dunning, Jr., F. W. Bobrowicz, and R. C. Raffenetti, J. Chem. Phys. **72**, 406 (1980).
164. T. H. Dunning, S. P. Walch, and A. F. Wagner, in D. G. Truhlar (ed.) Potential Energy Surfaces and Dynamics Calculations, Plenum, New York, 1981, p 329.
165. G. C. Schatz and S. P. Walch, J. Chem. Phys. **72**, 776 (1980).
166. G. C. Schatz and H. Elgersma, in D. G. Truhlar (ed.) Potential Energy Surfaces and Dynamics Calculations, Plenum, New York, 1981, p 311.
167. G. C. Schatz, J. Chem. Phys. **74**, 1133 (1981).
168. O. Rashed and N. J. Brown, J. Chem. Phys. **82**, 5506 (1985).

169. A. D. Isaacson, M. T. Sund, S. N. Rai, and D. G. Truhlar, J. Chem. Phys. **82**, 1338 (1985).
170. G. Herzberg, Molecular Spectra and Molecular Structure. I. Spectra of Diatomic Molecules, Van Nostrand, Princeton, 1950.
171. R. J. Bartlett, I. Shavitt, and G. D. Purvis, J. Chem. Phys. **71**, 281 (1979).
172. E. Kraka and T. H. Dunning, Jr., to be published.
173. T. H. Dunning, Jr., private communication.
174. D. G. Truhlar, unpublished calculations.
175. E. Kraka and T. H. Dunning, Jr., private communication.
176. A. D. Isaacson and S.-C. Hung, J. Chem. Phys. **101**, 3928 (1994).
177. G. Simons, R. Parr, and J. M. Finlan, J. Chem. Phys. **59**, 3229 (1973); G. Simons, *ibid.* **61**, 369 (1974).
178. L. B. Harding and W. C. Ermler, J. Comput. Chem. **6**, 13 (1985); W. C. Ermler, H. C. Hsieh, and L. B. Harding, Comput. Phys. Commun. **51**, 257 (1988).
179. Statistical Analysis System, Release 6.06, SAS Institute, Inc., Cary, NC, 1989.
180. W. L. Hase, G. Mrowka, R. J. Brudzynski, and C. S. Sloane, J. Chem. Phys. **69**, 3548 (1978).
181. W. L. Hase and K. C. Bhalla, J. Chem. Phys. **75**, 2807 (1981).
182. R. J. Duchovic, W. L. Hase, and H. B. Schlegel, J. Phys. Chem. **88**, 1339 (1984).
183. Z. Latajka and S. Scheiner, Int. J. Quantum. Chem. **29**, 285 (1986).
184. S. Scheiner and Z. Latajka, J. Phys. Chem. **91**, 724 (1987).
185. W. J. Hehre, R. Ditchfield, and J. A. Pople, J. Chem. Phys. **56**, 2257 (1972).
186. M. J. Frisch, M. Head-Gordon, H. B. Schlegel, K. Raghavachari, J. S. Binkley, C. Gonzalez, D. J. DeFrees, D. J. Fox, R. A. Whiteside, R. Seeger, C. F. Melius, J. Baker, R. Martin, L. R. Kahn, J. J. P. Stewart, E. M. Fluder, S. Topiol, and J. A. Pople, GAUSSIAN 88, Gaussian, Inc., Pittsburgh, PA, 1988.
187. D. G. Truhlar, N. J. Kilpatrick, and B. C. Garrett, J. Chem. Phys. **78**, 2438 (1983).
188. A. Gonzalez-Lafont, T. N. Truong, and D. G. Truhlar, J. Chem. Phys. **95**, 8875 (1991).

DIRECT DYNAMICS METHOD FOR THE CALCULATION OF REACTION RATES

DONALD G. TRUHLAR
Department of Chemistry and Supercomputer Institute
University of Minnesota
Minneapolis, Minnesota 55455-0431

1. Introduction

Accurate quantum dynamics calculations for atom-diatom reactions have advanced to the stage where the nuclear-motion Schrödinger equation can be solved essentially exactly for a given potential energy surface [1]. For example, we recently reported accurate quantum mechanical rate constants for the reaction $D + H_2 \rightarrow HD + H$ over a wide temperature range [2]. In this case the potential energy surface is very well known, and the dynamical results for the most accurate potential energy surface [3] agree with experiment [4] within 12% (maximum deviation) over the 200–900K temperature interval, with slightly large errors at higher T (16% at 1300 K, 22% at 1500 K). This is quite satisfying for a totally *ab initio* calculation of a chemical reaction rate.

Unfortunately, accurate quantum dynamics calculations are beyond the state of the art for systems with 5 or more atoms. But the atom-diatom calculations can be used to test the accuracy of more practical theories. For example, a 1986 calculation [5] based on variational transition state theory (VTST) with semiclassical tunneling (ST) contributions agrees with the accurate quantum dynamical results[2] within 10–20% over the whole 300–1500 K range of T. This is particularly encouraging for two reasons: (i) VTST/ST calculations provide an easily understood and classically visualizable picture of dynamical bottlenecks and tunneling paths [6]; (ii) VTST/ST calculations are practical for much larger systems [7].

VTST [8] provides a generalization of conventional transition state theory [6a,9] (TST). A conventional TST calculation is equivalent to calculating the one-way flux through a phase space hypersurface that divides reactants from products and passes, perpendicular to the imaginary-frequency normal mode, through the highest-energy saddle point of the potential energy on the lowest-energy path from reactants to products [6c]. In VTST one still calculates a one-way flux through a hypersurface that separates reactants from products, but now the location and orientation of the dividing surface are chosen, in

229

D. Heidrich (ed.), The Reaction Path in Chemistry:
Current Approaches and Perspectives, 229–255.
© 1995 *Kluwer Academic Publishers. Printed in the Netherlands.*

accordance with the fact that this flux provides a classical upper bound to the equilibrium reaction rate, to minimize the one-way flux [6c,10]. Where the variation is carried out for a thermal rate constant, i.e., when the flux is calculated for a canonical ensemble, this is called canonical variational transition-state theory (CVT). A practical way to accomplish this variation of the dividing surface is to consider a one-parameter sequence of dividing surfaces locally orthogonal to a reaction path, with the parameter being the distance s along the path at which the surface intersects the path [8j,8k,10,11].

To carry out a VTST calculation, one first defines a reaction coordinate s. In our work we define this as the union of the paths of steepest descents from the saddle point (which is the conventional transition state) towards reactants and towards products, where these paths are computed in an isoinertial coordinate system [12], e.g., normal coordinates of the saddle point [12a,12d,12e]. (An isoinertial coordinate system is any system in the kinetic energy has the form $\sum\limits_{i=1}^{F} p_i^2/2\mu$, where μ is a common reduced mass for all degrees of freedom.) The union of these paths is called [12d,12e] the minimum-energy path (MEP); some workers call it the intrinsic reaction path, or less suitably, the intrinsic reaction coordinate [12f].

The calculation of the one-way classical flux through the dividing surface at s reduces to

$$\text{Flux} = k[X][Y] \tag{1}$$

where k is the rate constant, $[X]$ and $[Y]$ are concentrations of the bimolecular reactants (for a unimolecular reaction one gets the same results without $[Y]$), and [6c,8k]

$$k = \frac{k_B T}{h} K_C^{GT}(s) \tag{2}$$

where k_B is Boltzmann's constant, T is temperature, h is Planck's constant, and $K_C^{GT}(s)$ is the classical equilibrium constant for

$$X + Y \longleftrightarrow GTS(s) \tag{3}$$

where GTS(s) is a species in which one degree of freedom is suppressed by constraining the value of the reaction coordinate to be s. Although GTS(s) is not an ordinary species, it is conventional, dating back to Eyring's 1935 paper [13], to replace $K_C^{GT}(s)$ by a quantal equilibrium constant, in which the levels of X, Y, and GTS(s) are quantized. This was justified as the leading term in a semiclassical series by Wigner [14], but it is neither rigorous nor unique. Recently though it has been shown [15] that short-lived resonances occur at maxima in vibrationally adiabatic potentials, with a spectrum corresponding very closely to the quantized energy levels usually assumed for GTS(s). These states are broadened in accordance with their finite lifetimes, and the broadening may also be interpreted as a consequence of tunneling [15]. The level widths for H + H$_2$ [15a] are several $k_B T$ below the transition state theory thresholds, which leads to large rate enhancements.

For the D + H$_2$ reaction, the VTST/ST calculations mentioned above [5] indicate that tunneling increases the rate constant by factors of 3, 7, and 60 at 400, 300, and 200 K, respectively. This confirms that tunneling is an important ingredient in accurate reaction rate calculations. Table 1 shows similar rate enhancements for several other gas-phase hydrogen-atom transfer reactions we have studied in recent years, namely, the H + H$_2$ reaction [16], the 1,5 sigmatropic shift in *cis*-pentadiene [7b], the CF$_3$ + CH$_4$ reaction, [17] and the OH + CH$_4$ reaction [18]. I anticipate similarly large below-threshold tunneling contributions for many H, H$^+$, and H$^-$ transfer reactions in both the gas phase and solution.

We have found that several ingredients are required for an accurate semiclassical calculation of tunneling contributions to chemical reaction rates [7b,10a,16,17,19], and we have developed multidimensional semiclassical tunneling methods to calculate such

Table 1. Rate constant including tunneling divided by rate constant without tunneling for four reactions at various temperatures.

T(K)	k(VTST/ST) ÷ k(VTST)
H + H$_2$	
250	60
300	20
500	3
1,5 sigmatropic shift in *cis*-pentadiene	
500	6
CF$_3$ + CH$_4$	
300	200
400	20
500	7
OH + CH$_4$	
250	20
300	7
400	$2\frac{1}{2}$

tunneling contributions [7b,10a,16,17]. A critical element in making such calculations practical is that for motion in the vicinity in the minimum-energy path we base the tunneling calculation

on the ground-state vibrationally adiabatic [12b,12d,12e,20] potential energy curve, called $V_a^G(s)$. The calculation of $V_a^G(s)$ assumes that vibrational motions transverse the reaction path adjust adiabatically to reaction-path motion and hence retain their quantum number (not their energy) [12b,12d,12e]. This is justified by the reaction-coordinate motion being in a threshold region. Analysis of accurate quantal calculations confirms, furthermore, that at the lowest energies at which reaction occurs, the flux emanating from several low-energy reactant states all tends to pass through the transition state [15b,21] region in the ground vibrational state. The kinetic energy operator to associate with $V_a^G(s)$ is not, however just $-(\hbar^2/2\mu)d^2/ds^2$. Because the MEP is curved, there is an internal centrifugal effect. Unlike the classical centrifugal force, which points to the convex side of a curved path (bobsled effect [12b]), in quantal tunneling regions the centrifugal force is toward the concave side of the path, and it causes corner cutting [12b,19a,22].

For small curvature of the reaction path, we have found, generalizing Marcus and Coltrin's treatment [23] of collinear H + H_2, that these negative internal centrifugal effects may be treated accurately [19c,19e,19f] even for multidimensional systems in terms of a single dominant tunneling path [19a]. We developed a quantitative multidimensional treatment based on an implicit effective path that cuts the corner to no greater extent in any transverse mode than the distance from the MEP to that mode's concave-side ground-state turning point [24]. Thus we calculate an action integral along an implicit path at or inside of the concave-side vibrational turning point in the direction of the internal centrifugal force (this force picks out, for each s, a particular linear combination of vibrational mode direction along which corner cutting occurs) [7b,24]. The path is implicit because the quantity directly approximated is the effective reduced mass, not the tunneling path. The method is called the small-curvature tunneling (SCT) approximation for short or centrifugal dominant small-curvature semiclassical adiabatic (CD-SCSA or CD-SCSAG) approximation for completeness (the G in the latter acronym stands for ground-state, and is included when the CD-SCSA algorithm is used to calculate ground-state transmission coefficients).

When reaction-path curvature is large, the situation is more complicated [19b,19d,24,25]. In this case, there may be significant contributions for a range of tunneling paths, the tunneling in the exoergic direction of reaction may proceed directly into vibrationally excited states of the products, and the tunneling paths may proceed through regions farther from the MEP than the transverse vibrational turning point, the radius of curvature of the reaction path, or both. In the general case, points along the dominant tunneling paths cannot be described in terms of an expansion of the potential about the MEP. The wider region covered by significant tunneling paths in the large-curvature case is called the reaction swath. In our calculations we switch smoothly from reaction-path coordinates to local $3N$-dimensional Cartesians in this region. For large-curvature cases we calculate action integrals in the exoergic direction, including tunneling into quasidiabatic states of the product, along series of straight-line paths satisfying a resonance condition[19d,24,26]. The semiclassical algorithm for tunneling calculations in large-curvature cases is called the large-curvature tunneling (LCT) approximation for short or the large-curvature version-3 (LC3 or LCG3) approximation to be more precise (the G in the latter acronym stands for ground-state, and is included when the LC3 algorithm is used to calculate ground-state transmission coefficients).

Often it is not possible to tell a priori whether the SCT or LCT approximation is most appropriate. Typically a poor choice of tunneling path will underestimate the tunneling, a result which can be understood by identifying the optimum semiclassical tunneling path with the path of least imaginary action [26]. Thus we employ an approximation called the microcanonical optimized multidimensional tunneling approximation (μOMT) [17]. In this method, for each total energy we calculate the tunneling probability by both the SCT and LCT approximations, and we accept whichever gives the larger tunneling probability as the better result.

Whenever one includes tunneling one should also include nonclassical reflection [27]. Just as tunneling represents the quantum mechanical phenomenon by which a system with less energy than the maximum of the effective potential is nevertheless transmitted past it, nonclassical reflection is the quantum mechanical (diffraction) effect by which a system with more energy than the maximum of the effective potential is nevertheless reflected. Because these effects partially cancel, one should not include one without the other. Because the Boltzmann factor is bigger in the tunneling region, the result is that tunneling usually dominates nonclassical reflection, and it is common to mention only tunneling when speaking of the net effect. The net factor by which these two effects increase (or decrease) the reaction rate is called the transmission coefficient.

2. Dynamical methods

The dynamics methods we employ are reviewed above, and full details are presented elsewhere. In particular, the polyatomic variational transition state theory calculations are described briefly in the original journal article [28] and in full detail in a book chapter [10]. The SCT, LCT, and μOMT tunneling methods are also explained elsewhere [7b,17,24,25]. VTST and these multidimensional tunneling methods are also summarized in the chapter by Isaacson in the present volume.

The discussion so far has concentrated on the calculation of bimolecular rate constants for gas-phase reactions under thermal conditions. Many extensions, e.g., unimolecular reactions [10], microcanonical ensembles [8j,8l,10,29], and state-selected reactions [19c,30] are described elsewhere but are not reviewed here.

At this point we review the information needed for the dynamics calculations described above. The essential input to the dynamics calculations can all be calculated from the masses of the atoms and a potential energy surface (also called a potential energy function or the Born-Oppenheimer potential.)*

First we must find (in the language of electronic structure codes—optimize) the saddle point and calculate its energy (which is called $V^{\neq}$ when measured with respect to the motionless state of reactants) and hessian. From the geometry we calculate moments of inertia, and from the hessian we calculate vibrational frequencies. Similarly we calculate the geometry, energy, moments of inertia, and vibrational frequencies for reactants.

Then we calculate the MEP in the direction toward the reactant or reactants (negative s) and in the direction toward product or products (positive s). This may be done by following the negative gradient [8k,8l,10,20e] (Euler method), following the negative gradient with added stabilization steps [12d,32], or by various [32c,32d,32e,33] more sophisticated (though not necessarily more efficient) algorithms. Typically this requires a small step size, called the gradient step size δs. At a series of points along the MEP, spaced by the generalized normal mode stepsize Δs ($\geq \delta s$), we calculate the Hessian and carry out a generalized normal mode analysis to find the transverse vibrational directions and frequencies. These transverse vibrational modes are localized to the $(3N-7)$-dimensional space ($3N-6$ for a linear molecule) that is orthogonal to the MEP [11], as well as to overall translation and rotation, which is accomplished conveniently by restricting attention to the coordinate hyperplane perpendicular to the tangent to the reaction path [8j] or by diagonalizing a projected force constant matrix in isoinertial coordinates [34].

We estimate the electronic partition functions of reactants and products from the degeneracies and excitation energies of low-lying electronic states. Typically only the ground electronic state needs to be included for closed-shell species and only fine structure excited states (e.g., $^2P_{1/2}$ for halogens, $^2\Pi_{1/2}$ for OH) need be included for open-shell species.

In general reactant properties are denoted by a superscript R, for the supersystem of X and Y, or by separate superscripts X and Y, conventional transition state properties are denoted $\neq$, and generalized transition state properties are denoted by the superscript GT and/or by affixing the arguments s.

From the energies, moments of inertia, symmetry numbers, vibrational frequencies ω_i^R and $\omega_i(s)$, and reduced moments of inertia (the latter if one or more vibrations is

---

*The gradient field of the potential energy surface is the force field.

treated as a hindered internal rotor), we calculate the vibrationally adiabatic ground-state potential curve as [8l,10,20d,28]

$$V_a^G(s) = V_{MEP}(s) + \frac{1}{2}\hbar \left[\sum_m^{F^{GT}-1} \omega_m(s) - \sum_{m=1}^{F^X} \omega_m^X - \sum_{m=1}^{F^Y} \omega_m^Y \right] \qquad (4)$$

where F is the number of internal degrees of freedom ($3N$–6 for a general species and $3N$–5 for a liner species) The generalized standard-state free energy of activation profile is computed as [20e,35]

$$\Delta G^{GT,0}(s) = V_{MEP}(s) + G^{GT,0}(T,s) - G^{R,0}(T) \qquad (5)$$

$V_{MEP}(s)$ is the Born-Oppenheimer potential energy along the MEP. The standard-state free energy of reactants $G^{R,0}(T)$ is computed by well known formulas, and $G^{GT,0}(T,s)$ is computed similarly, but excluding contributions from the reaction coordinate s.

In calculating $G^{GT,0}(T,s)$ and $G^{R,0}$, vibrations that correspond to hindered internal rotations are singled out for special attention, and a correction for their anharmonicity is added using a one-dimensional model [17,24,36]. Multiple saddle points due to hindered internal rotation or symmetry are handled in calculating $G^{GT,0}(T,s)$ by standard use of symmetry numbers.

The canonical variational transition-state theory (CVT) rate constant is given by

$$k^{CVT}(T) = \frac{k_B T}{h} K^{\neq,0} \exp\left[-\max_s \Delta G^{GT,0}(T,s) \Big/ k_B T \right] \qquad (6)$$

where for bimolecular reactions $K^{\neq,0}$ is the reciprocal of the concentration in the standard state. (For unimolecular reactions it is unity).)

The final predicted rate constant, including tunneling, is given by

$$k^{CVT/T}(T) = \kappa(T) k^{CVT}(T) \qquad (7)$$

In the small-curvature tunneling approximation, $\kappa(T)$ requires, in addition to some of the information detailed above, the curvature components $C_m(s)$ of the curvature of the reaction path, where each curvature component measures the projection of the curvature vector on a particular generalized normal mode direction m. Calculation of $\kappa^{LCT}(T)$ or $\kappa^{\mu OMT}(T)$ requires, in addition, values of the Born-Oppenheimer potential V in the reaction swath, typically at points where it cannot be computed from the available harmonic expansion around the MEP.

3. Interfacing the dynamics calculations to the potential energy surface

We have employed four general strategies for obtaining potential energy surface information for reaction-path dynamics calculations, and these are discussed in this section in chronological order of their development.

3.1. ANALYTIC POTENTIAL ENERGY SURFACES

The most obvious way to specify a potential energy surface is as an analytic function of the coordinates. Typically such a function is obtained by semiempirical valence bond theory and/or a valence force field with bond breaking terms or by a fit to *ab initio* electronic structure calculations of the Born-Oppenheimer potential energy surface as a function of internuclear distances. This kind of analytic potential has been widely employed for systems with three atoms and less widely so for systems with 4–12 atoms [37]. Some recent examples include surfaces for the reactions $H + CH_4$ [38], $Cl^-(H_2O)_2$ $+ CHCl$ [7a,39], $F + H_2$ [40], and $H + HBr$ [41].

The difficulty with this approach is that it is very time consuming and typically unsystematic. For polyatomics, the functional forms may suffer from missing or inadequate stretch-stretch or stretch-bend couplings. Even for triatomics, the barrier shape may depend strongly on the chosen functional form. One requires many electronic structure calculations and considerable care to create an accurate (or even useful) surface with all the important internal-coordinate couplings and without artifacts or spurious features.

The methods presented next attempt to circumvent the laborious and painstaking fitting process in one way or another. These methods may all be called direct dynamics in that the dynamics calculations are based *directly* on the electronic structure data without the intermediary of a fit. The first of the methods (section 3.2) is called straight direct dynamics because it is an implementation of this approach in its purest and most straightforward form.

3.2. STRAIGHT DIRECT DYNAMICS

In a second approach, called direct dynamics, we carry out the dynamics calculation precisely the same way as when using an analytic potential energy function, but whenever a potential energy, gradient, or hessian is needed, it is calculated by a full electronic structure calculation, employing *ab initio* methods [42], *ab initio* plus scaled-correlation methods [43], density functional or tight-binding theories [44], or semiempirical neglect-of-differential overlap methods [45]. The direct dynamics approach was originally employed for classical mechanical dynamics calculations [46] in which context it has received considerable further development [47]. Direct dynamics methods have also been employed for trajectory surface hopping calculations [48]. The present chapter is concerned with semiclassical direct dynamics based on reaction paths or reaction swaths;

several examples of this kind of calculations have been reported, both in our group [7b,17,32d,49] and by others [50], and reviews [51] of some of this work are available as well.

One difficulty with straight direct dynamics is that the large number of electronic structure calculations required tends to mitigate against employing high levels of theory. Thus, for example, some direct dynamics calculations have employed minimum basis sets [32d] and semiempirical molecular orbital theory [7b].

One way to keep the cost of the calculations low but improve the accuracy is to use semiempirical molecular orbital calculations in which some of the parameters are fit to data for the specific reaction of interest or for a limited range of reactions. We call this approach SRP for specific reaction parameters or specific range parameters. In several applications we have combined the SRP approach with semiempirical molecular orbital theory employing the neglect of diatomic differential overlap (NDDO) approximation. This is called the NDDO-SRP approach [49].

The NDDO parameters may be adjusted using experimental exoergicities, activation energies, rate constants, and/or reagent geometries and frequencies or by using higher-level *ab initio* data at selected stationary points or other geometries. Alternatively one can use some combination of experimental and *ab initio* data, with the combination depending on availability of data and feasibility of high-level calculations. The parameters may be adjusted by trial and error [17,49a,49b,49c,49d] or by an optimization routine, for example by a genetic algorithm [49e].

The reaction $CF_3 + HCD_3 \rightarrow CF_3H + CD_3$ provides a prototype example of the use of the NDDO-SRP approach [17].

One difficulty with this approach is that, although it is not very difficult to adjust a few NDDO parameters to improve the predicted barrier height and exoergicity, one must be careful not to introduce spurious wells or spuriously deep wells in the potential energy surface or to make the predicted reactant and product geometries or vibrational frequencies unphysical. Furthermore, if one tries to adjust the NDDO parameters to also improve the saddle point and/or reagent frequencies, the task becomes considerably more difficult. As one practical solution to this problem, we have found [49e,52] that genetic algorithms [53] are a useful technique for this difficult parameter adjustment. (Genetic algorithms are the most well known members of the larger set of evolutionary strategies [54] for nonlinear optimization.)

3.3. INTERPOLATED VARIATIONAL TRANSITION STATE THEORY (IVTST)

We have developed another approach to direct dynamics that we call interpolated variational transition state theory (IVTST) [55]. In this approach we carry out electronic structure calculations, including energies, gradients, and hessians, at the reactants, products, saddle point, and zero, one, or two additional points near the saddle point. These additional points are obtained by taking a step along the MEP in the direction of reactants or products or both. All other information needed for VTST calculations with SCT transmission coefficients is obtained by interpolation from this data.

The IVTST method was tested [55] for the OH + H_2 and CH_3 + H_2 reactions and six isotopomeric analogs by using previously available [38,56] analytic potential energy functions for these systems. To carry out the tests we pretended that we know the potential energy surface data only at the three stationary points (reactions, products, and saddle point) and zero, one, or two additional points on the MEP near the saddle point. The rest of the surface information was interpolated. The tests were very encouraging at the level of variational transition state theory without tunneling or including zero-curvature tunneling. Small-curvature tunneling calculations were also carried out and were less accurate in some cases due primarily to the difficulty of interpolating the effective reduced mass quantitatively in regions where its shape is qualitatively correct.

The IVTST method is presently developed only for small-curvature tunneling. Since large-curvature tunneling often proceeds through regions far from the MEP, it is not clear if this could be interpolated reliably using only the data used for the presently developed interpolation schemes. Further work to extend the IVTST method to large-curvature systems would be desirable.

Even for small-curvature tunneling, a difficulty with the IVTST method as applied so far is that the nonstationary points on the MEP are very near the saddle point. This is primarily a matter of economics since it is less expensive to follow the MEP only a short distance. This could be circumvented in the future by various strategies. For example, one could take a few or several gradient-only points with $\delta s < \Delta s$ before calculating the additional nonstationary-point hessians, or one could generalize the method to use points off the MEP.

The IVTST method has been applied successfully to the reactions Cl + CH_4 [55], OH + CH_4 [18], OH + CD_4 [57], and OH + C_2H_6 [7d]. Because these applications were very successful, and because IVTST can be incorporated into electronic structure codes in a very systematic and general way, IVTST may often be the method of choice for future applications. Hence further development of the method would probably be very useful. Nevertheless we are currently even more enthusiastic about dual-level schemes, which are discussed next.

3.4. DUAL-LEVEL DYNAMICS

Our third generation direct dynamics scheme is called VTST with interpolated corrections (VTST-IC) or dual-level direct dynamics (DLDD) or triple-slash (////) dynamics. This approach involves three steps [49c]:

(1) First, using straight direct dynamics with a low level (LL) of electronic structure theory, we perform full VTST calculations with optimized multidimensional tunneling, including frequencies along the reaction path and large-curvature tunneling through the reaction swath.

(2) Then, using a high level (HL) of electronic structure theory, we carry out energy and frequency calculations at 3 or 4 stationary points.

(3) In the final step we use the HL calculations to interpolate corrections to the energies, frequencies, and moments of inertia of the LL ones.

If HL denotes the high level (e.g., QCISD(T)/aug-cc-pVTZ or QCIS(T)/aug-cc-pVTZ//MP2/aug-cc-pVTZ) and LL denotes the low level (e.g., NDDO-SRP or MP2/6-31G*), then the final result of these three steps is denoted HL///LL, which is a direct generalization of the // notation of electronic structure theory. In particular //LL means that stationary point geometries are calculated at level LL, whereas ///LL mans that the reaction path is calculated at level LL.

In step (1) we calculate geometries, reaction-path curvature, and energies along the MEP, we obtain the generalized normal modes orthogonal to the MEP, including s-dependent frequencies and mode directions, and for large-curvature tunneling we calculate energies in the corner-cutting swath. In our VTST-IC algorithm [49c] we correct all the energies and frequencies but only two aspects of the geometries, namely the moments of inertia for the rotational partition functions and the reduced moments of inertia for the hindered internal rotation(s), if present. The geometries, however, are an intrinsic source of the reaction-path curvature and the mode directions (which affect the tunneling calculations) and these are not corrected. Thus it is very important that the low-level calculation give reasonable geometries along the reaction path, including of course the saddle point geometry.

As a first step in the validation of the new approach, we have performed two critical tests. In one we tested /// direct dynamics against straight direct dynamics for the reaction CF_3 + $HCD_3 \rightarrow CF_3H + CD_3$ [49c]. This reaction was chosen because it is dominated by large-curvature tunneling [17], and so it tests the most difficult part of the correction algorithm. In this case we used our 6-parameter NDDO-SRP implicit potential energy function [17], based on changing six parameters of the general AM1 [45b,45c] parameterization of NDDO theory, as the high level of theory, and we created a purposely very inaccurate surface to serve as the low level in order to provide a severe test. The test was carried out by comparing a straight direct dynamics calculation at the high level to another calculation in which we assumed that we know the high level results through the hessian expansion at four stationary points (reactants, well between reactants and saddle point, saddle point itself, and well between saddle point and products) and used these results to "correct" the low-level calculation. For the low level, we varied only one parameter in the MNDO [45a] general parameterization of NDDO theory in order to make the exoergicity right (0.7 kcal exoergic). This yields a barrier height that is 6.6 kcal too high (21.2 kcal vs. the "high-level" 14.6). It is *very easy* to get a better low-level surface, but that was not our goal here. Rather this is designed to be a difficult test for dual-level direct dynamics. Table 2 shows some of the results [49c]. At 350 K, a straight direct dynamics calculation predicts a rate constant larger than conventional TST at 350 K by a factor of about 20. The table shows that we can get a factor of about 15 by using the high-level results through a hessian expansion about only four stationary points. The actual ratios computed using more significant figures than shown in Table 2 yield factors of 20 and 13, respectively, a discrepancy of 35%. This discrepancy is reduced to 15% at 400 K and to an average value of 28% at 600–1500 K. For $CF_3 + CD_3H \rightarrow CF_3D + CD_2H$, the average error from 350 K to 1500 K is only 25%. Recalling that the low-level calculation used was designed to provide a difficult test, the results are satisfactory.

A second test was carried out for the reaction $OH + CH_4 \rightarrow H_2O + CH_3$ [49c]. Results obtained at several levels for this reaction are shown in Table 3. The original AM1 parameters yield a barrier height of 11.1 kcal/mol and an exoergicity of 21.1. The cc-pVTZ basis set of Dunning [58] was adjusted to yield the correct (experimental) exoergicity of 13.3 kcal/mol at the MP2-SAC [43b] level; this calculation then predicts a barrier height of 7.4 kcal/mol and a conventional TST rate constant at 350 K of 0.4 ×

Table 2. Rate constants (10^{-20} cm^3 molecule^{-1} s^{-1}) calculated by conventional and variational transition state theory for $CF_3 + CHD_3 \rightarrow CF_3H + CD_3$.

Potential surface				Dynamical level	T (K)	k(T)
High level	Low level	Approach	HL points			
AM2-SRP(6)[a]	· · ·	conventional	2	TST	350	0.2
					400	2
AM1-SRP(6)	· · ·	single-level	600	CVT/μOMT	350	4
					400	20
MNDO-SRP(1)	· · ·	"	600	CVT/μOMT	350	0.0003
					400	0.004
AM1-SRP(6)	MNDO-SRP(1)	dual-level	4	CVT/μOMT	350	3
					400	17

[a]Number in parentheses is number of SRP parameters.

Table 3. Rate constants (10^{-14} cm^3 molecule^{-1} s^{-1}) calculated by various approaches and with various sources of potential data for the reaction OH + CH$_4$ $\rightarrow$ H$_2$O + CD$_3$.

Potential surface				Dynamical level	k (T = 350 K)
High level	Low level	Approach	HL points		
AM1-SRP[a]	$\cdots$	conventional	2[b]	TST	1.1
AM1-SRP	$\cdots$	VTST	210	CVT/μOMT	0.9
MP2-SAC/cc-pVTZ[c]	$\cdots$	conventional	2[b]	TST	0.4
"	$\cdots$	IVTST	3	CVT/SCT	2.0
"	$\cdots$	"	5	CVT/SCT	1.1
"	AM1-SRP	VTST-IC	3	CVT/SCT	1.1
"	"	"	3	CVT/μOMT	1.1
				Experiment	1.6

[a] two parameters reoptimized
[b] reactant and saddle point only
[c] basis adjusted for correlation balance as explained in Ref. [18]

10^{-14} cm^3 molecule^{-1} s^{-1}. Using the reactant and saddle point properties required for this calculation plus a calculation at the product geometry (which was needed to adjust the basis set to predict the correct exoergicity) along with the IVTST method raises the predicted rate constant to 2.0×10^{-14} kcal/mol, and adding two more MP2-SAC points lowers this to 1.1×10^{-14} [18]. Recall that the IVTST method, at least as currently formulated, cannot treat large-curvature tunneling so these results are based on the SCT approximation. The estimation of the reaction-path curvature throughout the whole region that is important for tunneling on the basis of only three points in the vicinity of the saddle point is the least trustworthy part of this calculation.

Repeating this calculation at the VTST-IC level (/// dynamics) allows us to make a more reliable estimate of the reaction-path curvature and of the shapes of the various generalized transition state theory frequencies as well as to include large-curvature tunneling. Thus we adjusted the AM1 parameters to reproduce the MP2-SAC barrier height and the experimental exoergicity, and we used this surface as a low-level surface for the dual-level VTST-IC approach [49c]. Table 3 shows that, even using the results of the MP2-SAC calculation for only 3 points (reactant, saddle point, and product) yields a rate constant of 1.1×10^{-14}, in excellent agreement with the IVTST level.

The first application of the VTST-IC method to a new problem was to the reaction OH + NH$_3$ $\rightarrow$ H$_2$O + NH$_2$ [49d]. In this case we used two different low-level theories.

The first low-level theory is an NDDO-SRP calculation in which the functional form of the PM3 version of NDDO theory was generalized to allow a larger number of resonance integral parameters. In the original AM1 and PM3 parameterizations, there are five resonance parameters for a system composed of H, C, and N, namely β_{Hs}, β_{Cs}, β_{Cp}, β_{Ns}, and β_{Np}. Resonance parameters $\beta_{X\ell X'\ell'}$.for the interaction of an ℓ-type orbital on atom X with an ℓ'-type orbital on atom X are then approximated by

$$\beta_{X\ell X'\ell'} = \frac{\beta_{X\ell} + \beta_{X'\ell'}}{2} \tag{18}$$

In our treatment we took all nine unique pairs (only nine because there are no O-O or N-N pairs in the OH + NH$_3$ system) of X ℓ and X′ ℓ' as independent. The values of β_{HsHs}, β_{NsOs}, β_{NsOp}, and β_{NpOs} were kept the same as in PM3 (based on eq. 8 with PM3 values for the β_{Xl}), but we independently re-optimized β_{HsNs}, β_{HsNp}, β_{HsOs}, β_{HsOp}, and β_{NpOp}.

For the alternative low level, we used an *ab initio* direct dynamics calculation at the MP2/6-31G** [42] level.

For the high level we used the MP2/aug-cc-pVTZ level for stationary point optimizations, the QCISD(T)/aug-cc-pVTZ level for single point calculations of the energy at the resulting optimized geometries, and the MP2/aug-cc-pVDZ level for hessians at the corresponding stationary points. (The notation for correlation levels and basis sets is standard [42,578,59].)

Table 4 compares the calculated barrier heights at various levels of theory. This table clearly illustrates the advantages of dual-level and interpolatory techniques in that very large basis sets and high levels of treating electron correlation are required to obtain an accurate barrier height. If one were forced to use such a computationally demanding electronic structure calculations for all steps of the dynamics calculation, direct dynamics without interpolation would be impractical, whereas if one restricted oneself to the more affordable lower level, the barrier and frequencies would not be nearly as accurate as in the dual-level calculation.

Figure 1 compares the profile of the most strongly varying generalized normal mode frequency as calculated four different ways. The frequency shown in Figure 1 correlates with N–H stretching in the reactants and with O–H stretching in the products. This curve shows a shape very characteristic of atom-transfer reactions in general, in that it is typical for the frequency associated with the breaking bond to decrease rapidly when the bond switching occurs in the strong interaction region, where there are two half bonds, and then to increase as the associated vibrational mode transforms itself into a product stretch [8k,81,35,60]. Interpolation of this bond-switching-mode frequency is very challenging because of its rapidly varying character, but it is critical to capture this variation because it is a high-frequency mode, and high-frequency modes carry a considerable amount of zero point energy. (For example, the difference in zero point energy between a 1500 cm^{-1} mode in the bond-switching region and a 3500 cm^{-1} mode in the reactant region is 2.9 kcal/mol, and even the difference in zero point energy between a 1500 cm^{-1} mode and a 2000 cm^{-1} mode at two different points in the bond-switching region is 0.7 kcal/mol.) Figure 1 shows that the two corrected calculations give identical frequencies at $s = 0$, as they should since they are both corrected to the same high level at this point, but the reactive mode profile differs elsewhere because of differences between the two low-level calculations. The s value around which the minimum of the bond-switching-mode frequency tends to be centered depends strongly on the geometry of the saddle point, and this is a primary reason why it is important for the low-level calculation to yield a reasonably accurate saddle point geometry. Since saddle point geometry tends to correlate with reaction exoergicity [37a,61], we believe it is important for the low-level calculation to have the correct (or close to the correct) exoergicity, and this is one reason that we have emphasized the exoergicity in deriving SRP parameters. Figure 1 also illustrates the advantage of VTST-IC over IVTST, namely that reasonable low-level surfaces are excellent *chemical* interpolators because they have the chemical variation of the frequencies "built in." Thus dual-level approaches yield more accurate vibrational profiles than interpolations based on *mathematically* motivated forms.

Table 5 gives calculated rate constants with the chosen higher level and the alternative lower levels. Table 5 shows excellent agreement between calculations carried out with the two very different lower levels, up through CVT/SCT dynamics. It would be very expensive to calculate large-curvature tunneling with an *ab initio* lower level, but here the dual-level approaches with an NDDO-SRP lower level really prove their mettle. Large-curvature effects increase the calculated rate constant by a factor of 2.0 at 250 K, bringing the result into excellent agreement with experiment. The agreement with experiment for

Table 4. Number of basis functions and calculated barrier height (kcal/mol) for OH + NH$_3$ → H$_2$O + NH$_2$ at various levels of electronic structure theory.

Level	Number	V$^{\neq}$
PM3	12	13.7
PM3-SRP	12	5.9
MP2/6-31G**	50	9.0
MP2/aug-cc-pVDZ	82	5.8
MP2/aug-cc-pVTZ	184	4.5
QCISD(T)/aug-cc-pVTZ//MP2	184	3.6

[a]X/Y//X´ means a single-point energy calculated at level X/Y at a geometry optimized at level X´/Y.

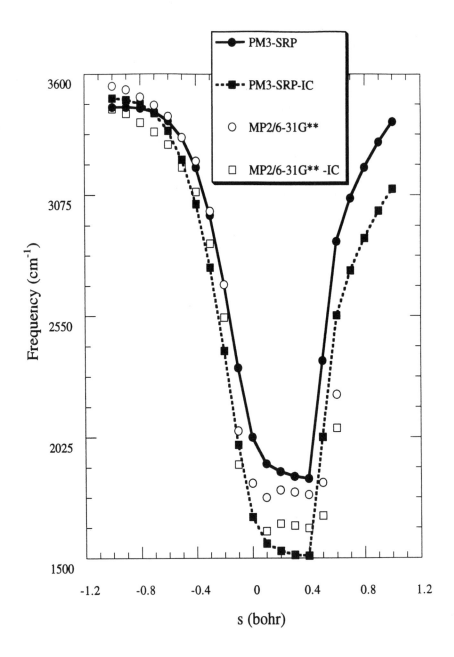

Figure 1. The generalized-normal-mode frequency for the bond-switching mode as a function of reaction coordinate for $OH + NH_3 \rightarrow H_2O + NH_2$.

the OH + NH$_3$ reaction is especially impressive because of the large Arrhenius curvature. Table 6 gives Arrhenius activation energies over three temperature ranges showing the dramatic lowering at low temperature.

Further improvements in these methods should allow a wide variety of interesting processes to be studied.

Table 5. Calculated VTST-IC rate constants (10^{-13} cm^3 molecule^{-1} s^{-1}) at the QCISD(T)/aug-cc-pVTZ//MP2[MP2/aug-cc-pVDZ]///LL level of electronic structure theory.[a]

Dynamical level	T (K)	LL	
		MP2/6-31G**	NDDO-SRP
TST	250	0.18	0.18
	300	0.42	0.42
	1000	25	25
CVT/SCT	250	0.66	0.68
	300	1.03	1.06
	1000	26	25
CVT/μOMT	250	· · ·	1.36
	300	· · ·	1.69
	1000	· · ·	26
Experiment	250	1.2	1.2
	300	2.0	2.0
	1000	31	31

[a]HL[HessL]///LL denotes a higher level of HL, except for the hessians which are calculated at level HessL, and a lower level of LL in a dual-level direct dynamics calculation.

Table 6. Arrhenius activation energies from fits to calculated rate constants at pairs of temperatures in the low, middle, and high-temperature regimes, based on NDDO-SRP as the low level and CVT/μOMT dynamics.

Temperature interval (K)	E_a (kcal/mol)
250–300	0.65
400–600	2.1
800–1000	4.6

4. Vibrational coordinates

Another area where technical advances are in progress is the treatment of the vibrations perpendicular to the reaction path. A central assumption of all reaction-path methods, including variational transition state theory and small-curvature tunneling methods, is the separation of the coordinates into three sets: external coordinates describing the overall translation and rotation, a reaction coordinate describing the motion of the system along some direct route from reactants to products, and the remaining coordinates, which will be called the bound vibrational coordinates.

For linear displacements from a stationary point, separable coordinates are uniquely defined for small displacements by normal mode coordinates [62], which simultaneously diagonalize the kinetic energy to infinite order and the potential energy to second order (i.e., through quadratic terms in the potential). Thus, to the extent that one stays in a region where the quadratic expansion of the potential is trustworthy, these coordinates separate the physical motion, and they are not just an artificial mathematical imposition. Not only is the motion separable in normal coordinates, but the coordinates themselves are very convenient for calculations, since they are rectilinear.

Such comfortingly natural and convenient coordinates do not exist though if we add the requirement that one of the coordinates describes the global motion of the system along some direct route from reactants to products. In general this cannot be accomplished in a convenient way with rectilinear coordinates, and one must introduce at least one curvilinear coordinate. The basic formalism for such a treatment was first advanced in the context of the large-amplitude vibrational motion of small molecules by various workers [63] and in the context of bimolecular reactions by Hofacker [20b], Marcus [12b,12c,20c,22], and others (for an extensive set of references, see Refs. 1–34 in [31a]). The approach used for large-amplitude vibrational motion was further developed for the case where the large-amplitude "coordinate" is the minimum energy path by Miller *et al.* [34] and by Natanson [64]. In these approaches, the reaction coordinate is curvilinear but the coordinates transverse to the reaction path are Cartesian. Hence we call this kind of treatment the Cartesian vibration (CV) approximation.

For variational transition state theory, one requires the vibrational Hamiltonian only in a hypersurface perpendicular to the reaction coordinate. For such calculations the global CV treatments for chemical reactions reduce to an earlier method [8j] in which a local Cartesian system is erected at a point on the MEP, with one of the axes aligned with the tangent to the path. However, for reaction-path-based tunneling calculations, one must consider the globally curved nature of reaction coordinate.

Both for VTST and for tunneling calculations, though, the CV coordinate system is often unsatisfactory, for two reasons. The first problem is that it has the same disadvantages as rectilinear coordinates for vibrational calculations on bound systems—in particular it does not provide a physically natural picture of valence forces, and so there are large anharmonic cross terms [65]. This problem is discussed elsewhere in the context of reaction-path dynamics [66]. Although this problem is serious in both bound-state theory (spectroscopy) and reaction-path dynamics, an even more serious problem occurs in the latter because even the harmonic frequencies are not independent of the coordinate system used for the generalized normal mode analysis at a non-stationary point [31,64]. As a consequence the generalized normal mode frequencies may become unphysically imaginary (as if the reaction path were a ridge) even when the reaction path sits at the bottom of a nonbifurcating valley [31a,67]. This happens frequently in the CV treatment because the Cartesian vibrational coordinates are nonphysical.

A better approach to modeling the vibrations transverse to the reaction path is to use curvilinear vibrational coordinates based on bond stretches, valence angle bends, bbond torsions (defined by dihedral angle between the ABC plane and the BCD plane in a group of atoms with the sequential bonding pattern A–B–C–D), and sometimes "improper dihedrals" (the angle between a bond and a plane). These curvilinear coordinates should provide a better description of generalized transition state vibrations for the same reasons that they are preferable to rectilinear coordinates for treating vibrations of bound molecules, namely that they tend to follow natural contours of the potential energy surface. Techniques for treating the transverse vibrations in curvilinear coordinates have been developed [31] and should be very useful for future work.

5. Summary

This chapter provides an account of our recent efforts to interface dynamics calculations based on reaction-path potentials and tunneling, including tunneling through the large-curvature reaction swath, with electronic structure theory.

6. Acknowledgments

I am grateful to Kim Baldridge, Jose C. Corchado, Joaquin Espinosa-Garcia, Bruce C. Garrett, Angels Gonzalez-Lafont, Mark S. Gordon, Michael Gu, Wei-Ping Hu, Charles Jackels, Yi-Ping Liu, Da-hong Lu, Gillian C. Lynch, Vasilios Melissas, Grigory Natanson, Kiet Nguyen, Ivan Rossi, Rozeanne Steckler, and Thanh Truong for their collaboration on the development and implementation of direct dynamics and internal coordinates techniques for chemical kinetics. The author's research on variational transition state theory and semiclassical tunneling method is supported by the U.S. Department of Energy, Office of Basic Energy Sciences. The author's research on high-energy chemical reactions is supported by NASA.

References

[1]   (a) D. G. Truhlar, D. W. Schwenke, and D. J. Kouri, J. Phys. Chem. **94**, 7346 (1990). (b) D. E. Manolopoulos and R. E. Wyatt, J. Chem. Phys. **92**, 810 (1990). (c) W. H. Miller, Annu. Rev. Phys. Chem. **41**, 245 (1991). (d) J. M.

Launay, Theor. Chim. Acta **79**, 183 (1991). (e) R. T Pack, E. A. Butcher, and G. A. Parker, J. Chem. Phys. **99**, 9310 (1993).

[2]   S. L. Mielke, G. C. Lynch, D. G. Truhlar, and D. W. Schwenke, J. Phys. Chem. **98**, 8000 (1994).

[3]   A. J. C. Varandas, F. B. Brown, C. A. Mead, D. G. Truhlar, and N. C. Blais, J. Chem. Phys. **86**, 6258 (1987).

[4]   J. V. Michael and J. R. Fisher, J. Phys. Chem. **94**, 3318 (1990).

[5]   B. C. Garrett, D. G. Truhlar, A. J. C. Varandas, and N. C. Blais, Int. J. Chem. Kinet. **18**, 1065 (1986).

[6]   (a) M. M. Kreevoy and D. G. Truhlar, in Investigation of Rates and Mechanisms of Reaction, 4th edition, C. F. Bernasconi (ed.), John Wiley & Sons, New York, 1986, Pt. I, p. 13. (b) D. G. Truhlar and B. C. Garrett, J. Chim. Phys. **84**, 365 (1987). (c) S. C. Tucker and D. G. Truhlar, in New Theoretical Concepts for Understanding Organic Reactions, J. Bertrán and I. G. Csizmadia (eds.), Kluwer, Dordrecht, 1989, p. 291.

[7]   (a) S. C. Tucker and D. G. Truhlar, J. Amer. Chem. Soc. **112**, 3347 (1990). (b) Y.-P. Liu, G. C. Lynch, T. N. Truong, D.-h. Lu, D. G. Truhlar, and B. C. Garrett, J. Amer. Chem. Soc. **115**, 2408 (1993). (c) S. E. Wonchoba and D. G. Truhlar, J. Chem. Phys. **99**, 9637 (1993). (d) V. S. Melissas and D. G. Truhlar, J. Phys. Chem. **98**, 875 (1994).

[8]   (a) E. Wigner, J. Chem. Phys. **5**, 720 (1937). (b) J. Horiuti, Bull. Chem. Soc. Japan **13**, 210 (1938). (c) M. A. Eliason and J. O. Hirschfelder, J. Chem. Phys. **30**, 1426 (1959). (d) J. C. Keck, J. Chem. Phys. **32**, 1035 (1960). (e) J. C. Keck, Advan. Chem. Phys. **13**, 85 (1967). (f) D. G. Truhlar, J. Chem. Phys. **53**, 2041 (1970). (g) A. Tweedale and K. J. Laidler, J. Chem. Phys. **53**, 2045 (1970). (h) G. W. Koeppl, J. Amer. Chem. Soc. **96**, 6539 (1974). (i) W. H. Miller, J. Chem. Phys. **61**, 1823 (1974). (j) B. C. Garrett and D. G .Truhlar, J. Chem. Phys. **70**, 1593 (1979). (k) B. C. Garrett and D. G. Truhlar, J. Phys. Chem. **83**, 1052 (1979). (l) B. C. Garrett and D. G. Truhlar, J. Phys. Chem. **83**, 1079 (1979).

[9]   (a) S. Glasstone, K. Laidler, and H. Eyring, The Theory of Rate Processes, McGraw-Hill, New York 1941. (b) K. J. Laidler, Chemical Kinetics, 3rd ed., Harper & Row, New York, 1987.

[10]   D. G. Truhlar, A. D. Isaacson, and B. C. Garrett, in Theory of Chemical Reaction Dynamics, M. Baer (ed.), CRC Press, Boca Raton, 1985, Vol. 4, p. 65.

[11]   D. G. Truhlar and B. C. Garrett, Accounts Chem. Res. **13**, 440 (1980).

[12]   (a) I. Shavitt, J. Chem. Phys. **49**, 4048 (1968). (b) R. A. Marcus, J. Chem. Phys. **45**, 4493 (1966). (c) R. A. Marcus, J. Chem. Phys. **49**, 2610 (1968). (d) D. G. Truhlar and A. Kuppermann, J. Amer. Chem. Soc. **93**, 1840 (1970). (e) D. G. Truhlar and A. Kuppermann, J. Chem. Phys. **56**, 2232 (1972). (f) K. Fukui, in The World of Quantum Chemistry, R. Daudel and B Pullman (eds.), Reidel, Dordrecht, 1974, p. 113.

[13]   H. Eyring, J. Chem. Phys. **3**, 107 (1935).

[14]   E. P. Wigner, Z. Phys. Chem., Abt. B **19**, 203 (1932).

[15]   (a) D. C. Chatfield, R. S. Friedman, D. G. Truhlar, B. C. Garrett, and D. W. Schwenke, J. Amer. Chem. Soc. **113**, 486 (1991). (b) D. C. Chatfield, R. S. Friedman, D. G. Truhlar, and D. W. Schwenke, Faraday Discussions Chem. Soc. **91**, 289 (1991). (c) D. C. Chatfield, R. S. Friedman, G. C. Lynch, D. G. Truhlar, and D. W. Schwenke, J. Chem. Phys. **98**, 342 (1993). (d) D. C. Chatfield, R. S. Friedman, D. W. Schwenke, and D. G. Truhlar, J. Phys. Chem. **96**, 2414 (1992).

[16]   B. C. Garrett and D. G. Truhlar, Proc. Natl. Acad. Sci. USA **76**, 4755 (1979).

[17]   Y.-P. Liu, D.-h. Lu, A. Gonzalez-Lafont, D. G. Truhlar, and B. C. Garrett, J. Amer. Chem. Soc. **115**, 7806 (1993).

[18]   V. S. Melissas and D. G. Truhlar, J. Chem. Phys. **99**, 1013 (1993).

[19]   (a) R. T. Skodje, D. G. Truhlar, and B. C. Garrett, J. Chem. Phys. **77**, 5955 (1982). (b) B. C. Garrett, D. G. Truhlar, A. F. Wagner, and T. H. Dunning, Jr., J. Chem. Phys. **76**, 1380 (1982). (c) B. C. Garrett and D. G. Truhlar, J. Chem. Phys. **81**, 309 (1984). (d) B. C. Garrett, N. Abusalbi, D. J. Kouri, and D. G. Truhlar, J. Chem. Phys. **83**, 2252 (1985). (e) B. C. Garrett, D. G. Truhlar, and G. C. Schatz, J. Amer. Chem. Soc. **108**, 2876 (1986). (f) G. C. Lynch, D. G. Truhlar, and B. C. Garrett, J. Chem. Phys. **90**, 3102 (1989); erratum: **91**, 3280 (1989). (g) B. C. Garrett, T. Joseph, T. N. Truong, and D. G. Truhlar, Chem. Phys. **136**, 271 (1989); erratum: **140**, 207 (1990). (h) D. G. Truhlar, J. Chem. Soc. Faraday Trans. **90**, 1740 (1994).

[20]   (a) M. A. Eliason and J. O. Hirschfelder, J. Chem. Phys. **30**, 1426 (1959). (b) L. Hofacker, Z. Naturforsch. **18a**, 607 (1963). (c) R. A. Marcus, Discussions Faraday Soc. **44**, 7 (1967). (d) D. G. Truhlar, J. Chem. Phys. **53**, 2041 (1970). (e) B. C. Garrett, D. G. Truhlar, R. S. Grev, and A. W. Magnuson, J. Phys. Chem. **84**, 1730 (1984).

[21]   G. C. Lynch, P. Halvick, D. G. Truhlar, B. C. Garrett, D. W. Schwenke, and D. J. Kouri, Z. Naturforsch. **44a**, 427 (1989).

[22]   R. A. Marcus, J. Chem. Phys. **49**, 2617 (1969).

[23]   R. A. Marcus and M. E. Coltrin, J. Chem. Phys. **67**, 2609 (1977).

[24]   D.-h. Lu, T. N. Truong, V. S. Melissas, G. C. Lynch, Y.-P. Liu, B. C. Garrett, R. Steckler, A. D. Isaacson, S. N. Rai, G. C. Hancock, J. G. Lauderdale, T. Joseph, and D. G. Truhlar, Computer Phys. Commun. **71**, 235 (1992).

[25]   T. N. Truong, D.-h. Lu, G. C. Lynch, Y.-P. Liu, V. S. Melissas, J. J. P. Stewart, R. Steckler, B. C. Garrett, A. D. Isaacson, A. Gonzalez-Lafont, S. N. Rai, G. C. Hancock, T. Joseph, and D. G. Truhlar, Computer Phys. Commun. **75**, 143 (1993).

[26]   B. C. Garrett and D. G. Truhlar, J. Chem. Phys. **79**, 4931 (1983).

[27]   B. C. Garrett and D. G. Truhlar, J. Phys. Chem. **83**, 2921 (1979).

[28]   A. D. Isaacson and D. G. Truhlar, J. Chem. Phys. **76**, 1380 (1982).

[29]   A. D. Isaacson, M. T. Sund, S. N. Rai, and D. G. Truhlar, J. Chem. Phys. **82**, 1338 (1985).

[30]   (a) D. G. Truhlar and A. D. Isaacson, J. Chem. Phys. **77**, 3516 (1982). (b) B. C. Garrett and D. G. Truhlar, J. Phys. Chem. **89**, 2204 (1985). (c) B. C. Garrett, N. Abusalbi, D. J. Kouri, and D. G. Truhlar, J. Chem. Phys. **83**, 2252 (1985). (d) B. C. Garrett and D. G. Truhlar, Int. J. Quantum Chem. **29**, 1463 (1986). (e) R. Steckler, D. G. Truhlar, and B. C. Garrett, J. Chem. Phys. **84**, 6712 (1986). (f) B. C. Garrett, D. G. Truhlar, A. J. C. Varandas, and N. C. Blais, Int. J. Chem. Kinet. **18**, 1065 (1986). (g) B. C. Garrett, D. G. Truhlar, J. M. Bowman, and A. F. Wagner, J. Phys. Chem. **90**, 4305 (1986). (h) J. Z. H. Zhang, Y. Zhang, D. J. Kouri, B. C. Garrett, K. Haug, D. W. Schwenke, and D. G. Truhlar, Faraday Discussions Chem. Soc. **84**, 371 (1987).

[31]   (a) G. A. Natanson, B. C. Garrett, T. N. Truong, T. Joseph, and D. G. Truhlar, J. Chem. Phys. **94**, 7875 (1991). (b) C. F. Jackels, M. Z. Gu, and D. G. Truhlar, J. Chem. Phys., in press.

[32]   (a) K. Ishida, K. Morokuma, and A. Komornicki, J. Chem. Phys. **66**, 2153 (1977). (b) M. W. Schmidt, M. S. Gordon, and M. J. Dupuis, J. Amer. Chem. Soc. **107**, 2585 (1985). (c) B. C. Garrett, M. J. Redmon, R. Steckler, D. G. Truhlar, K. K. Baldridge, D. Bartel, M. W. Schmidt, and M. S. Gordon, J.

Phys. Chem. **92**, 1476 (1988). (d) K. K. Baldridge, M. S. Gordon, R. Steckler, and D. G. Truhlar, J. Phys. Chem. **93**, 5107 (1989). (e) V. S. Melissas, D. G. Truhlar, and B. C. Garrett, J. Chem. Phys. **96**, 5758 (1992).

[33]   (a) M. Page and J. W. McIver, Jr., J. Chem. Phys. **88**, 992 (1988). (b) C. Gonzalez and H. B. Schlegel, J. Phys. Chem. **94**, 5523 (1990). (c) J.-Q. Sun and K. Ruedenberg, J. Chem. Phys. **99**, 5257 (1993).

[34]   W. H. Miller, N. C. Handy, and J. E. Adams, J. Chem. Phys. **72**, 99 (1980).

[35]   (a) B. C. Garrett and D. G. Truhlar, J. Amer. Chem. Soc. **101**, 4534 (1979). (b) B. C. Garrett and D. G. Truhlar, J. Amer. Chem. Soc. **101**, 5207 (1979).

[36]   D. G. Truhlar, J. Comp. Chem. **12**, 266 (1991).

[37]   (a) C. A. Parr and D. G. Truhlar, J. Phys. Chem. **75**, 1844 (1971). (b) D. G. Truhlar and R. E. Wyatt, Adv. Chem. Phys. **36**, 141 (1977). (c) W. L. Hase and R. J. Wolf, in Potential Energy Surfaces and Dynamics Calculations, D. G. Truhlar (ed.), Plenum, New York, 1981, p. 37. (d) D. G. Truhlar, R. Steckler, and M. S. Gordon, Chem. Rev. **87**, 217 (1987). (e) G. C. Schatz, Rev. Mod. Phys. **61**, 669 (1989).

[38]   T. Joseph, R. Steckler, and D. G. Truhlar, J. Chem. Phys. **87**, 7036 (1987).

[39]   S. C. Tucker and D. G. Truhlar, J. Amer. Chem. Soc. **112**, 3338 (1990).

[40]   S. L. Mielke, G. C. Lynch, D. G. Truhlar, and D. W. Schwenke, Chem. Phys. Lett. **213**, 10 (1993); erratum: **217**, 173 (1994).

[41]   (a) G. C. Lynch, D. G. Truhlar, F. B. Brown, and J.-g. Zhao, Hyperfine Interactions **87**, 885 (1984). (b) G. C. Lynch, D. G. Truhlar, F. B. Brown, and J.-g. Zhao, J. Phys. Chem. **99**, 207 (1995).

[42]   W. J. Hehre, L. Radom, P. v. R. Schleyer, and J. A. Pople, Ab Initio Molecular Orbital Theory, Wiley, New York, 1986.

[43]   (a) F. B. Brown and D. G. Truhlar, Chem. Phys. Lett. **117**, 307 (1985). (b) M. S. Gordon and D. G. Truhlar, J. Amer. Chem. Soc. **108**, 542 (1986).

[44]   (a) W. Kohn and L. Sham, Phys. Rev. A **140**, 1133 (1955). (b) W. A. Harrison, Electronic Structure and the Properties of Solids, Freeman, San Francisco, 1980. (c) D. Tománek and M. A. Schluter, Phys. Rev. B **36**, 1208 (1987). (d) T. Ziegler, Chem. Rev. **91** (1991) 651.

[45]   (a) M. J. S. Dewar and W. Thiel, J. Amer. Chem. Soc. **99**, 4899, 4907 (1977).
        (b) M. J. S. Dewar, E. G. Zoebisch, E. F. Healy, and J. J. P. Stewart, J. Amer.
        Chem. Soc. **107**, 3902 (1985). (c) M. J. S. Dewar and E. G. Zoebisch,
        Theochem **180**, 1 (1988). (d) J. J. P. Stewart, J. Comp. Chem. **10**, 221 (1989).
        (e) J. J. P. Stewart, J. Computer-Aided Mol. Design **4**, 1 (1990).

[46]   (a) I. Wang and M. Karplus, J. Amer. Chem. Soc. **95**, 8160 (1973). (b) D. J.
        Malcome-Lawes, J. Chem. Soc. Faraday Trans. II **71**, 1183 (1975). (c) C.
        Leforestier, J. Chem. Phys. **68**, 4406 (1978).

[47]   (a) R. Car and M. Parrinello, Phys. Rev. Lett. **55**, 2471 (1985). (b) O. F.
        Sankey and R. E. Allen, Phys. Rev. B **33**, 7164 (1986). (c) M. R. Pederson, B.
        M. Klein, and J. Q. Broughton, Phys. Rev. B **38**, 3825 (1988). (d) C. Z.
        Wang, C. T. Chen, and K. M. Ho, Phys. Rev. B **39**, 8586 (1989). (e) O. F.
        Sankey and D. J. Niklewski, Phys. Rev. B **40**, 3979 (1989). (f) J. Harris and
        D. Hohl, J. Phys.: Condens. Matter **2**, 5161 (1990). (g) D. A. Drabold, P. A.
        Fedders, O. F. Sankey, and J. O. Dow, Phys. Rev. B **42**, 5135 (1990). (h) G.
        Galli and M. Parrinello, in Computer Simulations in Materials Science, M. Meyer
        and V. Pontikis (eds.), Kluwer, Dordrecht, 1991, p. 283. (i) R. N. Barnett, U.
        Landman, A. Nitzan, and G. Rajagopal, J. Chem. Phys. **94**, 608 (1991). (j) R.
        M. Wentzcovitch and J. L. Martins, Solid State Commun. **78**, 831 (1991). (k) F.
        S. Khan and J. Q. Broughton, Phys. Rev. B **43**, 11754 (1991). (l) M. C.
        Payne, M. P. Teter, D. C. Allen, T. A. Arias, and J. D. Joannopoulos, Rev.
        Mod. Phys. **64**, 1045 (1992). (m) B. Hartke and E. A. Carter, Chem. Phys.
        Lett. **189**, 358 (1992). (n) R. M. Wentzcovitch, J. L. Martins, and G. D. Price,
        Phys. Rev. Lett. **70**, 3947 (1993). (o) D. A. Gibson and E. A. Carter, J. Phys.
        Chem. **97**, 13429 (1993). (p) R. N. Barnett and U. Landman, Phys. Rev. B **48**,
        2081 (1993). (q) M. J. Field, in Computer Simulations of Biomolecular Systems,
        Vol. 2, W. F. Van Gunsteren, P. K. Weiner, and A. J. Wilkonson (eds.),
        ESCOM, Leiden, 1993, p. 82. (r) D. L. Lynch, N. Troullier, J. D. Kress, and L.
        A. Collins, J. Chem. Phys. **101**, 7048 (1994). (s) W. Chen. W. L. Hase, and
        H. B. Schlegel, Chem. Phys. Lett. **228**, 436 (1994). (t) C. Lee, X. Long, I.
        Carpenter, S. Smithline, and G. Fitzgerald, Int. J. Quantum Chem. **47**, 527
        (1994).

[48]   (a) A. Warshel and M. Karplus, Chem. Phys. Lett. **32**, 11 (1975). (b) D. G.
        Truhlar, J. W. Duff, N. C. Blais, J. C. Tully, and B. C. Garrett, J. Chem. Phys.
        77, 764 (1982).

[49]   (a) A. Gonzalez-Lafont, T. N. Truong, and D. G. Truhlar, J. Phys. Chem. **95**,
        4618 (1991). (b) A. A. Viggiano, J. Paschkewitz, R A. Morris, J. F. Paulson,
        A. Gonzalez-Lafont, and D. G. Truhlar, J. Amer. Chem. Soc. **113**, 9404 (1991).
        (c) W.-P. Hu, Y.-P. Liu, and D. G. Truhlar, J. Chem. Soc. Faraday Trans. **90**,
        1715 (1994). (d) J. C. Corchado, J. Espinosa-García, W.-P. Hu, I. Rossi, and

D. G. Truhlar, J. Phys. Chem. **99**, 687 (1995). (e) W.-P. Hu, I. Rossi, and D. G. Truhlar, work in progress.

[50] (a) A. Tachibana, I. Ukazaki, M. Koizumi, K. Hori, and T. Yomabe, J. Amer. Chem. Soc. **107**, 1190 (1985). (b) S. M. Colwell and N. C. Handy, J. Chem. Phys. **82**, 1281 (1985). (c) C. Doubleday, Jr., J. W. McIver, Jr., and M. Page, J. Phys. Chem. **92**, 4367 (1988). (d) B. C. Garrett, M. L. Koszykowski, C. F. Melius, and M. Page, J. Phys. Chem. **94**, 7096 (1990). (e) T. N. Truong and J. A. McCammon, J. Amer. Chem. Soc. **113**, 7504 (1991). (f) B. C. Garrett and C. F. Melius, in Theoretical and Computational Models for Organic Chemistry, S. J. Formosinho, I. G. Csizmadia, and L. G. Arnaut (eds.), Kluwer, Dordrecht, 1991, p. 35. (g) J. W. Storer and K. N. Houk, J. Amer. Chem. Soc. **115**, 10426 (1993). (h) T. N. Truong, J. Chem. Phys. **100**, 8014 (1993). (i) J. Espinosa-García and J. C. Corchado, J. Chem. Phys. **101**, 1333 (1994). (j) T. N. Truong and W. Duncan, J. Chem. Phys. **101**, 7408 (1994). (k) J. Espinosa-García and J. C. Corchado, J. Chem. Phys. **101**, 8700 (1994). (l) T. N. Truong and T. J. Evans, J. Chem. Phys. to be published. (m) R. L. Bell and T. N. Truong, J. Chem. Phys., to be published.

[51] (a) D. G. Truhlar and M. S. Gordon, Science **249**, 491 (1990). (b) M. Page, Computer Phys. Commun. **84**, 115 (1994).

[52] I. Rossi and D. G. Truhlar, Chem. Phys. Lett. **233**, 231 (1995).

[53] (a) D. E. Goldberg, Genetic Algorithms in Search, Optimization, and Machine Learning, Addison-Wesley, Reading, MA, 1989. (b) J. H. Holland, Scientific American, July, 1992, p. 66. (c) E. J. Anderson and M. J. Ferris, in Parallel Processing for Scientific Computing, J. Dongarra, P. Messina, D. C. Sorensen, and R. G. Voigt (eds.), SIAM, Philadelphia, 1989, p. 137. (d) T. Bäck and F. Hoffmeister, in Proceedings of the Fourth International Conference on Genetic Algorithms, R. K. Belew and L. B. Booker (eds.), Morgan Kaufmann, San Mateo, CA, 1991, p. 92. (e) R. S. Judson, E. P. Jaeger, A. M. Treasurywala, and M. L. Peterson, J. Comp. Chem. **14**, 1407 (1993). (f) D. B. McGarrah and R. S. Judson, J. Comp. Chem. **14**, 1385 (1993). (g) Y. L. Xiao and D. E. Williams, J. Phys. Chem. **98**, 7191 (1994).

[54] (a) T. Bäck, F. Hoffmeister, and H.-P. Schwefel, in Proceedings of the Fourth International Conference on Genetic Algorithms, R. K. Belew and L. B. Booker (eds.), Morgan Kaufmann, San Mateo, CA, 1991, p. 2. (b) J. J. Grefenstette, in Proceedings of the Fourth International Conference on Genetic Algorithms, R. K. Belew and L. B. Booker (eds.), Morgan Kaufmann, San Mateo, CA, 1991, p.303. (c) K. DeJong and W. Spears, in Proceedings of the Fifth International Conference on Genetic Algorithms, S. Forrest (ed.), Morgan Kaufmann, San Mateo, CA, 1993, p. 618. (d) W. M. Spears, K. A. De Jong, T. Bäch, D. Fogel,

and H. de Garis, in Machine Learning, ECML-93, P. B. Bradzil (ed.), Springer, Berlin, 1993, p. 442. (e) T. Bäch and H.-P. Schwefel, Evolutionary Computation **1**, 1 (1993).

[55]   A. Gonzalez-Lafont, T. N. Truong, and D. G. Truhlar, J. Chem. Phys. **95**, 8875 (1991).

[56]   (a) S. P. Walch and T. H. Dunning, J. Chem. Phys. **72**, 1303 (1980). (b) G. C. Schatz and H. H. Elgersma, Chem. Phys. Lett. **73**, 21 (1980).

[57]   V. S. Melissas and D. G. Truhlar, J. Chem. Phys. **99**, 3542 (1993).

[58]   T. H. Dunning, Jr., J. Chem. Phys. **90**, 1007 (1989).

[59]   J. A. Pople and M. A. Head-Gordon, J. Chem. Phys. **82**, 284 (1985).

[60]   (a) H. S. Johnston, Gas Phase Reaction Rate Theory, Ronald Press, New York, 1966. (b) B. C. Garrett, D. G. Truhlar, and R. S. Grev, in Potential Energy Surfaces and Dynamics Calculations, D. G. Truhlar (ed.), Plenum, New York, 1981, p. 587.

[61]   G. S. Hammond, J. Amer. Chem. Soc. **77**, 334 (1955).

[62]   E. B. Wilson, Jr., J. C. Decius, and P. C. Cross, Molecular Vibrations, McGraw-Hill, New York, 1955.

[63]   (a) J. B. Howard, J. Chem. Phys. **5**, 442 (1937). (b) J. B. Howard, J. Chem. Phys. **5**, 451 (1937). (c) B. Kirtman, J. Chem. Phys. **37**, 2516 (1962). (d) B. Kirtman, J. Chem. Phys. **41**, 775 (1964). (e) J. T. Hougen, Can. J. Phys. **42**, 1920 (1964). (f) P. R. Bunker, J. Chem. Phys. **47**, 718 (1967), **48**, 2832(E) (1968). (g) J. T. Hougen, P. R. Bunker, and J. W. C. Johns, J. Mol. Spectrosc. **34**, 136 (1970). (h) D. C. Moule and C. V. S. R. Rao, J. Mol. Spectrosc. **45**, 120 (1973). (i) G. Dellepiane, M. Gussoni, and J. T. Hougen, J. Mol. Spectrosc. **47**, 575 (1973). (j) D. Papoušek, J. M. R. Stone, and V. Špirko, J. Mol. Spectrosc. **48**, 17 (1973). (k) A. R. Hoy and P. R. Bunker, J. Mol. Spectrosc. **52**, 439 (1974).

[64]   G. A. Natanson, Mol. Phys. **46**, 481 (1982).

[65]   A. D. Isaacson, D. G. Truhlar, K. Scanlon, and J. Overend, J. Chem. Phys. **75**, 3017 (1981).

[66] D. G. Truhlar, F. B. Brown, R. Steckler, and A. Isaacson, in The Theory of Chemical Reaction Dynamics, D. C. Clary (ed.), Reidel, Dordrecht, 1986, p. 285.

[67] D. G. Truhlar, J. Chem. Soc. Faraday Trans. **90**, 1608 (1994).

AB INITIO STUDIES OF REACTION PATHS IN EXCITED-STATE HYDROGEN-TRANSFER PROCESSES

Andrzej L. Sobolewski

Institute of Physics, Polish Academy of Sciences, 02 668 Warsaw, Poland

and

Wolfgang Domcke

Institute of Physical and Theoretical Chemistry, Technical University of Munich, 85748 Garching, Germany

1. Introduction

The reaction path concept is nowadays widely employed for the qualitative characterization of chemical reaction dynamics, see [1-3] for reviews. The availability of analytic energy gradients in *ab initio* calculations, in particular at the self-consistent-field (SCF) and second-order Moller- Plesset (MP2) levels, has allowed the systematic determination of reaction paths even for relatively large polyatomic systems [2]. Nearly all of these applications have been concerned with the potential-energy (PE) surface of the electronic ground state.

When considering reaction paths on the PE surfaces of excited states, as required for the rationalization of photochemistry [4], two major additional complications arise. First, reliable *ab initio* energy calculations for excited states are typically much more involved than ground-state calculations. Secondly, multi-dimensional surface crossings are the rule rather than the exception for excited electronic states. The concept of an isolated Born-Oppenheimer(BO) surface, which is usually assumed from the outset in reaction-path theory, is thus not appropriate for excited-state dynamics. At surface crossings (so-called conical intersections [5-7]) the adiabatic PE surfaces exhibit non-differentiable cusps, which preclude the application of the established methods of mathematical reaction-path theory [1,3]. As an alternative to non-differentiable adiabatic PE surfaces, so-called diabatic surfaces [8] may be introduced, which are smooth functions of the nuclear coordinates. However, the definition of these diabatic surfaces and associated wave functions is not unique and involves some subtleties [9-11].

Despite the lack of a comprehensive reaction-path theory for excited- state surfaces, the reaction path concept can fruitfully be applied at a pragmatic level to make progress in the understanding of photochemical dynamics. In the present Chapter

D. Heidrich (ed.), The Reaction Path in Chemistry:
Current Approaches and Perspectives, 257–282.
© *1995 Kluwer Academic Publishers. Printed in the Netherlands.*

we review recent *ab initio* work on photoinduced excited-state intramolecular proton (or hydrogen) transfer (ESIPT) processes, in which the reaction-path concept has been employed to explore relevant regions of the multi-dimensional PE surfaces, in particular barriers and surface crossings. These applications illustrate the value of the reaction path concept for overcoming the dimensionality problem of polyatomic PE surfaces, as has previously been demonstrated for many ground-state reactions. In addition, the results shed light on the possibilities and limitations of present-day *ab initio* technology in photochemistry and may thus be of some interest for a broader readership.

The ESIPT process is one of the simplest and most fundamental photochemical reactions and it has attracted much interest for many years. The most recent and perhaps a most comprehensive review of the vast literature concerning this field has been given by Formosinho and Arnaut [13] (for a review of selected topics see also Refs. [14-17]).

In general, the ESIPT process manifests itself in the rise and decay of the strongly Stokes-shifted fluorescence of the tautomer after absorption to the lowest excited singlet manifold of the primary molecular species. In most cases the fluorescence of the primary form does not appear at all. Following the radiative or radiationless decay of the excited tautomer, there is a reverse isomerization on the ground-state surface, so the entire process is cyclic. Time-resolved spectroscopy has revealed the following facts characteristic to ESIPT: (i) the dynamics of the excited-state proton transfer encompasses a large range of time scale, ranging from the femtosecond to the microsecond regime, although the majority of the relevant processes occur in the picosecond domain, (ii) the dynamics of the process can be strongly dependent on the solvent, particularly with respect to the polarity and the formation of intermolecular hydrogen bonds, and (iii) there is a temperature- (or excess-energy-) dependent very efficient non-radiative deactivation process competitive with ESIPT.

The large range of time scales for ESIPT indicates that quantum-mechanical tunneling is the dominant process in an intrinsic ESIPT process on the lowest excited singlet state PE surface. In non-alcoholic solvents the yield of non-radiative decay is independent of the viscosity, thus indicating the intramolecular nature of the process. The transfer of a proton between two groups of an aromatic molecule is associated with significant changes in dipole moments and a quite large ($\leq 10000 cm^{-1}$) fluorescence shift. This indicates large structural and electronic rearrangements of the system during the ESIPT process. In effect, the occurrence of the ESIPT reaction depends on a delicate balance between all the above mentioned factors. The large number of studies of ESIPT processes have lead to significant progress in the understanding of the basic mechanism of the process, but the complexity of the problem still precludes a unifying view of the ESIPT phenomenon.

The large size of the systems which are experimentally investigated with respect to the ESIPT process has precluded for a long time any advanced *ab initio* study of the reaction. Most of the earlier attempts to characterize the PE surfaces relevant for the

ESIPT process have used either semi-empirical or simple *ab initio* methods (see for instance [18-27]). Only recently significant progress in the theoretical description of the electronic structure of excited states of larger polyatomic systems has been made [28,29], which has allowed a more quantitative *ab initio* treatment of the excited-state PE surfaces relevant for this process [30-37]. In the following we discuss the more important conclusions concerning ESIPT reactions which emerge from these studies.

2. Theoretical methodology

2.1. Concerted vs. minimum energy reaction paths

The description of the reaction dynamics of a polyatomic molecule requires generally a full exploration of the 3N-6 dimensional (N being the number of atoms) PE surface. For a qualitative characterization of reaction, however, one can follow a simplified approach based on the concept of the reaction path (for a comprehensive review see [1-3,40] and references therein). Among others the concerted reaction (CR) path approach based on a straight line, least motion intramolecular coordinate, that interpolates linearly between the equilibrium geometries of the reactant and the product has often been used [27,30,31,36-38]. The geometries can either be defined in internal or Cartesian coordinates. Although the formulation of the CR path is straightforward, its definition in terms of Cartesian coordinates rises some problems because the distance in mass-weighted Cartesian coordinates between adjacent points is altered by rotation or translation of their respective reference axis. It has been shown that in its Cartesian definition all couplings in the kinetic energy part of the molecular Hamiltonian can be eliminated and transferred to the potential energy part [38]. This can significantly simplify the theoretical treatment of the molecular dynamics. On the other hand, if one is interested in a qualitative characterization of the PE function along the reaction coordinate, the CR path approach has a serious disadvantage: it does generally not pass through the saddle point which separates the minima of reactants and products [3].

With respect to a more quantitative characterization of the PE function along the reaction coordinate, in terms of local minima and barriers between them, the minimum energy reaction (MER) path concept is more promising. The MER path approach, also called the intrinsic reaction path approach [39], is defined as the steepest descent path from the transition state (a saddle point on the PE surface) down to the local minima that are equilibrium geometries of reactant and product. It has been shown [40] how to express the Hamiltonian of an N atom molecular system in terms of the reaction coordinate, and this approach has been used successfully to describe a variety of processes in polyatomic reaction dynamics. It is also well known that for many reactions (PT process among them) the MER path is very sharply curved, so that the relevant dynamical motion deviates far from it. This is not a particularly important point for our present purpose of a qualitative characterization of the PE surfaces relevant for the PT reaction, as far as we do not consider the dynamics of this process. Therefore we may adopt a widely used simplified version of the MER path formulation. In this approach one defines one of the 3N-6 intramolecular degrees of freedom as the reaction coordinate, while the remaining (3N-7) coordinates are optimized at each step of the reaction. There are no strict rules for choosing the reaction coordinate. In principle, this can be any of the 3N-6 intramolecular degrees of freedom. In practice, this should be the coordinate which changes the most when the reaction proceeds.

The description of the PT reaction typically involves a reaction coordinate of the type $X - H...Y$, where the light hydrogen nucleus is moving between two heavier (X and Y) heteroatoms. At the extremal points of the coordinate the hydrogen atom becomes chemically bonded to one of the atoms and forms a hydrogen bond with the other, and *vice versa*. Possible choices of the reaction coordinate are: the $X - H$ (or $H...Y$) distance, the difference between the $X - H$ and $H...Y$ distances, or the angular position of the H-atom with respect to the rest of molecular moiety. In the following presentation the $X - H$ distance is used to define the MER coordinate, although the difference between the $X - H$ and $H...Y$ distances would probably give a more "symmetrical" description of the reaction. One should also be aware that in some cases (of sharp curvature) the reaction valley may not be uniquely defined for a given reaction coordinate, so the transition state can be missed [3].

2.2 Role of molecular symmetry

Most of the molecular systems studied with respect to the ESIPT process remain planar or almost planar in the ground state. Almost exclusively the stable tautomeric form of these compounds consists of aromatic or quasi-aromatic ring(s) with some substituents (hydroxy-, amino-, carbonyl-, *etc.*) connected to the ring via single chemical bonds [13-17]. The nuclear geometry of these substituents with respect to the aromatic ring is often decisive for the overall symmetry of the system. The planarity of the molecular system is particularly important with respect to the excited-state calculations. If the nuclear frame of the system conserves the symmetry plane, its Hamiltonian possesses (at least) C_s symmetry. Within this point group, molecular orbitals of the system can be classified in terms of σ, n, and π orbitals, and the lowest electronically excited states generally result from the $\pi\pi^*$ and $n\pi^*$ excitations. As such they fall into two distinct symmetry representations, A' and A", respectively, in the C_s point group. Any out-of-plane deformation of the nuclear frame destroys the C_s symmetry of the system and mixes the states together. It is then often practically impossible to get a converged solution for the higher of the two lowest close lying excited electronic states of the same symmetry (A) and multiplicity, which in the C_s symmetry are distinct (*i.e.* A'($\pi\pi^*$) and A"($n\pi^*$)) and rather easy to converge.

While the stabilization of the ground state due to the out-of-plane deformation of the substituents is generally very small, on the energy scale relevant for the PT reaction, this may not be the case for the excited states. Two important sources of out-of-plane deformation of an aromatic (or aza-aromatic) ring after electronic excitation can be expected. First, the rigidity of the ring is weakened due to the promotion of one of the electrons to an antibonding π^*-orbital, and second, there are generally closely spaced $A'(\pi\pi^*)$ and $A''(n\pi^*)$ states which are non-adiabatically coupled to each other due to out-of-plane (A'' symmetry) vibrations. The so-called pseudo-Jahn-Teller interaction [12] can distort the nuclear geometry of some of the excited electronic states along the vibrational (out-of-plane) coordinates of the coupling modes. It is thus convenient, for practical reasons, to restrict the molecular system to C_s symmetry in calculations of the excited-state PE functions along the PT reaction coordinate and to consider any distortion from the C_s symmetry as resulting from vibronic interactions. This approach, of course, makes sense if the distortion from the C_s symmetry has not a serious influence on the energy. This point must be carefully checked for each individual system.

2.3 *Ab initio* methods used in the present studies

The molecular geometries of most of molecular systems considered in this study were optimized at the Hartree-Fock (HF) level for the ground state, whereas in the optimization of the excited-state geometries the configuration interaction scheme with single excitations from the HF reference (CIS) were used. These simple methods are very efficient and relatively cheap which is an important factor in the search for minimum-energy nuclear configurations of large polyatomic systems. They have, however, an important disadvantage: they do not properly describe dissociation of the molecular system. For a correct description of dissociation and also for a proper description of some of the excited-states one needs a multi-reference electronic wave function. This can be provided by the complete-active-space self-consistent-field (CASSCF) approximation, *i.e.* by performing a full configuration-interaction (CI) calculation within a limited orbital space [28]. It is well established that the CASSCF method is especially useful in describing a process where the electronic structure varies strongly as a function of nuclear geometry, as for instance along the PT reaction coordinate. For larger molecules, however, the size of the CI space quickly becomes too large to be tractable. The method of choice, when dealing with the excited states of large polyatomic systems, is to perform the CASSCF calculations with a selected orbital space for a given electronic state of interest, thus treating variationally all the near-degeneracy effects in the multi-configuration electronic wave function. The remaining dynamic correlation effects are then added in a subsequent step with the use of second-order perturbation theory with the CASSCF wave function as the reference (CASPT2) [29]. For the (closed-shell) ground state, the wave function of which is essentially given by one determinant, the CASPT2 approximation should give results similar to the much cheaper MP2 theory. For more details concerning the computational methods, see [30-37].

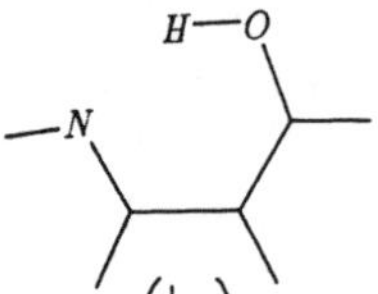
(Ia)

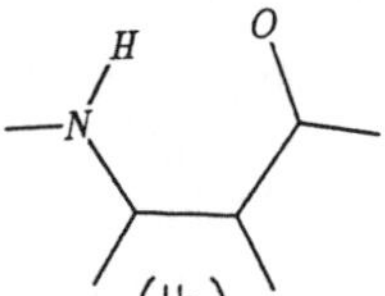
(Ib)

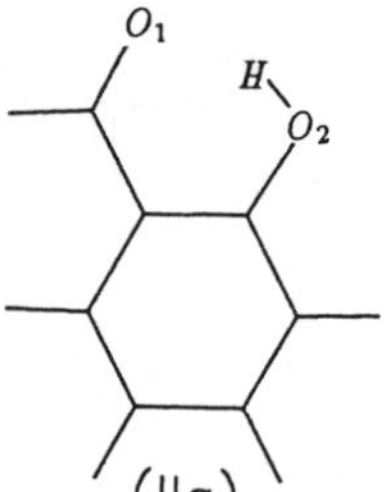
(IIa)

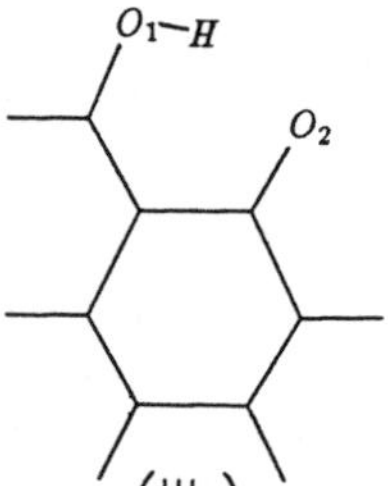
(IIb)

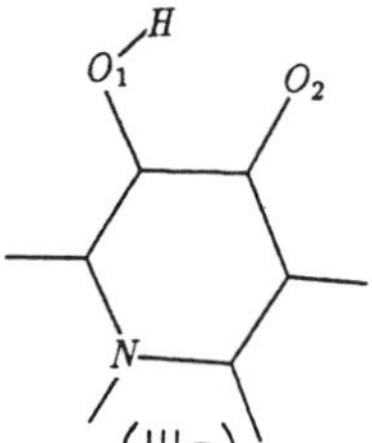
(IIIa)

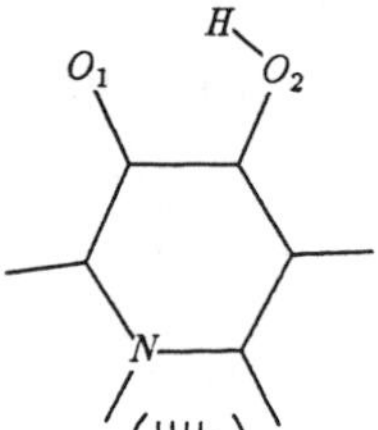
(IIIb)

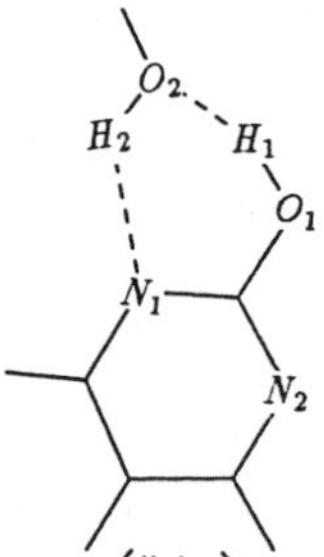
(IVa)

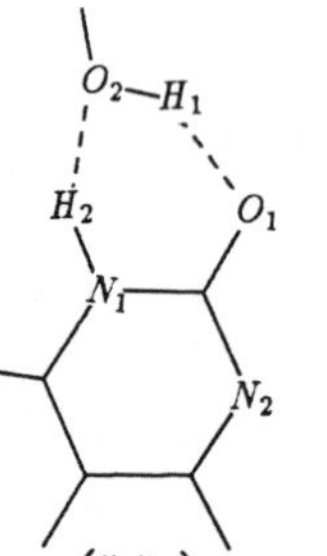
(IVb)

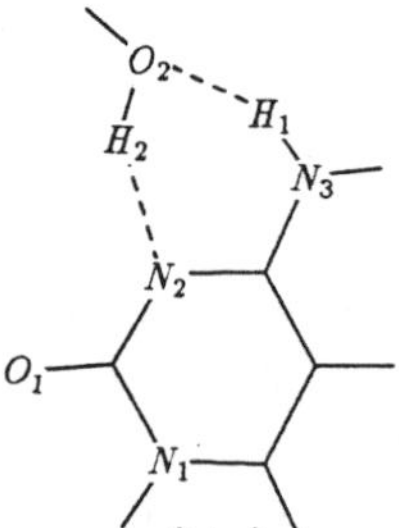
(Va)

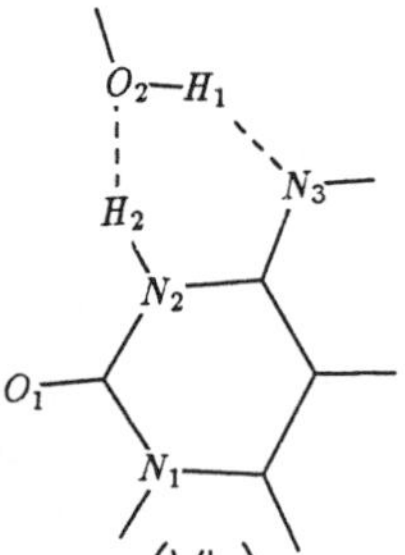
(Vb)

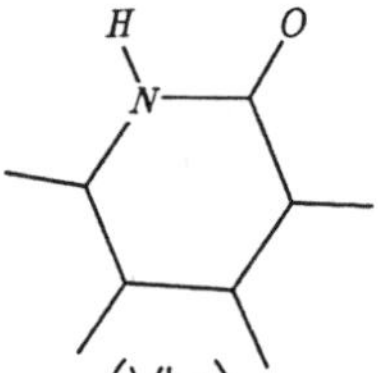
(VIa)

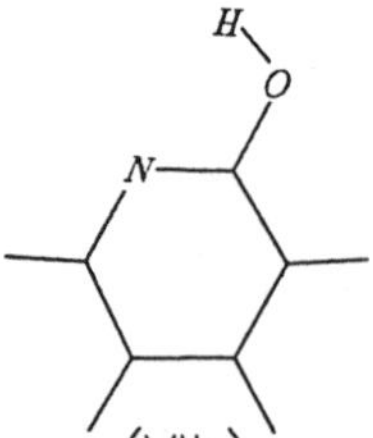
(VIb)

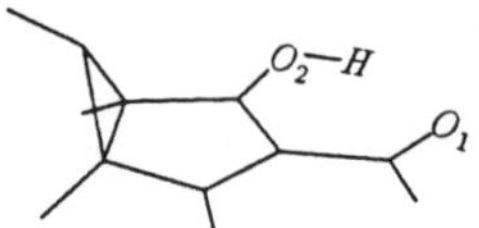
(VII)

3. Theoretical results.

The proton transferring compounds can be divided into two distinct categories (for a more detailed classification see [41]): (A) molecules with an intramolecular hydrogen bond(s) (compounds I-III), and (B) molecules without an intramolecular hydrogen bond(s) (compounds IV-VI). If, however, the compounds of the latter category form hydrogen bonded molecular complexes with protic solvents (water, alcohol, *etc.*, for instance), the PT reaction in such systems which changes one tautomeric form into another can be considered as a concerted double-proton transfer along pre-existing hydrogen bonds (systems IV and V). There is also experimental evidence of an optically induced PT reaction in series of isolated compounds of B-category [42,43], so theoretical results for this "exotic" (from the point of view of the standard PT reaction) compound (VI) will also be presented.

A typical proton transferring system contains a molecular fragment of the type X-H...Y, where X and Y are heteroatoms which can form hydrogen bonds. The length of the XH (or YH) chemical bond is of the order of 1.0Å. The length of the hydrogen bond can vary over a wide range from 1.5Å up to 2.0Å, depending mostly on the nuclear geometry of the X-H...Y molecular fragment and on the strength of the hydrogen bonded complex. The Y...H distance in the isolated systems of the B-category is much larger, for instance in 2-pyridone it is of the order of 2.5Å. From an quantitative point of view, all the above mentioned systems differ in the distance over which the hydrogen is transferred in the reaction. Does this have an influence on the basic mechanism responsible for the ESIPT process? In the following presentation we try to answer this question.

3.1 Systems with an intramolecular hydrogen bond

In a molecular system with an intramolecular hydrogen bond the X and Y heteroatoms constitute a part of molecular system which together with a "transferring" hydrogen atom form typically a six- (in some cases a five-) membered atomic ring closed by the hydrogen bond (I-III). The simplest system which can serve as a model for this class of compounds is malonaldehyde [44]. The two tautomeric forms of this molecule are identical, and there is no physical property that can monitor in real time the transfer of the H atom. The tunneling splitting, however, is present in the molecular spectrum and indicates the occurrence of the chemical process. If one replaces one of the oxygen atoms of malonaldehyde by a nitrogen atom, for instance, the two tautomeric structures of this compound (I) correspond to different molecular species, and consequently there is an asymmetry in the double-minimum PE functions along the PT reaction coordinate. As far as ground- and excited-state PE functions are concerned, two situations are possible: the asymmetry is of the same kind in both states (common asymmetry), or it is reversed in the excited-state (reversed asymmetry) [45]. Only the latter situation may be relevant for a typical PT cycle. Another basic question of the ESIPT process concerns the existence of a double or single minimum in the electronic states involved in the process. If the latter situation prevails, tautomerization is simply achieved by a vibrational relaxation process. The absence of an energy barrier would lead to PT rates in the subpicosecond domain.

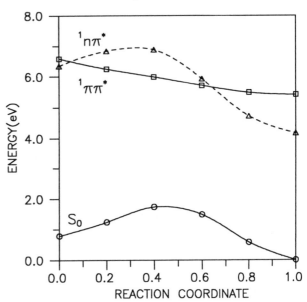

Figure 1. CASSCF potential energy functions calculated along the PT reaction coordinate for the ground- (circles), and for the lowest $\pi\pi^*$ (squares) and $n\pi^*$ (triangles) excited singlet states of APN ($Q_{PT}=0$ -form Ia, $Q_{PT}=1$ -form Ib).

As a first example, 1-amino-3-propenal (APN) has been considered in Ref.[30] as a model system for the photo-induced ESIPT process. The PE surfaces of the ground and the lowest excited singlet electronic states were characterized by *ab initio* CASSCF calculations. The concerted reaction path for PT was defined in internal coordinates for the (Ia) and (Ib) nuclear structures of APN optimized in the ground state. The resulting PE functions are presented in Figure 1. Upon inspection of this figure one sees that first of all both tautomeric forms represent, as expected, local minima on the ground-state PE surface and are separated from each other by a barrier along the PT reaction coordinate. Unfortunately, the keto-form (Ib), which is usually the product form in the ESIPT process, is lower in the energy on the ground-state PE surface of the considered model compound. Thus the picture presented in Figure 1 represents the "common" asymmetry example. The ground state PE surface is not, however, the main subject of interest in this Chapter, so let us concentrate our attention on the lowest excited-state PE surfaces. As it has been already anticipated in Section 3.2, the two lowest excited singlet states for such a planar system, $^1A''(n\pi^*)$ and $^1A'(\pi\pi^*)$, differ in both the orbital symmetry and in the electronic nature. The wave function of the first state is dominated by the $n\pi^*$ configuration (16a')(4a''), while the wave function of the second singlet is dominated by the HOMO-LUMO $\pi\pi^*$ configuration (3a'')(4a''). As expected, the $\pi\pi^*$ excited singlet state has significant oscillator strength for the radiative transitions from/to the ground state ($\simeq 0.3$) while the oscillator strength of the $n\pi^*$ excited state is very small $(10^{-2} - 10^{-5}$ depending on the tautomeric form [30]). In other words, the $^1A'(\pi\pi^*)$ state is the "bright" state which can be directly populated by absorption from the ground state and it should also be the "emitting" state for the fluorescence. The $^1A''(n\pi^*)$ state is an essentially "dark" state for absorption from the ground state and it also has a negligibly small probability of spontaneous emission.

The PE functions of both excited states are also very different along the PT reaction coordinate (Figure 1). While the PE function of the $^1A''(n\pi^*)$ state shows an qualitatively similar behavior as the ground-state PE function with a significant barrier for the PT reaction, the PE function of the $^1A''(\pi\pi^*)$ state is a barrierless along the reaction coordinate, providing a pathway for an ultrafast ESIPT reaction. Moreover, in this particular molecular system, the PE functions of both excited singlet states cross each other twice along the PT reaction coordinate. The two PE surfaces intersect only when the C_s symmetry of the nuclear frame is conserved. Any out-of-plane vibrational motion will mix the $^1A''(n\pi^*)$ and $^1A'(\pi\pi^*)$ states, resulting in a multidimensional conical intersection of their PE surfaces. The strong non-adiabatic effects which operate at such conical intersections [46,47] will cause a complex time-dependent dynamics of the wave packet on the coupled PE surfaces. Generally one can say that the "bright" $^1A''(\pi\pi^*)$ state is the proton transferring state, whereas the "dark" $^1A''(n\pi^*)$ state is an intruder state which tends to inhibit the PT reaction. This interference is one of the key factors which can govern the dynamics of the reaction. The most important conclusion which emerges from the calculations of Ref.[30] is that the ESIPT reaction cannot be generally considered as a process occurring on a single adiabatic PE surface. Consideration of $n\pi^* - \pi\pi^*$ vibronic coupling is essential for the understanding of the dynamics of the process. The energy separation of the lowest excited $n\pi^*$ and $\pi\pi^*$ singlet states is sensitive to the size of molecular system as well as to the environment in which the ESIPT reaction is investigated. Whereas the effect of the environment on the excited state PE surfaces of polyatomic molecules is still beyond any reasonable *ab initio* treatment, the effect of molecular size can to some extent be investigated. In Ref.[31,32] PE surfaces relevant for the ESIPT process in the "real" proton transferring systems o-benzaldehyde (OHBA) and N-substituted-3-hydroxypyridinones have been studied. The resulting PE functions of OHBA (II) along the concerted PT reaction coordinate (the tautomeric structures were optimized in the ground state) calculated at the CASPT2 levels of theory are presented in Figure 2. One sees that the PE functions of the lowest excited singlet states are qualitatively similar to those presented in Figure 1 for the model system of APN. They also reveal the importance of dynamic electron correlation effects which are neglected at the CASSCF level of approximation. The inclusion of dynamic correlation effects at the CASPT2 level eliminates completely the barriers for PT on the $^1A'(\pi\pi^*)$ and on the ground-state PE surfaces present in the CASSCF approximation (see Ref.[31]), while the height of the barrier on the $^1A''(n\pi^*)$ PE surface is reduced. One should be careful in the interpretation of the CASPT2 results (see also discussion in Ref.[32]) since in this approach the dynamic electron correlation effects are treated perturbatively. Nevertheless, one should emphasize that the qualitative picture of ESIPT dynamics in OHBA which emerges from the CASPT2 results presented in Figure 2 is in full accord with experimental observations, where neither the existence of a barrier for PT on the $S_1(\pi\pi^*)$ nor on the S_0 PE surfaces has been reported [48-50].

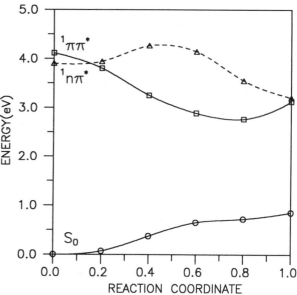

Figure 2. CASPT2 potential energy functions calculated along the PT reaction coordinate for the ground- (circles), and for the lowest $\pi\pi^*$ (squares) and $n\pi^*$ (triangles) excited singlet states of OHBA ($Q_{PT}=0$ -form IIa, $Q_{PT}=1$ -form IIb).

The PE functions considered along the concerted reaction coordinate, discussed so far, do not provide any information concerning the importance of geometrical (or vibrational) relaxation along the reaction path for PT, because by definition all intramolecular coordinates are changing linearly (proportionally) along the reaction path. The possible role of the out-of-plane vibrations for pseudo-Jahn-Teller distortions has been mentioned above. Some information with this respect can be obtained from the minimum energy reaction (MER) approach. This treatment has been applied in investigations of the PE functions relevant for ESIPT in 3-hydroxy-4-pyridinone (3HPD), structure III, and its methyl derivative 3-hydroxy-2-methyl-4-pyridinone (3H2MPD) [32] for which the occurrence of ESIPT process has recently been reported [51]. In the MER path investigations of Ref.[32] the O_1H distance has been defined as the reaction coordinate, and the rest of the remaining intramolecular coordinates has been optimized at each step. A full optimization of the reaction path has been performed for 3HPD and 3H2MPD in the ground and in the lowest spectroscopically relevant excited $A'(\pi\pi^*)$ singlet state. The obtained variations of some of the intramolecular coordinates involved in the PT reaction are qualitatively similar for both states, so in the following we present (Figure 3) and discuss those corresponding to the excited $^1A'(\pi\pi^*)$ state of 3HPD.

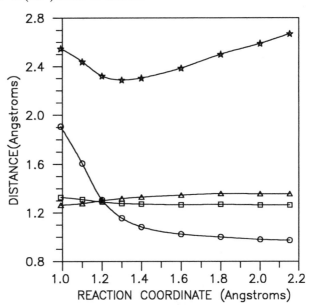

Figure 3. Variation of some of intramolecular distances of 3HPD along the PT reaction coordinate (the O_1H distance): O_2H -(circles), O_1O_2 -(stars), CO_1 -(squares), and CO_2 -(triangles).

The first conclusion resulting from inspection of Figure 3 is that the variation of the intramolecular coordinates along the reaction path is far from being linear. This reflects the essence of the PT reaction mechanism. Most sensitive to the reaction coordinate (O_1H) is the distance of the "mobile" hydrogen atom from the second oxygen atom (O_2H). It decreases abruptly with increase of the O_1H distance, up to a certain point. It is the point where the hydrogen atom is almost equally separated from both oxygen atoms. After passing this point the hydrogen forms a chemical bond with the second oxygen atom (O_2), and its bond length remains practically constant vs. the reaction coordinate. The hydrogen jumping (PT reaction) is accompanied by a significant decrease of the O_1O_2 distance, which reaches its minimum at the transition point. This results from a simultaneous decrease of the CCO_1 and CCO_2 bond angles (not indicated in the figure, see original paper for details). The most important contribution which the decrease of the distance between the oxygen atoms has on the PT process is the lowering of the barrier for the PT reaction. The other intramolecular coordinates do not show such drastic changes along the reaction path. Perhaps the most noticeable one is interchange of the single/double bond character between the CO_1 and CO_2 bonds expressed in the variation of their lengths.

Let us note that the variations of the intramolecular coordinates along the PT reaction path depend on the definition of the reaction coordinate. The only general trend which has already been discussed in Section 2.2 and which was observed for all the systems discussed in this Chapter is that near the transient point for PT the two heteroatoms X and Y between which the hydrogen atom is transferred are getting closer each other and this decreases the barrier for PT.

3.2. Hydrogen bonded complexes

In compounds in which the hydrogen atom is far away from the acceptor atom (Y), an intramolecular (intrinsic) proton transfer is impossible because of the lack of a hydrogen bond and generally the PT process requires a mediator (for an exception, see the next Section). The simplest mediating system is the water molecule which can form two hydrogen bonds with the compound of interest and the PT reaction which transforms the tautomeric forms into each other can be viewed as a concerted biprotonic transfer (IV,V). Theoretical investigations of the PE functions relevant for ESIPT in hydrogen bonded complexes of 2-pyrimidinone (2PMD) and cytosine (CYT) with water have been reported in Refs. [33] and [34], respectively.

In a low-temperature matrix one tautomer, the hydroxy-form of 2PMD, is predominating [52]. Complexation with water stabilizes more its oxo-form (IVb), and both forms (hydroxy- and oxo-) of the complex are observed [53]. Among possible tautomers of cytosine, three: the amino-hydroxy-, the amino-oxo- and the imino-oxo- are observed in vacuum [54a] and in low-temperature matrices [42b,54b]. In aqueous solution the amino-oxo (the native form of DNA) is found to be the dominant tautomer [55]. These experimental findings concerning the ground state have been confirmed by the theoretical results. Thus the oxo-form of 2PMD is higher in energy by $0.19eV$ than the hydroxy-form, but complexation with water (IV) decrease the difference to only $0.05eV$ [33]. Similarly, the amino-oxo-form of cytosine is by only $0.03eV$ more stable than the imino-oxo-form, but complexation with water (V) increases the difference up to $0.14eV$ [34].

In the calculations of the reaction path for the PT reaction in hydrogen bonded complexes $2PMD{:}H_2O$ and $CYT{:}H_2O$ the MER path concept has been applied. As the reaction coordinate the O_1H_1 and the N_3H_1 bond lengths, respectively, were chosen. The rest of the intramolecular degrees of freedom was optimized at each step of the reaction. Both systems are generally non-planar with one hydrogen atom of water tilted out of the molecular plane. This has, however, a minor effect on the ground-state energy of both systems. The relative energy difference between two tautomers is only by $110cm^{-1}$ and $130cm^{-1}$ changed by imposing C_s symmetry for $2PMD{:}H_2O$ and $CYT{:}H_2O$ systems, respectively. This has been taken as a justification for optimizing the excited-state geometry within the C_s point group. In both systems the $A''(n\pi^*)$ singlet state is significantly higher in energy (by about $0.5eV$) than the $A'(\pi\pi^*)$ lowest excited singlet state, thus being unimportant for the molecular dynamics on the optically excited PE surface. The variations of intramolecular coordinates involved in the PT reaction are qualitatively similar for both systems in the considered electronic states. In Figure 4 the results of Ref.[33] for the $^1A'(\pi\pi^*)$ state of $2PMD{:}H_2O$ system are presented as a typical example. The first conclusion from inspection of this figure is that, similarly to the case of an intramolecular PT process considered in preceding Section, the variations of the intramolecular coordinates along the reaction path are far from being linear. The most characteristic effect is the variation of the O_2H_2 bond length of the water molecule. It remains practically unchanged vs. the reaction coordinate (the O_1H_1 bond length) up to a certain point. When hydrogen atom (H_1) becomes almost equally spaced from both oxygens, its length increases abruptly and, simultaneously the O_2H_1 and the N_1H_2 distance shorten and new chemical bonds are formed. Both, the O_2H_1 and N_1H_2, distances remain practically constant for the remaining part of the reaction coordinate. The hydrogen jumping is accompanied by a significant decrease of the distance between both components of the complex at the transient point. The mechanism of the double-PT reaction in hydrogen bonded complexes of $2PMD$ (as well as of cytosine) with water is apparently the same as discussed above for an intrinsic intramolecular PT reaction.

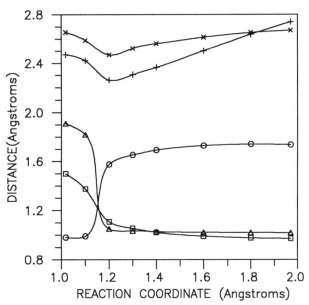

Figure 4. Variation of some of intra- and inter- molecular distances of the 3PMD:H_2O complex along the PT reaction coordinate (the O_1H_1 distance): O_2H_2 -(circles), O_2H_1 -(squares), N_1H_2 -(triangles), O_1O_2 -(+), and N_1O_2 -(x).

The PE functions obtained from the optimization procedure along the PT reaction path (Figure 4) represent a rather crude estimation since simple levels of theory (HF or CIS approximations) with a relatively small basis set (3-21G) were employed. In Figures 5 and 6 the PE functions of $2PMD$:H_2O and CYT:H_2O, respectively, recalculated with the use of MP2 for the ground state and CASPT2 for the excited states are presented. Gaussian basis sets of double-zeta quality with polarization functions have been employed in these calculations. Figures 5 and 6 show that the ground state PE functions behave very similar vs. the PT reaction coordinate in both systems. The product tautomeric forms are slightly higher in energy than the initial forms and are separated by significant barriers ($\simeq 0.5eV$ for $2PMD$:H_2O and

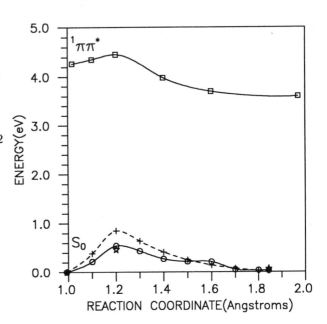

Figure 5. The PE functions of 3PMD:H_2O complex vs. the PT reaction coordinate (O_1H) preserving the C_s point group calculated by means of the MP2 method for the ground state - (circles) and by means of the CASPT2 method for the $\pi\pi^*$ excited singlet state -(squares). The crosses denote energy of the HF reference used in the MP2 calculations. The MP2 results for the ground state calculated at the C_1 optimized geometry are denoted by stars.

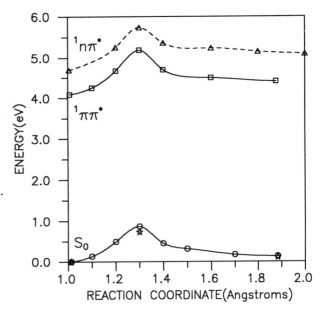

Figure 6. The PE functions of CYT:H_2O complex vs. the PT reaction coordinate (N_3H_1) preserving the C_s point group calculated by means of the MP2 method for the ground state - (circles) and by means of the CASPT2 method for the $\pi\pi^*$ -(squares) and for the $n\pi^*$ - (triangles) excited singlet states. The MP2 results for the ground state calculated at the C_1 optimized geometry are denoted by stars.

$\simeq 1.0eV$ for $CYT:H_2O$), so both systems are rather well protected against a thermally induced PT reaction. For a detailed discussion of the ground-state PE surfaces we refer to the original papers [33,34]. In the following we concentrate the attention on the excited-state PE functions.

In both complexes the excited $A''(n\pi^*)$ singlet states are significantly higher in energy than the $S_1(\pi\pi^*)$ states (see Figure 6), so we can limit discussion to the latter. The S_1 PE functions of both systems are significantly different. While the $S_1(\pi\pi^*)$ PE function of the $2PMD:H_2O$ complex has only a moderate barrier ($\simeq 0.19eV$) along the PT reaction coordinate and the reaction is by about $0.67eV$ exothermic, the amino-oxo-form of the $CYT:H_2O$ complex is even more stable in the excited S_1 state than in the ground state. In other words, the two systems considered here belong to two distinct categories with respect to the shape of their PE functions, *i.e.* to the reversed and to the common asymmetry case, respectively.

Concluding this paragraph, we can confirm that although the complex of cytosine with water seems to be stable with respect to any thermal- or photo-induced PT reaction, the $2PMD:H_2O$ complex may indeed be the candidate of choice for an experimental detection of ESIPT reaction in a hydrogen bonded molecular complex. The possibility of generation of "rare" tautomeric forms of cytosine in photoinduced bimolecular secondary processes has been theoretically considered in Ref.[35].

3.3. Systems without intramolecular hydrogen bonds

The majority of ESIPT reactions occur in systems where an intra- (or inter−) molecular hydrogen bond(s) between the transferred hydrogen atom and a heteroatomic proton acceptor facilitates the reaction. Generally, these molecular systems have six- (or five-) membered rings closed by hydrogen bond(s). There is an important class of compounds which can exist in two (or more) tautomeric structures, but there is no intramolecular hydrogen bond along which the proton can be transferred in a presumed PT reaction which transforms one tautomeric form into another. To this category belong, among others, 2-pyridone (VI) as well as isolated 2PMD (IV) and cytosine (V). Generally such systems have four-membered rings closed by an imagined line connecting the hydrogen atom to a heteroatomic proton acceptor. Such compounds did not attract special attention with respect to the ESIPT reaction until recently, when experimental evidence of the process has been reported [42(a)]. Up to date a whole class of compounds without an intramolecular hydrogen bond which transfer a proton after an optical excitation in low-temperature rare gas matrix has been characterized [42,43]. These systems have a peculiar feature which discriminates them from typical ESIPT systems. Namely, the proton is transferred from a ring nitrogen atom to an oxo- or tio- group, whereas in ESIPT systems the reverse process generally occurs. This indicates that the physical mechanisms responsible for these phenomena might be fundamentally different.

Among others the photo-induced tautomerization of 2-pyridone (2PY) has been

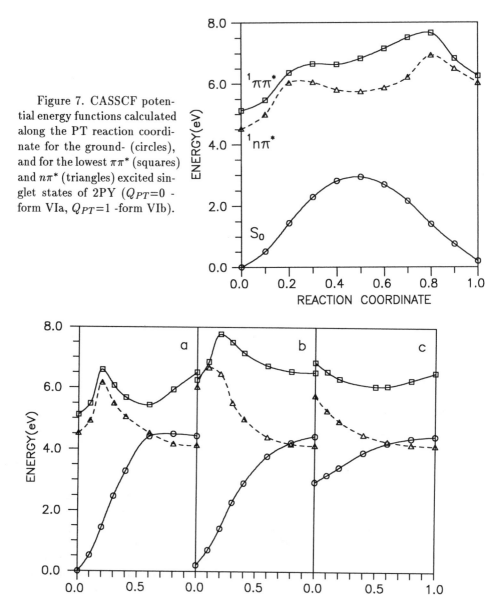

Figure 7. CASSCF potential energy functions calculated along the PT reaction coordinate for the ground- (circles), and for the lowest $\pi\pi^*$ (squares) and $n\pi^*$ (triangles) excited singlet states of 2PY ($Q_{PT}=0$ - form VIa, $Q_{PT}=1$ -form VIb).

Figure 8. CASSCF potential energy functions of the ground (circles) and the lowest excited singlet states, $\pi\pi^*$ (squares) and $n\pi^*$ (triangles), of 2PY calculated along the reaction coordinate to "mobile' hydrogen atom dissociation: (a) from equilibrium geometry of form VIa, (b) from equilibrium geometry of form VIb, and (c) from reaction coordinate $Q_{PT}=0.5$ of Figure 7.

experimentally detected [43]. The PE surfaces relevant for the ESIPT process in this system have been theoretically characterized in Ref.[36]. The two tautomeric forms of 2PY (VIa and VIb) were optimized in the ground state and the concerted reaction path along which one form is transformed into another was chosen as the reaction coordinate. The resulting PE functions of the ground- and the lowest excited-singlet states are presented in Figure 7. One sees that both tautomeric forms are separated by a significant barrier ($\simeq 3.0 eV$) along the PT reaction coordinate on the ground-state PE surface. The height of the barrier is significantly reduced at the real saddle-point for the reaction [56,57]. Similarly to typical ESIPT systems the aromatic form (VIb) of 2PY has the lowest excited singlet state located at higher energy than the non-aromatic form (VIa). The two tautomeric forms are also separated by barriers on the excited PE surfaces. The excited-state PE functions show, however, a behavior which is qualitatively different from that discussed above for typical ESIPT systems. It is particularly evident that the PE function of the $^1A"(n\pi^*)$ state has an additional (third) local minimum roughly at the position of the top of barrier in the ground state. An attempt to optimize the molecular geometry at the "third" minimum of the $^1A"(n\pi^*)$ PE surface lead to dissociation of the "mobile" hydrogen atom. Accepting one particular nuclear geometry along the path for dissociation where the hydrogen atom is almost equally spaced from both the nitrogen and oxygen atoms, the concerted reaction path for dissociation can be defined. In Figure 8 the PE functions of the ground and the two lowest excited singlet states calculated along the reaction coordinate for dissociation from the three distinct minima of the $^1A"(n\pi^*)$ PE function of Figure 7 are presented. One sees that apart from the "third" local minimum (Figure 8c) the two other minima at the geometries of stable tautomeric forms are protected by barriers against dissociation on all PE surfaces considered. Generally, one can say that there is no reaction coordinate connecting the two tautomeric forms of 2PY on the $^1A"(n\pi^*)$ PE surface which can avoid the dissociation of hydrogen atom. This is a direct consequence of lack of an intramolecular hydrogen bond along the PT reaction coordinate.

More recently Barone and Adamo [27] have theoretically re-investigated the PE functions of 2PY along the CR path for tautomerization. Instead of the "third" local minimum on the excited $^1A"(n\pi^*)$ PE surface found in Ref.[36] they have obtained only an energy plateau in the central part of the CR path. The difference in the slope of PE functions reported in both papers can partially result from the definition of the CR path defined in Cartesian and intramolecular coordinates, respectively. The Authors of Ref.[27] claim, however, that they do not see any convicing evidence of dissociation of the "mobile" hydrogen atom on the $^1A"(n\pi^*)$ PE surface. This simply results from their method of calculation of the excited-state energies (CIS). One cannot properly describe dissociation at the HF level of approximation. A multi-reference wave function (of a CASSCF type) is needed for proper description of this process. Let us mention that a quite similar behavior with respect to ESIPT reaction has recently been reported for formamide [37].

The possible mechanism of ESIPT process which emerges from the theoretical study is discussed in Section 4.2.

3.4. Non-radiative decay channel

As has already been mentioned in the Introduction, it is a common observation that there exists an energetical threshold above which an efficient non-radiative deactivation process takes place in ESIPT systems. The process can be safely identified as internal conversion to the ground state. The extremely high efficiency of the process $(k_{nr} \simeq 10^{12} s^{-1})$ at the threshold indicates the presence of strong non-adiabatic interactions between the excited-state and the ground-state PE surfaces. Such strong non-adiabatic effects are usually expected near conical intersections of adiabatic PE surfaces [46,47]. For a closed-shell system the lowest-energy intersections of such a kind usually occur along the reaction path leading to the formation of radicalic of biradicalic forms [4]. This has been well documented in the case of the so-called "channel three" of benzene and pyrazine [58-61] in which such intersections occur along the reaction path to the so-called prefulvenic form. The energetical barrier for the nonradiative process of the order of few kcal/mol observed for many of ESIPT systems [13-17] resembles to a large extent the situation observed in benzene (see Ref.[61] and references therein). The tendency of formation of the prefulvenic structures from the first $\pi\pi^*$ excited singlet state seems to be a characteristic feature of simple aromatic systems [4,59]. Along the reaction coordinate to the prefulvenic form the aromatic ring is out-of-plane deformed and a new bond between the pair of meta-atoms of the ring is formed (c.f structure VII, for instance). In the case of benzene and pyrazine the reaction coordinate retains C_s symmetry, and thus there is a real intersection of the $S_0(A')$ and $S_1(A")$ states along the reaction path [58-61]. In the case of ESIPT systems considered in this Chapter, any out-of-plane deformation of the aromatic ring reduces the molecular symmetry to C_1 . In the C_1 symmetry all states belong to the same irreducible representation and the PE functions should avoid crossing.

The PE functions along the CR path leading to prefulvenic forms of OHBA and cytosine have been reported in Refs.[31,34]. The calculation of the excited-state energy for a system with no symmetry poses a serious computational problem, because the near-degeneracy of different states of the same symmetry leads to convergence problems. This can to some extent be avoided by state-averaged calculations of the CASSCF wave function which can also serve as a reference for the CASPT2 treatment. The state-averaged CASSCF PE functions of the ground and the lowest excited singlet states of OHBA (III) along the CR path to the prefulvenic form (VII) [31] are presented in Figure 9.

Inspection of Figure 9 shows a close correspondence between these PE functions and those determined along a similar reaction coordinate for benzene and pyrazine [60]. The only qualitative difference is that the real curve-crossings found for pyrazine and benzene are replaced by avoided crossings of the PE functions as an effect of the reduction of symmetry. The non-adiabatic coupling which is present at the avoided crossing of the excited- and ground-state PE surfaces [31,61] will cause efficient internal conversion to the ground state [62].

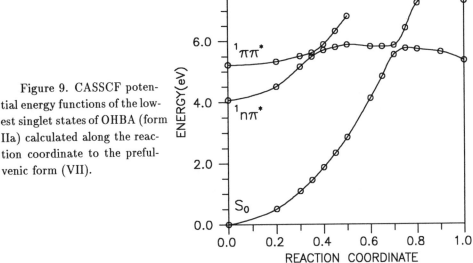

Figure 9. CASSCF potential energy functions of the lowest singlet states of OHBA (form IIa) calculated along the reaction coordinate to the prefulvenic form (VII).

4. Discussion

The results of extensive *ab initio* calculations of the PE surfaces reported above provide the quantitative basis for a qualitative interpretation of the complex photophysics of ESIPT systems. On the basis of the results one can distinguish two basic mechanisms responsible for the ESIPT process and its dynamics, depending on the geometry of the proton transferring center.

4.1. Mechanism of ESIPT process in systems with hydrogen bonds

In typical ESIPT systems, *i.e.* in molecules with intramolecular hydrogen bond(s) or in hydrogen bonded complexes, the transferred hydrogen atom(s) follows the pre-existing hydrogen bond(s). The results of theoretical calculations indicate that the proton can be only transferred on the $^1\pi\pi^*$ excited-state PE surface which can be either barrierless (for strong intramolecular H-bonds) [30,31] or with some barrier (for H-bonded complexes) [33,34]. The optical excitation prepares the system on the "bright" $^1A'(\pi\pi^*)$ PE surface of the initial tautomeric form (stable in the ground state). The optically prepared wave packet experiences a gradient along the PT re-action coordinate and will move and spread rapidly in this direction. If the $^1A"(n\pi^*)$ PE surface were absent or lying at much higher energy, we would observe a barri-erless direct ESIPT reaction on femtosecond time scales, a kind of intramolecular vibrational relaxation process. An important characteristic of the ESIPT systems resulting from the theoretical results discussed above is that in these systems the $A'(\pi\pi^*)$ and $A"(n\pi^*)$ singlet states are often very close each other at the geometry of vertical excitation from the ground state. One should have in mind that the ESIPT reaction is not a one-dimensional motion of a proton. It rather requires displacements along a number of intramolecular coordinates (*c.f.* Figures 3 and 4, for instance). The "intruder" $^1A"(n\pi^*)$ state, being almost degenerate with the "proton transferring" $^1A'(\pi\pi^*)$ state, increases the average density of vibrational states, over which the initial wave-packet can spread-out, but among them there is not any "good" (barrier-less) PT reaction coordinate. In other words, the $^1A"(n\pi^*)$ state provides a dilution effect for the "good" reaction coordinate on the $^1A'(\pi\pi^*)$ PE surface. This can be the key for the explanation of the anomalously slow PT rate constant observed for OHBA ($k_{PT} \simeq 10^9 s^{-1}$) [48] as compared to other ESIPT systems [13-17]. In the very similar system of salicylidenaniline (SA), where the carbonyl group oxygen of OHBA is replaced by an NH group, the PT rate constant ($k_{PT} \simeq 2 \times 10^{11} s^{-1}$) [63] is by more than two orders of magnitude larger than in OHBA. The calculations of Ref.[31] pre-dict a separation of $0.74eV$ between the $^1A"(n\pi^*)$ and $^1A'(\pi\pi^*)$ states of SA at the geometry of vertical excitation from the ground state instead of $-0.21eV$ predicted for OHBA. Such a separation prevents any serious interference between the excited PE surfaces, and thus ESIPT in SA can be considered to occur on a single adiabatic (and essentially barrierless) $^1A'(\pi\pi^*)$ PE surface. At higher excess of excitation en-ergy (under collisionless conditions) or at higher temperatures (in bulk conditions) a non-radiative deactivation channel opens. This can effectively compete with the PT process and with the radiative deactivation of the system.

An important conclusion which emerges from the present results is that the ES-IPT reaction may generally result from a delicate balance between PT dynamics on a single (and essentially barrierless) adiabatic $^1\pi\pi^*$ PE surface and a vibronically induced interaction with the close lying "interfering" $^1n\pi^*$ states as well as with the ground state, causing the non-radiative decay. The balance between these factors depends on the particular system, but the most crucial factor is the molecular size. All the ESIPT systems contain (by definition) heteroatoms and thus they have $n\pi^*$ excited electronic configurations. The energy of the lowest excited $n\pi^*$ singlet states does not depend strongly on the number of heteroatoms in the system and is of the order of about $4.5eV$ for the systems considered. The energy of the lowest $\pi\pi^*$ singlet state is rather sensitive to the size of the aromatic moiety and for larger systems it can lie in the visible part of the spectrum. The energetical separation between these states is expected to increase with the increase of molecular size and for larger poly-atomic systems the ESIPT process is expected to occur on a single $\pi\pi^*$ adiabatic PE surface. For smaller systems, however, one can expect a strong interference between both excited PE surfaces which leads to a complex dynamics on multi-dimensional interacting PE surfaces. This can significantly decrease the rate constant for PT and eventually prohibit the reaction.

The second "interfering" factor, the non-radiative decay channel, is also dependent on the molecular size. If it is true that the internal conversion in these systems results from the strong non-adiabatic interaction between the excited state and the ground state along the reaction path to the prefulvenic forms, this mechanism will operate only for aromatic (or azaaromatic) systems. The formation of prefulvenic structures is intrinsically connected with the promotion of one of the electrons from a bonding π-orbital to an anti-bonding π^*-orbital. This effect has first been found in ethylene, where promotion of one of the two π-electrons on an anti-bonding π^*-orbital causes intramolecular rotation around the CC bond by 90 degrees [4]. The larger the aromatic system is, the more rigid it is with respect to out-of-plane deformation. Thus one can expect a higher barrier in larger molecules which protects the planar system from the strong non-adiabatic region along the prefulvenic path.

The theoretical results and their discussion presented above provides a qualitative basis for the explanation of the experimental observation that the ESIPT process occurs effectively in rather large aromatic systems with intramolecular hydrogen bonds.

4.2. Mechanism of the ESIPT process in systems without hydrogen bonds

The results of the theoretical calculations of the excited-state PE surfaces of 2PY presented in Section 3.3 provide the basis for the definition of a qualitatively new picture of the ESIPT reaction in systems without intramolecular hydrogen bonds. The mechanism of the ESIPT process that emerges from the calculations is basically different from that discussed in the preceding Section for systems with pre-existing intramolecular hydrogen bond(s) along the reaction coordinate. The only common feature of both systems is that at the equilibrium geometries of the tautomeric forms the lowest singlet states have $\pi\pi^*$ and $n\pi^*$ characters, with the former carrying almost exclusively the oscillator strength for optical excitation from S_0. In other words, in both models of the ESIPT process, the optically "bright" state, $^1A'(\pi\pi*)$, can vibronically interact with the "dark" $^1A'(n\pi^*)$ state along the PT reaction path. By analogy with pyrazine molecule [60,64], we can expect that in 2PY there are intramolecular coordinates leading to low-energy intersections between the $^1\pi\pi^*$ and $^1n\pi^*$ PE surfaces. The two surfaces are vibronically coupled via out-of-plane vibrational modes. This coupling converts the surface intersections (allowed by symmetry) into multidimensional conical intersections. Thus a non-Born-Oppenheimer description of excited-state hydrogen transfer is needed in both (with and without intramolecular hydrogen bond) models. The basic feature that discriminates both models is the lack of a barrierless excited-state PE surface along the PT coordinate in the non-hydrogen bond system. Instead a dissociation of the "mobile" hydrogen atom represents an intrinsic feature of the ESIPT reaction in such systems.

The following scenario of the ESIPT process in non-hydrogen bonded systems emerges from the results of theoretical studies: The optically prepared wave packet on the $^1A'(\pi\pi^*)$ PE surface will spread over the surface, and will transfer also to the $^1A'(n\pi^*)$ PE surface due to vibronic coupling between the quasi-degenerate levels. Part of the wave packet (perhaps the dominant one) will follow the gradients in the direction of the given tautomeric minimum in both S_1 and S_2 states. This "nonreactive" part of the wave packet will lead back to the S_0 state of given tautomeric form via fluorescence or via internal conversion. There is, however, a probability that part of the wave packet (the "reactive" one) can pass over the excited-state barrier or will tunnel through and can reach the "dissociative" part of PE surface of the $^1A''(n\pi^*)$ state. Let us notice that the barrier heights of Figure 8 were not optimized and it is also expected that inclusion of dynamic electron correlation by a post-CASSCF method like CASPT2 may significantly decrease the barriers. However, the excited-state geometries were also not optimized, thus the barriers on the excited PE surfaces of 2PY may appear to be prohibitively large for a direct access to the "dissociative" region of the $^1A''(n\pi^*)$ PE surface. This means that in the reported experiments [42,43] the "dissociative" region is rather reached in a stepwise process, with the lowest singlet, or triplet states as intermediates. In this respect the results presented in Figure 8 provide only a qualitative characterization of the system. The "reactive" part of wave packet will feel a significant gradient in the direction of dissociation of the hydrogen atom. This can eventually lead to real dissociation of the "mobile" hydrogen. On the other hand, the wave packet moving on the dissociative part of the $^1A''(n\pi^*)$ PE surface can convert to the ground state due to the conical intersection between the two states as indicated in Figure 8. In condensed phases (like in solvents or in low-temperature matrices) the system can be effectively cooled down to one of the ground state local minima. As the net effect of the scenario sketched above there is a finite probability for the system to reach the ground state minimum different than the initial one. In the other words, PT reaction can take place via photon-induced dissociation- association (PIDA) mechanism.

At this level of theory it seems that the PT process via the PIDA mechanism is symmetric with respect to the initial tautomeric form. Thus, at longer wavelength of exciting light the keto to enol reaction will be induced, while at shorter wavelengths the reverse process can also occur. Whereas there is a number of experimental works reporting the keto to enol ESIPT in 2PY and in similar systems [42,43], there is up to date no evidence for the reverse reaction to occur. A possible explanation is that an efficient radiationless process to the ground state takes place after excitation of the enol (aromatic) form within the lowest singlet manifold. This may be the same mechanism as discussed above for the ESIPT systems with an intramolecular hydrogen bonds, namely, the non-adiabatic interactions between the ground and excited states along the reaction coordinate to the prefulvenic form. Thus the aromatic form cannot live long enough for stepwise absorption to higher states to take place. The keto-form of 2PY (Ia) is not an aromatic form and it cannot form such biradicalic species.

Finally we conclude that the physical mechanism responsible for ESIPT reactions in systems with and without intramolecular hydrogen bonds is basically different. The novel PIDA mechanism proposed in Ref.[36] for ESIPT in non-hydrogen bonded systems has to wait for experimental verification by direct observation of atomic hydrogen migration after optical excitation of suitable molecular systems.

5. Concluding remarks

We have considered in this Chapter the ESIPT process as a simple but nevertheless representative example of a photochemical reaction. The reaction-path concept has been used to construct *ab initio* PE functions for the PT reaction. On the basis of these quantitative theoretical data, a qualitative picture of the reaction mechanisms of different ESIPT systems has been elaborated. We hope to have shown that this particular kind of electronic structure calculations is useful for the rationalization of the vast amount of spectroscopic and kinetic data on ESIPT systems.

In all our calculations we have found that fairly high-level *ab initio* techniques are required for a reliable characterization of photochemically relevant PE functions. It seems that a CASSCF calculation with adequately chosen active space is required for the proper description of quasi-degeneracy effects (non-dynamical electron correlation). The inclusion of dynamical electron correlation effects is also essential. Second-order perturbation theory with the CASSCF wave function as reference state (CASPT2) seems to be an adequate method for excited-state reaction path studies, although more experience with this relatively new method is required before definitive conclusions can be drawn. Unfortunately the optimization of the reaction path itself at the CASPT2 level is not feasible at present, and one has to rely on geometry information obtained with simpler methods. In many cases we have found that the HF approximation yields even qualitatively wrong energy profiles, and also the CASSCF approximation is insufficient for obtaining qualitatively correct results for reaction barriers. In a few cases, where we have been able to compare the *ab initio* results with published results from semi-empirical calculations, the latter have been found to be totally unreliable for the prediction of reaction energy barriers.

Another general conclusion from the present studies is the relevance of near-degeneracies and multi-dimensional surface crossings of excited-state PE surfaces. It seems that in many ESIPT systems the reaction dynamics is largely determined by the interaction of a barrierless $\pi\pi^*$ surface with a $n\pi^*$ surface, which exhibits a significant barrier for PT. The established paradigm of reaction dynamics on a single isolated BO surface is thus not appropriate for ESIPT systems. The rigorous definition of reaction-path Hamiltonians for intersecting and interacting surfaces and the treatment of chemical dynamics represent a challenge for future theoretical work.

References:

1. P. G. Mezey, Potential Energy Hypersurfaces, in Studies in Physical and Theoretical Chemistry, Vol. 53, Elsevier, Amsterdam, 1987

2. H. B. Schlegel, Adv. Chem. Phys. **67**, 249 (1987)

3. D. Heidrich, W. Kliesch and W. Quapp, Properties of Chemically Interesting Potential Energy Surfaces, in Lecture Notes in Chemistry, Vol. **56**, Springer, Berlin, 1991

4. J. Michl and V. Bondacic-Koutecky, in Electronic Aspects of Organic Photochemistry, Willey, New York, 1990

5. G. Herzberg and H. C. Longuet-Higgins, Discuss. Faraday Soc. **35**, 77 (1963)

6. T. Carrington, Acc. Chem. Res. **7**, 20 (1974)

7. L. Salem, J. Am. Chem. Soc. **96**, 3486 (1974)

8. F. T. Smith, Phys. Rev. **179**, 111 (1969)

9. C. A. Mead and D. G. Truhlar, J. Chem. Phys. **70**, 2284 (1979) *ibid* **77**, 6090 (1982)

10. C. A. Mead, Rev. Mod. Phys. **64**, 51 (1992)

11. T. Pacher, L. S. Cederbaum and H. Köppel, Adv. Chem. Phys. **84**, 293 (1993)

12. I. B. Bersuker and V. Z. Polinger, Vibronic Interactions in Molecules and Crystals, Wiley, New York, 1989

13. S. J. Formosinho and L. G. Arnaut, J. Photochem. Photobiol. **A75**, 21 (1993)

14. E. M. Kosower and D. Huppert, Ann. Rev. Phys. Chem. **37**, 127 (1986)

15. M. Kasha, J. Chem. Soc. Faraday Trans. II **82**, 2379 (1986)

16. P. F. Barbara and H. P. Trommsdorff, eds. Special Issue on Spectroscopy and Dynamics of Elementary Proton Transfer in Polyatomic Systems, Chem. Phys. **136**, 152-360 (1989)

17. Special Issue (M. Kasha Festschrift) J. Phys. Chem. **95**, 10220-10524 (1991)

18. J. Calatan and J. I. Fernandez-Alonso, J. Mol. Struct. **27**, 59 (1975)

19. W. Orttung, G. W. Scott and D. Vosooghi, J. Mol. Struct. **109**, 161 (1984)

20. K. L. Kunze and J. R. de la Vega, J. Am. Chem. Soc. **106**, 6528 (1984)

21. H. Tanaka and K. Nishimoto, J. Phys. Chem. **88**, 1052 (1988)

22. J. Waluk, H. Bulska, A. Grabowska and A. Mordzinski, New J. Chem. **10**, 413 (1086)

23. S. Nagaoka and U. Nagashima, Chem. Phys. **136**, 152 (1989)

24. B. Dick, J. Phys. Chem. **94**, 5752 (1990)

25. N. P. Ernsting, Th. Arthen-Engeland, M. A. Rodriguez and W. Thiel, J. Chem. Phys. **97**, 3914 (1992)

26. X. Duan and S. Scheiner, Chem. Phys. Lett. **204**, 36 (1993)

27. V. Barone and C. Adamo, Chem. Phys. Lett. **226**, 399 (1994)

28. B. O. Roos, Adv. Quantum Chem. **69**, 399 (1987)

29. (a) K. Andersson, P.-A. Malmqvist, B. O. Roos, A. J. Sadlej and K. J. Wolinski, J. Phys. Chem. **94**, 5483 (1990); (b) K. Andersson, P.-A. Malmqvist and B. O. Roos, J. Chem. Phys. **96**, 1218 (1992)

30. A. L. Sobolewski and W. Domcke, Chem. Phys. Lett. **211**, 82 (1993)

31. A. L. Sobolewski and W. Domcke, Chem. Phys. **184**, 115 (1994)

32. A. L. Sobolewski and L. Adamowicz, Chem. Phys., in press.

33. A. L. Sobolewski and L. Adamowicz, J. Phys. Chem., submitted.

34. A. L. Sobolewski and L. Adamowicz, J. Chem. Phys., in press.

35. A. L. Sobolewski and L. Adamowicz, Chem. Phys. Lett., in press.

36. A. L. Sobolewski, Chem. Phys. Lett. **211**, 293 (1993)

37. A. L. Sobolewski, J. Photochem. Photobiol., in press

38. W. H. Miller, B. A. Ruf and Y. -T. Chang, J. Chem. Phys. **89**, 6298 (1988)

39. K. Fukui, Acc. Chem. Res. **14**, 363 (1981)

40. W. H. Miller, N. C. Handy and J. E. Adams, J. Chem. Phys. **72**, 94 (1980)

41. J. Heldt, D. Gormin and M. Kasha, Chem. Phys. **136**, 321 (1989)

42. (a) M. J. Nowak, J. Fulara and L. Lapinski, J. Mol. Str. **175**, 91 (1988); (b) M.J. Nowak, L. Lapinski and J. Fulara, Spectrochimica Acta, **45A**, 229 (1989); (c) L. Lapinski, J. Fulara and M.J. Nowak, Spectrochimica Acta , **46A**, 61 (1990); (d) L. Lapinski, J. Fulara, R. Czerminski and M.J. Nowak, Spectrochimica Acta, **46A**, 1087 (1990); (e) L. Lapinski, M.J. Nowak, J. Fulara, A. Les and L. Adamowicz, J. Physical Chemistry, **94**, 6555 (1990); (f) M.J. Nowak, L. Lapinski, H. Rostkowska, A. Les and A. Adamowicz, J. Phys Chem., **94**, 7406 (1990); (g) M.J. Nowak, L. Lapinski, J. Fulara, A. Les and L. Adamowicz, J. Phys. Chem., **95**, 2404 (1991); (h) H. Vranken, J. Smets, G. Maes, L. Lapinski, M.J. Nowak and L. Adamowicz, Spectrochimica Acta **50A**, 875 (1994);

43. M. J. Nowak, L. Lapinski, J. Fulara, A. Les and L. Adamowicz, J. Chem. Phys. **96**, 1562 (1992)

44. T. Carrington and W. H. Miller, J. Chem. Phys. **84**, 4364 (1986)

45. R. Rossetti, R. Rayford, R. C. Haddon and L. E. Brus, J. Am. Chem. Soc. **103**, 4303 (1981)

46. H. C. Longuet-Higgins, in Advances in Spectroscopy, Vol.2, H. W. Thompson (ed.), Interscience, New York, 1961, p.429

47. H. Köppel, W. Domcke and L. S. Cederbaum, Adv. Chem. Phys. **57**, 59 (1984)

48. S. Nagaoka, N. Hirota, M. Sumitami and K. Yoshihara, J. Am. Chem. Soc. **94**, 488 (1983)

49. S. Nagaoka, U. Nagashima, Chem. Phys. **136**, 153 (1989)

50. J. Catalan, F. Toribio and A. U. Acuna, J. Phys. Chem. **86**, 303 (1982)

51. P. -T. Chou, M. Chao, J. H. Clements, M. L. Martinez and C. -P. Chang, Chem. Phys. Lett. **220**, 229 (1994)

52. L. Lapinski, R. Czerminski, M. J. Nowak and J. Fulara, J. Mol. Struct. **220**, 147 (1990)

53. J. Smets, PhD thesis, Leuven, 1993.

54. (a) R. D. Brown, P. D. Godfrey, D. McNaughton and A. P. Pierlot, J. Am. Chem. Soc. **111**, 2308 (1988); (b) M. Szczesniak, K. Szczepaniak, J. S. Kwiatkowski, K. KuBulat and W. B. Person, *ibid* **110**, 8319 (1988)

55. M. Dreyfus, O. Bensaude, G. Dodin and J. E. Dubois, J. Am. Chem. Soc. **98**, 6338 (1976)

56. M. J. Scanlan and I. H. Hiller, Chem. Phys. Lett. **107**, 330 (1984)

57. M. Moreno and W. H. Miller, Chem. Phys. Lett. **171**, 475 (1990)

58. S. Kato, J. Chem. Phys. 88 (1988) 3045.

59. A. L. Sobolewski and W. Domcke, Chem. Phys. Lett. **180**, 381 (1991)

60. A. L. Sobolewski, C. Woywod and W. Domcke, J. Chem. Phys. **98**, 5627 (1993)

61. W. Domcke, A. L. Sobolewski and C. Woywod, Chem. Phys. Lett. **203**, 220 (1993)

62. A. L. Sobolewski, J. Chem. Phys. **93**, 6433 (1990)

63. P. F. Barbara, P. M. Rentzepis and L. E. Brus, J. Am. Chem. Soc. **102**, 2786 (1980)

64. C. Woywod, W. Domcke, A. L. Sobolewski and H.-J. Werner, J. Chem. Phys. **100**, 1400 (1994)

VIEWING THE REACTION PATH WITH THE HELP OF TIME-RESOLVED FEMTOSECOND SPECTROSCOPY

CHRISTOPH MEIER
Fakultät für Physik
Albert-Ludwigs-Universität
D-79104 Freiburg i. Br., Germany

VOLKER ENGEL
Inst. für Physikalische Chemie
Universität Würzburg
D-97070 Würzburg, Germany

1. Introduction

The spectroscopy of the *transition state*, which separates products from educts in a chemical reaction has been the subject of numerous studies [1]. A beautiful idea, which by now has been applied to a great variety of molecular systems is to use electron detachment spectroscopy to directly resolve resonance features which are the fingerprints of quasi-bound states of the molecule [2]. The latter are localized in the transition-state region. Nevertheless this kind of spectroscopy does not tell us about the history of the system when it evolves from this intermediate region towards the product channels. In this chapter we want to line out, from a theoretical point of view, how time-resolved spectroscopy can be used to monitor the evolution of a system which is prepared in the transition state, until it ends up in several product channels. This idea has been promoted by Zewail and coworkers over the last years and exciting experiments were performed. It is not the purpose of this contribution to review those experiments (see [3]). Rather, we will describe the principles of time-resolved spectroscopy using quantum mechanical and classical theory. In particular it will be seen that the motion along the reaction path can be directly taken from an experimental signal.

The article is organized as follows: in Sec. 2 we outline the scheme of a pump/probe experiment which is performed with two ultrashort laser pulses. Then the excitation mechanism which prepares the system in a coherent superposition of eigenstates in the transition-state region of the respective potential surface is described (Sec. 3). This localized wave packet evolves along the reaction path, as described in Sec. 4. In Sec. 5 it is shown how the wave-packet motion can be probed. A short summary concludes this chapter.

D. Heidrich (ed.), The Reaction Path in Chemistry:
Current Approaches and Perspectives, 283–294.
© 1995 *Kluwer Academic Publishers. Printed in the Netherlands.*

2. Pump/Probe Spectroscopy

The spectroscopic tool to be considered here is *femtosecond pump/probe spectroscopy*. This experimental technique uses two ultrashort laser pulses which are time-delayed with respect to each other. They are sent into a molecular sample and a signal is recorded as a function of the delay-time between the pulses. To be more specific, we assume the molecule to be in an inital state $|\psi_0\rangle|0\rangle$. Here $|\psi_0\rangle$ denotes the wave function for the nuclear motion and $|0\rangle$ the wave function of the electrons (the adiabatic separation of nuclear and electronic motion is assumed throughout). The pump pulse induces a transition and the resulting wave function which describes the molecule after the interaction with the electric field may be assigned as $|\psi_1\rangle|1\rangle$. We treat electronic excitation so that the molecule is prepared in another electronic state $|1\rangle$. After the pump pulse passed the sample, the molecule evolves unperturbed until the probe pulse starts interacting. This interaction results in a second excitation to (in our case) a final electronic state $|2\rangle$ with the respective nuclear wave function $|\psi_2\rangle$. The scheme just described is depicted in **Figure 1** and illustrates the idea of many pump/probe experiments.

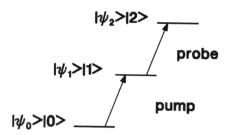

Figure 1: Illustration of a pump/probe scheme. $|\psi_n\rangle$ are nuclear molecular wave functions in different electronic states $n = 0, 1, 2$ and $|n\rangle$ denote the electronic part of the total wave function.

The measured signal can be e.g. the total fluorescence from state $|2\rangle$ to an energetically lower electronic state or an ion signal, as discussed below. The idea behind this kind of experiments is that the pump/probe signal which is connected to the wave function $|\psi_2\rangle|2\rangle$ depends on the time-evolution in the *intermediate* state $|1\rangle$, that is, on the nuclear dynamics described by the wave function $|\psi_1(t)\rangle$.
We will now look into the nature of the excitation process and then show which connection to the motion along a reaction path exists. A two-dimensional example is used for the purpose of illustration.

3. Wave-Packet Preparation

Consider a molecule in the initial state $|\psi_0\rangle|0\rangle$ as indicated in **Figure 1**. Assuming a weak electric field, the electronic excitation induced by the pump pulse can be described within time-dependent perturbation theory. The state obtained for a one-photon absorption process is (atomic units are used throughout) [4]

$$|\psi_1\rangle|1\rangle = i \int_0^t dt' U_1(t - t')\{|1\rangle W_{10}(t')\langle 0|\} U_0(t')|\psi_0\rangle|0\rangle. \tag{1}$$

Here U_1, U_0 are the time-evolution operators for the nuclear motion in the electronic states $|1\rangle$ and $|0\rangle$, respectively. The molecule-field interaction within the dipole approximation is

$$W_{10}(t) = \mu_\epsilon E_0 f(t) e^{-i\omega_1 t}, \qquad (2)$$

where μ_ϵ is the projection of the molecular dipole moment on the polarization vector of the field, ω_1 is the pump laser frequency, E_0 its field strength and $f(t)$ describes the pulse envelope.

Since we are concerned with femtosecond pulses the function $f(t)$ is assumed to have a temporal width less then 100 femtoseconds. If we take a Gaussian shape function of e.g. 50 fs width then the corresponding spectral destribution is about 70 meV broad. This energy interval is much broader than the separation between molecular (vibrational and rotational) states so that the pulse is able to excite several stationary eigenstates simultaneously. The resulting state is a coherent superposition of molecular eigenstates which is a *wave packet*. The most important property of this wave packet in the present context is that it is a spatially localized object. To illustrate this we will use the fragmentation of the HOD molecule as a numerical example. The process is initiated by laser excitation from the electronic ground state (X) to the first electronic state (A). The $X \to A$ transition of the water molecule is a "prototype for the direct dissociation of a triatomic molecule" [5]. Experimental and theoretical ab initio studies have been performed to a very high status of sophistication. Femtosecond studies have, to our knowledge, not been performed. One reason might be that the time scale for dissociation is so short, that the ultrashort pulses have to be in the order of 10 fs to monitor the hydrogen bond rupture. Nevertheless, we will use this example since the potential energy surface for the electronically excited state is well known and the system can be treated, to a very good approximation, as two-dimensional. Theoretical studies showed that the fragmentation process starting from the rotational/vibrational ground state in the X-state is nearly independend of the bonding angle (γ) [7]. This allows for the calculation of all properties for fixed bonding angle.

Energy contours of the potential energy surface [6] for the first excited state of water are plotted in **Figure 2** as a function of the OD distance r_D and OH distance r_H and for fixed angle $\gamma = 104°$ (the equilibium angle in the electronic ground state). The surface is purely repulsive so that the dissociation is mainly direct. The figure also shows contours of the probability density $|\psi_1(r_H, r_D, t)|^2$ at $t=6$ fs. The wave function was calculated using Eq.(1) with a 12 fs $\sin^2$-pulse (center at 6 fs). A frequency of $\omega_1 = 0.3$ au (8.6 eV) was used. Short pulses with this frequency are not yet available. Thus at least a two-photon excitation process is required so that the transition occurs. Since we want to study water as a model system to illustrate central ideas we will not further discuss this subject.

The time origin was fixed to be the center of the pulse , so that $t = 6$ fs is the time when the $\sin^2$-pulse has decayed to zero. The initial state, representing the vibrational ground state wave function was calculated with a two-dimensional potential [8]. This does not yield the correct ground state eigenenergy which is however not important here.

As can be taken from the figure a localized wave packet is prepared in the short pulse excitation process. Its location is close to the potential barrier which separates the two product channels $H + OD$ and $D + OH$, i.e. the packet is found in the vicinity of the transition-state region. Nevertheless parts of the initial wave packet have already moved towards the exit channels. Since the H-atom is lighter than the D-atom it

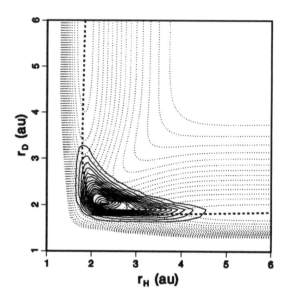

Figure 2: Contours of the potential energy surface of the A-state of HOD (dotted lines) in intervals of 0.01 au. The surface is displayed as a function of the distances between the H (D) atom and the oxygen and for fixed bending angle of $\gamma = 104^o$. Also shown is the minimum energy path (dashed line) and the probability density of the wave packet prepared by femtosecond excitation from the vibrational and electronic ground state of the molecule.

moves faster outward. This can be taken from the asymmetric wave packet which is streched more into the $H + OD$ channel.
Figure 2 also shows the minimum energy path connecting the two product channels (dashed line). The question which will be addressed in the following section is, how the wave packet will evolve along the potential energy surface. This motion will define the reaction path for the photochemically induced reaction on the two-dimensional potential energy surface, leading to the different products.
It has to be emphasized that only an ultrashort laserpulse can create a localized wave packet as displayed in the figure. The longer the pulse, the more the prepared state will be delocalized in coordinate space and thus resemble a single stationary scattering state of the molecule. The time evolution of such a state is given by a phase factor and thus the whole idea of pump/probe spectroscopy is lost.

4. Time evolution along the reaction path

As can be taken from **Figure 2**, the wave packet prepared by the pump pulse has most of its probability amplitude close to the potential barrier. Parts of it experience the gradient of the potential which points towards the $H + OD$ channel, other parts experience the gradient into the $D + OH$ channel. As a result, whence the packet

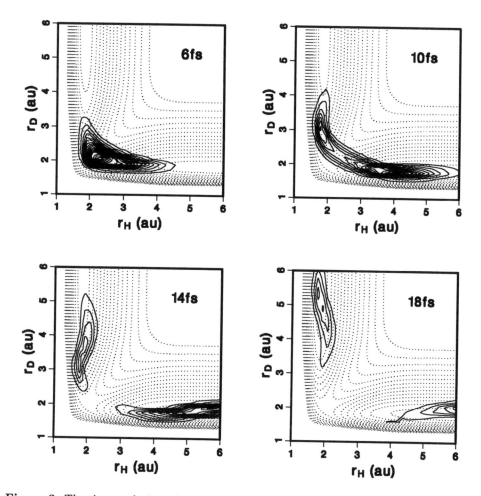

Figure 3: The time evolution of the wave packet prepared by the pump process. Contours of the wave packet probability density are displayed for different times.

starts moving, it bifurcates and splits into two disjunct parts. This is documented in **Figure 3** which contains snapshots of the wave-packet motion for different times t, as indicated.

What was mentioned in the previous section becomes clearer now: since the H dissociation is faster than the time it takes for the D atom to separate from the oxygen, the corresponding part of the wave packet enters the product channel earlier. This also implies that the $H + OD$ channel is preferentially populated [9]. The overall dissociation dynamics is very fast (about 14 fs). There is a small fraction of the wave packet which remains for a while in the transition state region. This "recurrence" of the packet gives rise to a weak structure of the $X - A$ absorption spectrum [10]. The centers of the wave packets oscillate around the minimum energy path. This motion documents that during the fragmentation OH and OD products are built which are vibrationally excited. The branching ratio for production of the two molecular products can be directly computed from the ratio of the norm of the wave packets in the different arrangement channels.

Figure 4 shows two typical trajectories which trace the quantum mechanical wave-packet motion. They are started on the symmetric stretch line with different initial momenta pointing into the exit channels. The two trajectories represent dissociation into $H + OD$ and $D + OH$, respectively. What was found for the quantum dynamics is more clearly demonstrated by the classical trjectories: the $H + OD$ dissociation is faster and the oscillations of both trajectories around the minimum energy path clearly shows the vibrational excitation of the fragments. Indeed, if we compute the time-evolution of the bondlength expectation values $\langle r_{H,D} \rangle$ from the bifurcated packets it is found that they resemble closely the classical trajectories (as can be expected from Ehrenfest's theorem). The wave-packet motion shows that the dissociation proceeds as can be anticipated classically.

This is of great importance since in our mind we picture a chemical reaction to proceed classically: the educts approach each other, following a trajcetory on a multidimensional potential energy surface. If the collision partners are close to each other, they interact and finally the separation into the products occurs.

Ultrashort pulses are able to prepare localized wave packets which (for short times) move in the average classically. This motion clearly defines a reaction path for the case that the reaction starts right at the transition state and the molecular fragments evolve into the arrangement channels. In this sense the laser induced process is a "half collision", since the first part of a full collision where the atomic and molecular species approach each other is missing.

The question now is how to trace the wave-packet motion and thus view the time evolution of the system during the reaction. This naturally yields information about the reaction path.

5. Probe of the molecular motion along the reaction path

As we have seen in the previous section the wave packet prepared by the pump-pulse moves along the reaction path towards all energetically accessible exit channels. It is the purpose of a pump/probe experiment to monitor this motion in real-time. Several detection schemes have been used and we will discuss the special case where the second laser pulse ionizes the molecule. In this case an ion-signal is measured as a function of the pump/probe delay. For a cornerstone experiment using pump/probe ionization spectroscopy see [11]. As before we will use perturbation theory to calculate the wave function which is prepared by the probe pulse, but now we have to describe an ionization continuum. This is done by expanding the total ionic wave function into a set of orthonormal electronic eigenstates $|E, 2\rangle$ which are labeled according to the energy of the photoelectron. Then one can derive the first order expression for the nuclear wave function in the final electronic state [12, 13]:

$$\psi_{2,E}(R, t; T) = i \int_T^t dt' \quad e^{-i(K+V_2)(t-(t'-T))} \quad \mu_{21}(E) f(t'; T)$$
$$e^{-i(\omega_2 - E)t'} \quad e^{-i(K+V_1)(t'-T)} \quad \psi_1(R, T). \tag{3}$$

K denotes the nuclear kinetic energy operator and V_1, V_2 are the potential energy operators in the electronic states $|1\rangle$ and $|2\rangle$, respectively. T is the delay time when

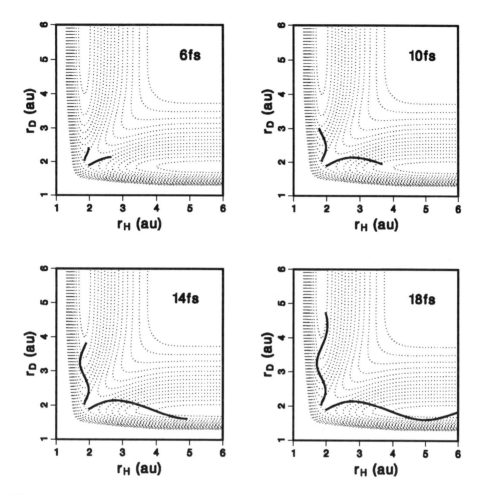

Figure 4: The time evolution of two trajectories starting with different intitial conditions in the region where the quantum mechanical wave packet prepared by the pump process is located intitally. The trajectories represent dissociation into $H + OD$ and $D + OH$, respectively.

the probe-pulse starts interacting with the molecule. The wave function in Eq.(3) is written as a function of all nuclear coordinates R. The above expression resembles the one in Eq.(1) but now explicitly depends on the delay time T and the energy of the ejected photoelectron.

We can now define the photoelectron spectra

$$P(E,T) = \lim_{t \to \infty} \langle \psi_{2,E}(R,t;T) | \psi_{2,E}(R,t;T) \rangle. \tag{4}$$

The total ion signal can be obtained by integrating $P(E,T)$ over the energy E.

To see how the dependence of the electron spectrum on the delay time comes about

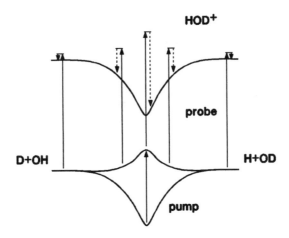

Figure 5: Cuts of the potential surfaces of the electronic ground state (X), the state in which the dissociation occurs (A) and the ground state of the ionic molecule along the minimum energy path in the A-state (see **Figure 2**). The straight lines correspond to the frequencies ω_1 and ω_2 of the pump and probe pulse. The dashed lines mark three different photoelectron energies.

we assume that the wave packet ψ_1 does not move essentially during the probe process. Then the kinetic energy operators in the propagators can be neglected. This assumption is fulfilled if very short pulses are used. Of course for the fast hydrogen dynamics the pulses have to be much shorter then in the case of a heavy atom. Eq.(3) now becomes

$$\psi_2(R, t; T) \sim \mu_{21}(E)\psi_1(R, T)I(t, T), \tag{5}$$

where we neglected unimportant phase factors and $I(t, T)$ is defined as

$$I(t; T) = \int_T^t dt' f(t'; T)e^{i(V_2 - V_1 - (\omega_2 - E))t'}. \tag{6}$$

We see that the wave function ψ_2 resembles the function ψ_1 (multiplied with the transition dipole moment) but is weighted with the Fourier-tranform $I(t; T)$ which determines the norm of the wave function. The integration is performed over a smooth envelope function $f(t, T)$ and an oscillating phase factor. According to the stationary phase approximation [14], the modulus of the integral will be largest at the stationary points R_s for which the exponent in Eq.(6) is zero:

$$V_2(R_s) - V_1(R_s) = \omega_2 - E \tag{7}$$

Note that the potentials depend on all nuclear coordinates so that R_s denote vectors. In our particular case we have $R_s = (r_{H,s}, r_{D,s})$. Eq.(7) tells us that the photoelectron spectrum will be largest, if the center of the wave packet ψ_1 happens to be close to a region of space where the energy difference $(\omega_2 - E)$ equals the difference between the potentials of the states $|1\rangle$ and $|2\rangle$. This is illustrated in **Figure 5**. Shown is the minimum energy path on the potential V_1. The two exit channels are indicated. Also shown are the potentials for HOD^+ and the electronic ground state of the neutral molecule along the same reaction coordinate. Three distinct cases are shown. The straight lines mark the pump and probe laser energies ω_1 and ω_2. and the dashed lines are three electron energies for which, according to Eq.(7) , one would expect a maximum in the photoelectron spectrum. In our calculation we used ω_2=0.45 au (12.4 eV), which is –as in the case of ω_1– an energy too high to be achieved by today's short pulse techniques. The largest photoelectron energy can be obtained if the wave packet is located around the potential energy barrier. As seen in **Figure 3** this happens only at short times. If the packets slide down the potential barrier the electron energy which fulfills the resonance condition Eq.(7) becomes lower and is lowest in the asymptotic region where the cut through the difference potential along the shown coordinate becomes constant. **Figure 6** contains the photoelectron spectra for three electron energies which correspond to the generic cases illustrated in **Figure 5**. Results are plotted for H_2O, HOD and D_2O.

In the case of E=0.32 au the signal maps the separation of the wave packets from the transition state region. Note that this separation is so fast that pump and probe pulse have to overlap in time to see this effect (they overlap until T=12 fs). The signal decays within 10 fs which is the characteristic time for the motion away from the potential barrier. If we monitor the electron spectrum for an energy of 0.18 au,

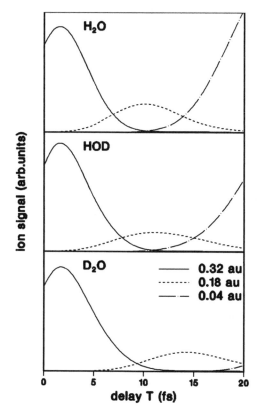

Figure 6: The photoelectron signal $P(E,T)$ for different energies E and delay times up to 20 fs. The signals are displayed for H_2O, HOD and D_2O respectively.

the signal is expected to be maximal if the packets are located farther out along the reaction path. Indeed we find the curves to peak at later times in all three cases. A comparison between H_2O and D_2O signals tells us that the H_2O dissociation proceeds faster. The HOD case resembles very much the H_2O case since the main dissociation channel is $H + OD$. Finally for the lowest energy E=0.04 au the electron spectrum reflects the population in the asymptotic exit channels. From what we just discussed it follows that, by monitoring the photoelectron spectrum for different energies as a function of delay time one is able to probe the wave packets at different locations on the potential energy surface and thus monitor the motion along the reaction path in real time. There is of course the ambiguity that if one has more than one wave packet there are always different points R_s for which Eq.(7) holds.

Instead of analyzing the photoelectron spectrum for fixed energy one may study the dependence of the full spectrum on the delay time. The result is shown in **Figure 7** (left panel) which contains $P(E,T)$ as a function of energy E and delay-time T for water and its isotopes.

These plots document a dynamical reflection principle: for a fixed time T the (multi-dimensional) wave packet is mapped on the energy scale via the resonance condition of Eq.(7). The variation of the delay-time T then allows for the observation of the wave-packet motion along the reaction path. Thus (e.g. in the case of HOD) we see a shift of the distribution from larger to lower energies which reflects the motion of the two packets from the transition state region towards the asymptotic product channels. Unfortunately the $D+OH$ channel is so weakly populated that the motion of the two parts of the initial wave packets can hardly be separated. Only a small shoulder which is best seen around T= 12 fs is due to the fact that one packet reaches the exit channel earlier than the other. To illustrate the reflection principle we just described one may directly correlate the classical motion depicted of **Figure 4** with the electron energies via Eq.(7). The result is displayed in **Figure 7** (right panel) which shows the energies

$$E(T) = \omega_2 - \{V_2(r_H(T), r_D(T)) - V_1(r_H(T), r_D(T))\}, \tag{8}$$

where $(r_H(T), r_D(T))$ are chosen to be the two trajectories in **Figure 4**. The time-evolution of $E(T)$ shows the shift of the maximum of the photoelectron kinetic energy distribution as the packets move along the reaction path. Again we observe the different fragmentation times leading to the two kind of fragments. **Figure 7** reveals the connection with the quantum dynamics.

6. Summary

In this contribution we have described the use of femtosecond pump/probe spectroscopy to monitor molecular wave-packet motion along the reaction path. The use of ultrashort laser pulses allows for the preparation of localized molecular wave packets which, at least for short times, move very similar to a classical particle. Electronic excitation from the ground to some electronically excited state which is antibonding and has several dissociation channels results in the preparation of a wave packet in the transition state region. In the course of the photochemical reaction, the wave packet will, under the influence of the potential energy between the different reaction partners, split and parts of the packet move along the reaction path into all fragment channels which are energetically allowed. To monitor this time evolution is the task of the pump and probe arrangement. According to a resonance condition the probe excitation, which is again initialized by a femtosecond pulse, is only effective if the packet

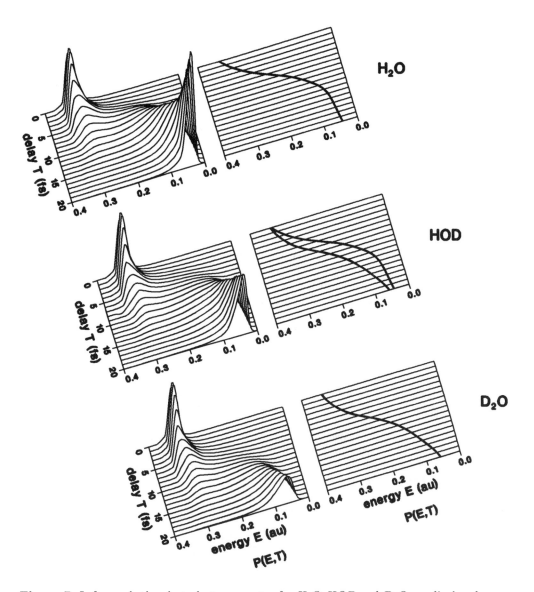

Figure 7: Left panel: the photoelectron spectra for H_2O, HOD and D_2O are displayed as a function of energy E and delay time T. Right panel: Mapping of the classical dynamics on the energy scale. The lines correspond to the shift of the maximum of the photoelectron kinetic energy distribution as a function of time. The mapping is performed with the classical version (Eq.8) of the resonance condition (Eq.7).

is localized in certain regions of coordinate space, i.e. at certain positions along the reaction path. In the case of a probe ionisation a dynamical reflection principle is encountered: the time dependence of the photoelectron energy distribution reflects the dynamics of the wave packets along the reaction coordinate. The mapping proceeds via the difference potential between the electronic states on which the reaction occurs and the final electronic state accessed by the probe transition. Although we have used a triatomic molecule for the purpose of illustration the central ideas outlined above are valid for larger systems as well. Again we would like to draw the readers attention on the exciting field of time-resolved measurements of molecular reactions [3].

References

[1] P. R. Brooks, Chem. Rev. **88**, 407 (1988).

[2] R. B. Metz, S. E. Bradforth, D. M. Neumark, Adv. Chem. Phys. **81**, 1 (1992), D. M. Neumark, Annu. Rev. Phys. Chem **43**, 153 (1992), Acc. Chem. Res. **26**, 33 (1993).

[3] A. H. Zewail, Faraday Discuss. Chem. Soc. **91**, 207 (1991), J. Phys. Chem. **97** 12427 (1993), Femtochemistry, Vols.1,2, World Scientific, Singapure, 1994.

[4] R. Loudon, The quantum theory of light, Clarendon, Oxford, 1983.

[5] V. Engel, P. Andresen, F. Crim, B. Hudson, R. Schinke, V. Staemmler, J. Phys. Chem. **96**, 3201 (1992).

[6] V. Staemmler and A. Palma, Chem. Phys. **93**, 63 (1985).

[7] R. Schinke, Photodissociation dynamics, Cambridge University Press, Cambridge, 1993.

[8] J. Zhang, D.G. Imre and J.H. Frederick, J. Phys. Chem. **98**, 1840 (1989)

[9] V. Engel and R. Schinke, J. Chem. Phys. **88**, 6831 (1988).

[10] V. Engel, R. Schinke and V. Staemmler, J. Chem. Phys. **88**, 129 (1988); J. Zhang and D.G. Imre, J. Chem. Phys. **90**, 1666 (1989).

[11] T. Baumert, M. Grosser, R. Thalweiser and G. Gerber, Phys. Rev. Lett. **67**, 3753 (1991) ;T. Baumert, B. Bühler, M. Grosser, R. Thalweiser, V. Weiss, E. Wiedenmann and G. Gerber, J. Phys. Chem. **95**, 8103 (1991)

[12] M. Seel and W. Domcke Chem. Phys. **151**, 59 (1991), J. Chem. Phys. **95**, 7806 (1991)

[13] Ch. Meier and V. Engel in Femtosecond Chemistry, J. Manz and L. Woeste (eds.), VCH, Heidelberg, 1994.

[14] C. Eckart, Rev. Mod. Phys. **20**, 399 (1948).

[15] Ch. Meier and V. Engel, Chem. Phys. Lett. **212**, 691 (1993); J. Chem. Phys. **101**, 2673 (1994).

INDEX

Understanding Chemical Reactivity

1. Z. Slanina: *Contemporary Theory of Chemical Isomerism*. 1986
   ISBN 90-277-1707-9
2. G. Náray-Szabó, P.R. Surján, J.G. Angyán: *Applied Quantum Chemistry*. 1987
   ISBN 90-277-1901-2
3. V.I. Minkin, L.P. Olekhnovich and Yu. A. Zhdanov: *Molecular Design of Tautomeric Compounds*. 1988
   ISBN 90-277-2478-4
4. E.S. Kryachko and E.V. Ludeña: *Energy Density Functional Theory of Many-Electron Systems*. 1990
   ISBN 0-7923-0641-4
5. P.G. Mezey (ed.): *New Developments in Molecular Chirality*. 1991
   ISBN 0-7923-1021-7
6. F. Ruette (ed.): *Quantum Chemistry Approaches to Chemisorption and Heterogeneous Catalysis*. 1992
   ISBN 0-7923-1543-X
7. J.D. Simon (ed.): *Ultrafast Dynamics of Chemical Systems*. 1994
   ISBN 0-7923-2489-7
8. R. Tycko (ed.): *Nuclear Magnetic Resonance Probes of Molecular Dynamics*. 1994
   ISBN 0-7923-2795-0
9. D. Bonchev and O. Mekenyan (eds.): *Graph Theoretical Approaches to Chemical Reactivity*. 1994
   ISBN 0-7923-2837-X
10. R. Kapral and K. Showalter (eds.): *Chemical Waves and Patterns*. 1995
    ISBN 0-7923-2899-X
11. P. Talkner and P. Hänggi (eds.): *New Trends in Kramers' Reaction Rate Theory*. 1995
    ISBN 0-7923-2940-6
12. D. Ellis (ed.): *Density Functional Theory of Molecules, Clusters, and Solids*. 1995
    ISBN 0-7923-3083-8
13. S.R. Langhoff (ed.): *Quantum Mechanical Electronic Structure Calculations with Chemical Accuracy*. 1995   ISBN 0-7923-3264-4
14. R. Carbó (ed.): *Molecular Similarity and Reactivity: From Quantum Chemical to Phenomenological Approaches*. 1995
    ISBN 0-7923-3309-8
15. B.S. Freiser (ed.): *Organometallic Ion Chemistry*. 1995
    ISBN 0-7923-3478-7
16. D. Heidrich (ed.): *The Reaction Path in Chemistry: Current Approaches and Perspectives*. 1995   ISBN 0-7923-3589-9

Kluwer Academic Publishers – Dordrecht / Boston / London